INJECTION-MOULD DESIGN FUNDAMENTALS

INJECTION-MOULD DESIGN FUNDAMENTALS

by

A. B. GLANVILL and E. N. DENTON

A.M.I.Mech.E. G.I.Mech.E.

AMERICAN EDITION

INDUSTRIAL PRESS INC., 200 MADISON AVENUE

NEW YORK, N.Y. 10016

13 15 17 18 16 14

PREFACE

The plastics industry has expanded at a phenomenal rate since 1945 and a substantial part of this increase has occurred in the field of injection moulding. This in turn has resulted in a very large increase in the number, size, precision and complexity of injection tools required, together with a corresponding increase in the number of engineers engaged in mould design and manufacture.

In the authors' view, this expansion in plastics mould engineering has resulted in practice running ahead of the standard authoritative technical literature available. Most serious perhaps, is the lack of publications on the fundamental principles involved in the design and performance of moulds. Such works that have been published are mostly of a general nature or have covered certain specific features.

For this reason young designers have largely had to acquire their technology through empirical knowledge and practical experience without the availability of adequate collected theoretical and fundamental data.

It has been the intention of the authors, therefore, primarily to set out the basic technical principles underlying mould design, to provide data on all the component details of the tool and to examine the fundamental construction and operation of the various types of mould. To this end it is hoped that the work may be of value to young engineers as a college text book and to practising designers and toolmakers, by providing the theoretical background to practical experience, and as a reference manual.

No attempt has been made to discuss highly complex specialist tools; most of these are designed for specific applications and are, in any event, generally extensions of a basic mould type.

In this respect, experienced designers never lose sight of the high static and dynamic loads prevailing in a running mould and arising from the flow and pressure of the injected plastic material. For this reason, light-weight tool components or complex mechanisms should be very carefully considered before their use is adopted. Failure of the mould in operation can be very expensive and, as a general rule, sound simplicity is safer than too-clever complexity.

Another objective of this book has also been to emphasize that not only must the mould form produce a dimensionally correct component, but that the effect of flow and pressure on the moulding itself must be equally considered, and these features are discussed in detail.

Finally, the authors would like to express their sincere appreciation to those who have in any way assisted with the preparation of the text or illustrations. Also, while acknowledgement is gladly made in the work to a number of concerns who have furnished information or illustrations, particular acknowledgement is due to Ekco Plastics Ltd.

While every effort has been made to ensure the accuracy of this publication no responsibility can be assumed.

A. B. GLANVILL

E. N. DENTON

* * *

ACKNOWLEDGEMENTS

EKCO PLASTICS LTD.
CINCINNATI MILLING MACHINES LTD.
E. H. JONES (MACHINE TOOLS) LTD.
ALFRED HERBERT LTD.
ROCKWELL MACHINE TOOL CO., LTD.
ENGIS LTD.
BERYLLIUM AND COPPER ALLOYS LTD.
DARWINS LTD.
IMPERIAL SMELTING CORPORATION LTD.
THE HOYT METAL CO. OF GREAT BRITAIN LTD.
LONDON AND SCANDINAVIAN METALLURGICAL CO., LTD.
IMPREGNATED DIAMOND PRODUCTS LTD.
WICKMAN LTD.
WESTON-EVANS & CO., LTD.
UDDEHOLM LTD.

REFERENCES

"Electroformed Moulds in Hard Nickel Alloy", *Canadian Plastics* September 1954.

P. Spiro, "The Production and Application of Hard Nickel Electroformed Moulds and Dies", *Plastics Institute Trans.*, Vol. XXIV, No. 57.

D. W. Rudorff, "Principles and Applications of Spark Machining", INSTITUTION OF MECHANICAL ENGINEERS.

* * *

\ttention is drawn to the fact that many of the devices and processes ibed and illustrated in this book are the subjects of valid patents at ne of writing, and many of the trade names mentioned are the ed trade marks of certain manufacturers and traders although al mention may not have been made.

CONTENTS

CHAPTER I

PLASTICS MOULD MANUFACTURING METHODS AND MATERIALS

It is intended in this chapter to review very briefly the principal methods and materials employed in the manufacture of plastics moulds, as a background of manufacturing facilities against which the tool design itself must be considered.

While it is a prime function of the mould design satisfactorily to produce plastics mouldings from the completed tool, it is clearly also paramount that the design adopted should be capable of economic and sound manufacture by the most suitable method and facilities available. It is thus essential that the mould design should be undertaken with a clear idea in mind not only of the overall production method to be adopted —e.g. die sinking, hobbing, electro-deposition, etc.—but also of the manufacturing procedure of every detail of the proposed design. Thus a knowledge and experience of actual toolmaking practice is almost essential if mould designs are to be soundly based. Detailed consideration of this aspect is beyond the scope of this book, but the following short review will serve to show the various methods of approach that are possible, and the chief materials employed.

Machining. By far the largest number of moulds are produced either by machining from the solid, or by building up from machined blocks. Even with forms produced by other methods, such as hobbing and casting, considerable further machining work is required for the final preparation of the form, together with such operations as mounting, and fitting of ejectors.

The steel to be used is either in the form of forged billet or bar stock which is cut approximately to size by means of a power hacksaw or bandsaw.

The initial operation on a mould to be completely produced by machining consists of preparing the steel that is to form the punch or cavity by squaring and bringing to size. For small or medium moulds, this may be done on a shaper, or, in the case of large billets, on a planing machine. Large blocks forming part of the tool construction such as backing plates and packing blocks are prepared in the same manner.

Additionally, when possible, the shaper is often employed for machining the actual form.

Cutting the form itself may be carried out on standard machine tools, either with or without special equipment. Such machines may comprise lathes and milling machines, and the special plant may be diesinking machines or duplicators. The actual method adopted will depend on the equipment provided in the particular shop, the machine capacity available, the complexity of the form, and the cost estimate to be met.

LATHES. The machining of circular or part-circular male tool or cavities can be carried out on a centre lathe, with the spherical or irregular contours being produced by the use of form cutters or profile templates. The method is rapid and straightforward; hence moulds of circular form are generally of substantially lower cost. For completely circular forms, all machining can be carried out in the lathe right through to final polishing, and for these, as for part-circular forms, the tool-post grinder can be used for finishing.

Besides its use for the machining of punches and cavities, the lathe is used for the production of many mould details such as guide pins and bushes, carrier pins, ejectors and push-back pins, core pins, and sprue bushes. It is also employed for the machining of threaded core pins that are used to produce moulded threaded holes. In this respect, it may be important that the pitch of the core pin is compensated to allow for the shrinkage of the plastics material, so that after cooling the moulded thread is of correct pitch. Some lathes have special screw-cutting compensation for this purpose. Such compensation is important where the moulded thread is to engage with a standard screw, but where very short threads are concerned or where the moulded thread is to engage another moulded thread (the shrinkages being equal) this compensation is sometimes ignored.

STANDARD MILLING MACHINES. These machines, generally of the vertical type, are very extensively used in the manufacture of moulds, not only for cutting the form, but also for machining details such as pads, cores, slots and runners. The milling of both male forms and cavities is common practice, but where an area of detail within a deep cavity would necessitate cutters of small diameter, careful consideration should be given to providing this area on a separate pad for subsequent fitting, in order to facilitate machining. With the use of a rotary table, arcs or circular forms may be machined. Other operations, such as milling ribs along a male form and spaced around its periphery may be executed with a dividing head. On a standard milling machine almost any form can be produced by means of the machine movements and the use of additional standard equipment, or by means of templates. Such methods, however, tend to require frequent resetting of the machine or work for changes of profile, direction and angle, or highly skilled attention if complex non-regular forms are to be machined; hence special-purpose machines are employed to advantage in these cases.

ROTARY-HEAD MILLING. The special Kearney and Trecker vertical milling machine illustrated in Figs. 1.1 and 1.2 has mounted upon a slide

Fig. 1.1. A Rotary-head Milling Machine which can be used for the Production of many Types of Mould without Resetting or the Use of Models or Templates. [*Courtesy: E. H. Jones (Machine Tools) Ltd.*]

a cutter which may be offset any distance up to 4 in. from true centre. The slide itself is mounted upon a rotary head which can be manually turned through 360 deg. or continuously power-rotated at speeds between $\frac{1}{6}$ and $2\frac{1}{2}$ r.p.m. By means of the normal table movements of the machine, it is possible to perform conventional cutting in longitudinal or transverse directions. In addition, by means of the rotary head, arcs, radii and circles can be machined, while, in combination with the offset-cutter slide, angular motion is produced. These movements represent the elements of geometric delineation in producing horizontal and vertical lines, circles, arcs and angles which comprise a major portion of many designs. The rotary-head machine, therefore, can reproduce these motions readily and

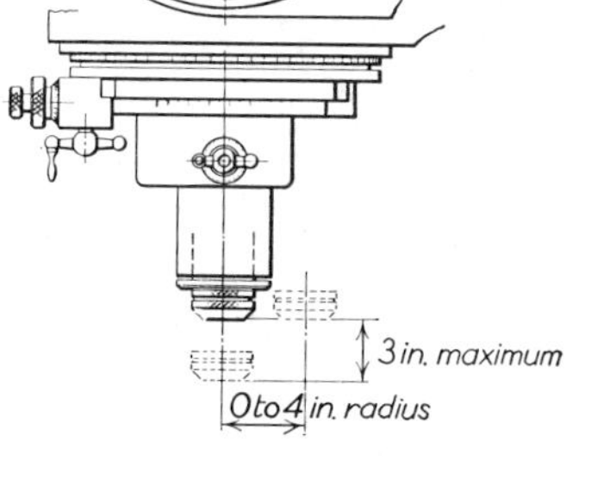

Fig. 1.2. The Rotary-head Movement, showing Method of setting the Spindle off-centre and the Vertical Adjustment of the Spindle on the 2D Kearney and Trecker Rotary-head Milling Machine. (*Courtesy: E. H. Jones (Machine Tools) Ltd.*]

much more rapidly than the multiple set-ups required on conventional milling machines. By the addition of a cherrying attachment controlling a combination of the horizontal spindle slide and vertical movement of the quill, a circular path is imparted to the cutter in the vertical plane. This attachment, together with the rotary head, is able to generate spherical shapes. Angular setting of the attachment can produce any angle in the vertical plane, or, again, by modification of the attachment and in combination with the rotary head, can produce conical forms.

Additional equipment such as rotary tables, dividing heads and attachments for universal and right-angle milling, slotting, and cornering, further extends the capabilities of the machine to produce complex forms without resetting.

DIE MILLING. An example of a special type of milling machine particularly suitable for mould manufacture is the Cincinnati Contourmaster tool

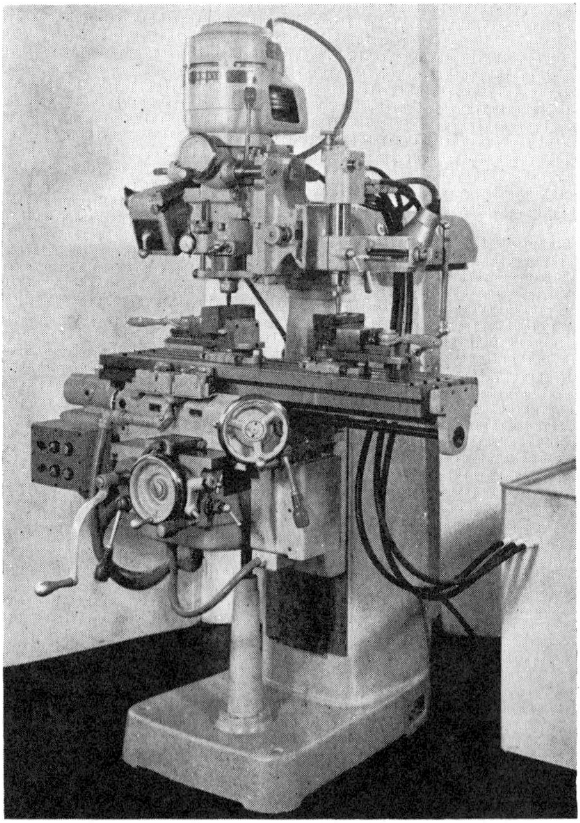

Fig. 1.3. Cincinnati Contourmaster Tool and Die-milling Machine. [*Courtesy: Cincinnati Milling Machines Ltd.*]

and die milling machine which has an unusually large range of movements. This machine is illustrated in Fig. 1.3.

It can be fitted with a tracer mechanism to operate from a model or template. Such equipment is mounted adjacent to the workhead and comprises a very lightly actuated stylus in contact with the model or template. The stylus operates a hydraulic valve which, in turn, actuates a hydraulic cylinder controlling the vertical movement of the cutter spindle. Thus, as the machine table is fed horizontally or transversely, the model or template is passed beneath the stylus, the form causing a rise or fall in the

Fig. 1.4. Duplicating with a Duplex Rotary-table Attachment. The Pair of Circular Tables carrying the Model and Workpiece rotate in Phase, while the Saddle is traversed to provide a Slow, Continuous Cross-feed. [*Courtesy: Cincinnati Milling Machines Ltd.*]

stylus spindle which, consequently, through the hydraulic circuit, causes a corresponding rise or fall in the cutter, which therefore machines a profile identical to that of the original model. Additional equipment includes a duplex rotary table which is intended primarily for circular moulds and is illustrated in use in Fig. 1.4. One table carries the work and the other the master, the work being under the cutter and the master under the tracer. The tables are rotated simultaneously in phase, while, at the same time, the saddle is automatically cross-fed at each revolution. Thus, by commencing the traverse at the outside perimeter, the entire surface of the master is traversed and the machine becomes, in effect, an automatic duplicator.

Tracer control of this type is also fitted to vertical milling machines of larger capacity, so that die sinking from a model or template can be performed on heavier moulds.

DUPLICATING ATTACHMENTS. One of the commonest methods in mould manufacture is that employing a duplicating attachment on a milling machine. Duplicating is particularly convenient when the form to be machined is too complex to be easily machined without frequent resetting on a standard machine, or is non-regular in form (e.g. double-curvature radii); or where the production of multiple forms is required from a single model. These attachments are used for the production of cavities, male tools and master hobs.

The duplicator may be used either with a 3-dimensional model to reproduce both the horizontal and vertical planes simultaneously, or with a template to reproduce in one or the other of these planes. Models may be of metal, plastics or hard plaster, or they may be duplicate impressions or moulded components.

Alternatively, several models can be used, portions of each being used as the work proceeds, to produce a composite form in the duplicated die.

One of the best known duplicators is the Gorton. This comprises a tracer head which is adjustable vertically and mounted parallel to the vertical cutter spindle of the milling machine, the downward feed of both being linked together by means of a shaft. The duplicator table itself is mounted upon the machine table and may be moved in all directions upon ball-bearing slides, under the control of a single lever. The model or template is secured upon an auxiliary table which is placed on the duplicator table below the tracer. The auxiliary table is independently adjustable in both directions by means of micrometer screws, for accurately positioning the model or for imparting movement to the model during duplicating: e.g. if the model is an original impression, it may be desired to incorporate slight changes in the new die. The workpiece itself is mounted upon the duplicating table beneath the machine cutter.

The operator controls the movement of the duplicator table by means of the hand-lever, so that the tracer stylus is passed across the model, or around the template, so that the workpiece attached to the table executes the same relative movement beneath the cutter. At the same time, with the other hand, the operator controls the vertical position of both cutter and tracer stylus. Thus, by constantly passing the stylus across the surface of the model and progressively altering the depth of cutter and stylus, an accurate duplicate of the master is eventually produced at the workpiece. In the case of a horizontal template, the cutter may remain at a set height, the stylus being traced around the template form.

The stylus is proportional in size and shape to the machine cutter being used, since essentially the duplicator produces work of the same size as the model. Slight changes in size can, however, be effected by altering the size of the stylus in relation to the size of the cutter. This may be advantageous if it is desired to cut a tool of slightly different size to the model, thus providing, for example, compensation for wall thickness.

The cutters used are normally single- or double-flute cutters, since

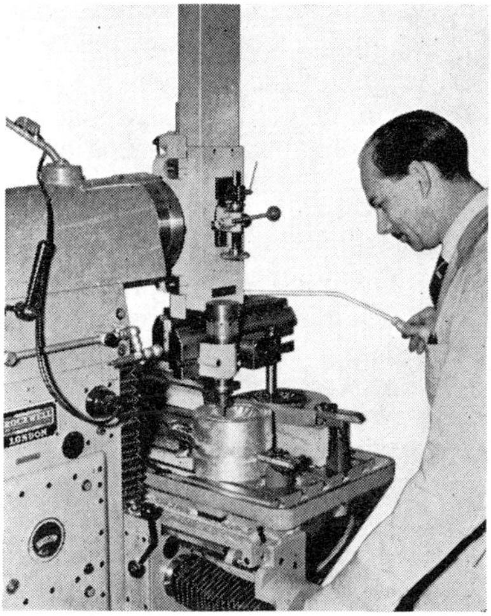

Fig. 1.5. The Thiel Silex 166 Copy-milling Attachment. [*Courtesy: Rockwell Machine Tool Co. Ltd.*]

these can be easily ground to the special sizes and shapes required to produce the desired form.

Another type of duplicating attachment is shown in Fig. 1.5. This is the Thiel Silex copy-milling attachment, and is simple and inexpensive. It carries its own cutter spindle which is mounted upon a slide together with the tracer, and the model is mounted alongside the work on the machine table. Traversing of work and model is effected manually, vertical movement being controlled by adjustable spring pressure.

Automatic Duplicating Machines. This type of equipment is extensively used for the sinking of cavities and the cutting of male forms, particularly of complex forms of medium-to-large size. Essentially, the machine is a horizontal milling machine with the work and model mounted upon a vertical platen. Mounted above the cutter spindle is an automatic tracer which controls the vertical motion of the cutter and the longitudinal and transverse actions of the machine table. By means of this equipment, almost any form can be duplicated automatically from a model or template.

Typical of this kind of equipment, and one of the best known, is the Pratt and Whitney Keller. The workpiece is secured to the vertical platen opposite the cutter, and the model or template is mounted above it, opposite the tracer.

It is essential that all corresponding points on the model and workpiece bear the same relationship as the position of the tracer to cutter, and, to facilitate setting up, the tracer may be adjusted in three dimensions, and the cutter spindle transversely.

If three-dimensional work, male or female, is to be duplicated, a model of any suitable material such as metal, plastics, wood or hard plaster is set up as the master, in conjunction with the automatic tracer head on the machine. The machine movements are set so that the tracer passes across the model in a series of parallel strokes either vertically or horizontally, with the cutter executing similar movements relative to the workpiece.

The length of the stroke is predetermined either by reversing dogs on the horizontal or vertical motion at uniform stroke, or by transverse dogs that cause reversal at the outline of the form or at predetermined depth. Combinations of these methods can be employed, or manual reversal can be effected by the operator. The feed increment occurs automatically at the completion of each stroke, the amount being preset. As the tracer passes over the model, its contact causes axial or side pressure on the stylus which in turn effects electrical contacts in the tracer head, these contacts controlling the various motions of the machine. Thus, as the form causes various changes of pressure on the stylus, the electrical controls operate magnetic clutches on the feed screws of the various machine slides to produce the motion called for by the tracer. The feed screws are operated through variable speed drives, the speed being controlled by the tracer head. Thus, when following a curved surface on the master, the tracer can set the relative speeds of movement on the compound slides to produce the required slope.

In profile work, instead of a solid model a template is employed as the master. The profile tracer is of a type responding to side pressure, which by deflection from the template outline makes or breaks electrical contacts in the head. There are four contacts disposed at 90 deg. corresponding to the right-and-left movement of the table and the up-and-down movement of the vertical slide. These contacts control the magnetic clutches on the machine-slide operating screws. As the direction of the template changes, the tracer makes or breaks these contacts, thereby varying the motions of the machine to reproduce the form. In profile duplicating, the cutter is set to a fixed predetermined depth.

A number of special attachments are also available for this machine. One of these is a two-plate attachment used in conjunction with the automatic and profile tracers in combination, the former following cross-section templates, and the latter an outline template.

With this equipment, three-dimensional work can be executed by means of templates instead of a solid model, although a series of templates must be employed if there is any change in the cross-sectional form. Another attachment allows the profile to be duplicated from a template while the cross-section is copied from a solid model of the cross-section, if the latter is varying and cross-section templates are impracticable.

ENGRAVING. Many moulds include decorative designs such as required on fancy goods, cosmetic containers, etc.; letters, numbers and insignia as used on radio knobs and scales, nameplates, etc.; or complex detailed forms such as exist on toys, models, escutcheons, etc. In many instances, these are conveniently machined on the mould by engraving. Where only a single pad, special cavity or master hob is required, it may be convenient to have this work done by a skilled hand engraver. However, where standard letters or figures can be used, or where a master template or model is readily provided, machine engraving is employed. This requires less skilled labour, is faster, and is generally more uniform.

ENGRAVING MACHINES. In their simplest form, engraving machines comprise a pantograph mechanism carrying a tracer, and a motor-driven cutter spindle. The tracer is manually guided around the profile of the model or master mounted upon the copyholder table, the cutter executing similar movements on the workpiece. By suitably adjusting the setting of the pantograph arms, the workpiece may be cut smaller or larger than the master.

The simplest machines are two-dimensional or letter-engraving machines in which the pantograph is free to operate in the horizontal plane. These are used for the engraving of raised or sunk characters, or designs from a master template or standard type. The masters are usually in brass or steel on an enlarged scale, but machines commonly operate from a 1 : 1 reproduction down to reductions of 1 : 50. Depth of cut is achieved by adjusting the cutter feed as work proceeds. The cutters used are usually of the single-lip type, of appropriate shape to produce the required profile; but straight and spiral-fluted end-mills and ball-end cutters are also employed.

With regard to curved surfaces, provision has to be made for the cutter to follow the curvature of the workpiece, in order to maintain an even depth of cut. This is achieved by mounting a stationary former guide directly above the tail end of the spindle, which is kept in upward contact with the former by means of a spring. The former is of opposite curvature to the master, and is positioned directly vertically in line with it. As the cutter is moved horizontally by the pantograph, the varying curvature of the former guide causes the cutter spindle to rise or fall and maintain an even depth of cut in the corresponding curvature of the workpiece.

The engraving of lettering and designs around cylinders and cones is also possible, but care must be exercised to ensure that such engraving is performed in the line of draw of the mould and does not produce undercuts in the moulding which would prevent the ejection of the component from the tool.

In addition to two-dimensional engravers, pantograph machines are also available which perform in three dimensions, so that the cutter duplicates also the vertical form of a model. This is achieved in different machines by various methods. One type employs a pivoted lever operating in the vertical plane and connecting the vertical movement of the tracer to the vertical movement of the cutter—the ratio relation between these movements being controlled by the setting of the bar. In another type, the whole horizontal pantograph can be tilted vertically from a pivot at the rear, carrying the tracer and cutter with it. Alternatively, the pantograph movement may be carried on vertical slides on the column of the machine, to rise and fall as required.

These three-dimensional machines are made in a variety of sizes, both for straightforward design engraving and for work on small-to-medium mould cavities. The well-known Deckel machine, for instance, will deal with moulds up to 550 lb. weight and is capable of reducing or enlarging work within the ratios of 1 : 1.5 and 1 : 10. The area covered by the pantograph depends upon the ratio being used, and on these machines varies from a rectangle of $5\frac{7}{8}$ by $13\frac{1}{4}$ in. at a 1 : 1.5 ratio, to a rectangle of

$2\frac{5}{32}$ in. by 2 in. at a 1 : 10 ratio. A three-dimensional pantograph machine is shown in Fig. 1.6.

This is a Deckel GK21 machine and, as shown, is fitted with the rough-milling attachment, in which the copying stylus is mounted through the adjustable slide on the bridge across the copy table. An auxillary cross-slide is mounted on the copy table itself beneath the bridge. The model is traced by the stylus by operation of the feed screws, and the work is rough-milled with fast-cutting end-mills. Finishing cuts are made with the normal fine-cutting engraving tools, by removing the bridge piece and operating the pantograph manually. Both two- and three-dimensional work can be performed.

PANTOGRAPH DIE-SINKING MACHINES. The three-dimensional pantograph engraving machine has been developed by increased massiveness

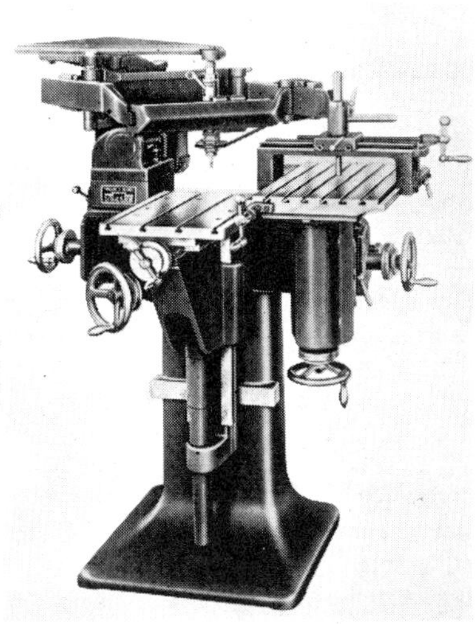

Fig. 1.6. Deckel GK21 3-dimensional Pantograph Engraving Machine shown fitted with the Rough-milling Attachment. [*Courtesy: Burton, Griffiths & Co. Ltd.*]

and additional operating features into a die-sinking machine. This type of equipment is capable of undertaking all normal cavity and punch machining—particularly that having complex detail—although it is often convenient to perform the initial rough-machining operations on standard machine tools and to complete the finishing and detailed cutting on the pantograph machine.

The Deckel KF 12 shown in Fig. 1.7 is a machine of this type and will accommodate moulds up to 1,300 lb. weight. The standard machine works on a normal ratio of 1 : 1 with a maximum cutting area of $19\frac{3}{4}$ by $15\frac{3}{4}$ in., but an enlarging and reducing attachment is available within the ratios of 1 : 1.5 and 1 : 4.

Additional equipment includes an optical tracing unit for rough milling with feed from the worktable screws, and a symmetrical attachment by which opposite-hand symmetrical duplicates of cavities or punches can be produced from a single master. Available with this machine also is a pattern-forming device by which many cylindrical, tapered, spherical,

Fig. 1.7. Deckel KF12 Pantograph Die Sinker duplicating a Die Cavity from a Model. [*Courtesy: Burton, Griffiths & Co. Ltd.*]

prismatic and plane surfaces can be generated directly without the use of a master. This equipment has a maximum cutting radius of 2.96 in. and 4 in. linear at one setting, and enables shapes within the capacity of the attachment to be produced direct from the drawing.

Steels for Machine-cut Moulds. The prime requirements in materials for plastics moulds are adequate strength, resistance to the action of the moulding material, and facility in working: cost may also be a consideration. A number of steels are now available for plastics moulds with varying properties, and whose selection is determined by the standard of performance required under the foregoing headings.

The strength requirements arise from the high pressures employed in moulding, with normal cavity pressures in injection moulding of from 5,000 to 20,000 lb. per sq. in. Closing tonnages are dependent on the press employed, but on very large dies may reach 3,000 tons. Resistance to these considerable forces is a function both of the mould design and of the inherent strength of the steel used. Not only is this strength consideration necessary for the main cavity or punch, but is just as important for cores and mould details, since these, too, are exposed to severe pressures. The necessity to withstand bending stresses and also the direct compressive loads requires that the core-strength of the material should be high while sufficient surface hardness should be retained.

Adequate surface hardness is important so that the following requirements can be met:

(a) Ability to withstand the erosion effects of stiff thermoplastic materials in tools where flow is restricted or obstructed.

(b) Resistance to wear in tools—particularly on long runs.

(c) Capability of the mould surface to take and retain a high polish—essential to the production of a fine finish on the moulding and to facilitate ejection.

In some moulding concerns, moulds other than those for use with nylon are not generally specifically hardened, except the moving parts. However, many other concerns harden all moulds, often for the additional reason of minimising damage during tool setting or operation.

The choice of method of hardening can vary with the type of steel employed, the hardness required for the particular application, and the intricacy and accuracy of the individual tool. Direct-hardening by air or oil is used and often carburizing can be employed. Air-hardening is generally employed on intricate tools, or to ensure minimum distortion, but the more elaborate case-hardening treatment, frequently the cyanide process, is used for maximum hardness.

It is essential that the steel should harden reasonably clean; particularly with intricate moulds, and in this connection, oil-hardening tends to produce less scale than air-hardening. Freedom from excessive scale is essential both to facilitate and to reduce subsequent polishing of the mould, and to maintain dimensional accuracy. Minimum distortion on hardening is similarly desirable. Finally, the ability of the steel to be readily and simply annealed is important, since this will enable any later modifications to the tools to be the more readily executed.

It is important that the mould material should possess good machining properties, and, to this end, the steel is generally supplied in the annealed condition (Brinell 170/286) for hardening after machining. However, many of the steels can be supplied in the heat-treated condition (Brinell 270/320) which still allows machining to be carried out and yet permits the immediate use of the mould without further processing. While this method provides greater difficulties in machining and a lower mould hardness, it does avoid the need to harden after machining, and thus eases the problem of distortion and polishing. For increased wear resistance, these moulds are frequently chromium-plated.

A final consideration is the ability of the steel to withstand chemical action of certain plastics materials, particularly P.V.C. This is often achieved by utilizing normal mould steels which are subsequently chromium-plated, but an alternative practice is to employ stainless steels. This latter procedure does have the advantage of allowing subsequent mould modifications to be carried out without first stripping off any plating, and overcomes the difficulty of plating in deep recesses. In general, however, stainless steels are more difficult to machine than other alloy steels.

The classes of steel normally employed for moulds are as follows:

(A) NICKEL-CHROMIUM STEEL AND NICKEL-CHROMIUM-MOLYBDENUM STEEL

These steels, whether direct-, oil- or air-hardening, or carburized, are probably the most widely used for injection moulds. They are supplied in billets and bar stock. Most steel producers supply alloys of this type specifically for mould manufacture, normally in the annealed condition, but also in the heat-treated condition for subsequent direct machining and immediate use. The 1½- and 2½-per-cent nickel-chromium steels have excellent strength with good fatigue resistance. The 1½-per-cent alloy is the more readily machined, but the 2½- and 3-per-cent varieties have increased toughness. The 4-to-4½-per-cent nickel-chromium steels have maximum strength and are probably the most widely used.

Hall and Pickles Ltd.—"Versatile", a 1½-per-cent nickel-chromium steel for oil or cyanide hardening and supplied annealed to a hardness of 197/255 Brinell.

English Steel Corporation Ltd.—"PLMB/1", a 1½-per-cent nickel-chromium-molybdenum steel, oil-hardening, and supplied heat-treated to a tensile of 60/65 tons for direct machining and immediate use; "PLMB/2", a 2½-per-cent nickel-chromium-molybdenum steel for oil hardening and supplied heat-treated to a tensile of 70/75 tons for direct machining and immediate use; "PLMC/1", a 2½-per-cent nickel-chromium-molybdenum steel for oil hardening, supplied annealed to 277 Brinell, and when hardened, quenched in oil and tempered has a hardness of 490/640 V.P.N.

Wm. Jessop & Sons, Ltd.: J. J. Saville & Co., Ltd.—"Jessop-Saville G.O.", a 3-per-cent nickel-chromium steel that is oil-hardening and carburizing. It is supplied annealed, or hardened and tempered to a tensile of 60 tons for direct machining and immediate use. The carburized case has a hardness of 60 Rockwell-C.

Edgar Allen & Co., Ltd. "A.13 Nickel-Chromium". This when oil-quenched and tempered yields a hardness of 520/540 VPN; when carburized and oil-quenched, a hardness of 700/760 VPN. It is supplied annealed to a hardness of 170/180 Brinell. "A.100 Nickel-Chromium-Molybdenum" which is supplied annealed to 190/220 Brinell. When oil-quenched and tempered, its hardness is 521/540 VPN; when carburized and oil-quenched, 700/740 VPN.

Hall & Pickles, Ltd. "Larport", a 4¼-per-cent nickel-chromium-molybdenum steel that oil- or air-hardens up to 53 Rockwell-C, and cyanide-carburizes up to 60 Rockwell-C. It is supplied annealed to 241/286 Brinell for subsequent heat treatment, or already treated to a hardness of 269/321 Brinell (60/70-ton tensile) for direct machining and immediate use.

English Steel Corporation. "PLMC/2", a 4-per-cent nickel-chromium-molybdenum steel that is air-hardening to 490/630 VPN, and is supplied annealed to 286 Brinell; "PLMC/3", a 4-per-cent nickel-chromium-molybdenum steel that case-hardens to 800 VPN, and is supplied annealed to 277 Brinell.

Richard W. Carr and Co., Ltd.—"Plutonic 552", a 4¼-per-cent nickel-chromium-molybdenum steel that is oil- or air-hardening to a hardness of 441/519 VPN or carburizes to 678/723 VPN.

Wm. Jessop & Sons, Ltd. : J. J. Saville & Co., Ltd.—"Jessop G.I. Special", and "Saville S.V.L.", 4.1-per-cent nickel-chromium-molybdenum steels that air-harden to 477 Brinell, oil-harden to 514 Brinell, case-harden to 60 Rockwell-C, and are supplied annealed to 269 Brinell.

(B) High-carbon High-chromium Steels

These steels with a general analysis of 1.25–2.5 per cent carbon and 12–14 per cent chromium are not so extensively used for machine-cut moulds as the nickel-chromium varieties but are suitable for moulds that require minimum distortion and high hardness for maximum abrasion resistance. They are not, however, so suitable for portable tools or others that are liable to shock. Air hardening is generally more applicable to small sections. Suppliers of these steels and their brands are as follows:

Hall and Pickles, Ltd.—"Vital", oil- or air-hardening to 57/63 Rockwell-C and supplied annealed to 217/255 Brinell.

Edgar Allen & Co., Ltd.—"Double Six", oil- or air-hardening to 720/860 VPN, and supplied to Brinell 230/250.

English Steel Corporation Ltd.—"Super C 12" and "CD", oil- or air-hardening to 650/860 VPN, and supplied annealed to 241/248 Brinell.

Richard W. Carr & Co., Ltd.—"Motor 235", oil- or air-hardening to 698/778 VPN, and supplied annealed.

Wm. Jessop & Sons Ltd. : J. J. Saville & Co., Ltd.—"Jessop Alloy C" and "Saville W.P.S.", oil- or air-hardening to 534/601 Brinell; supplied annealed.

(C) Corrosion-resisting and Stainless Steels

Edgar Allen & Co., Ltd.—"Imperial Stainless Mould Steel", oil-hardening to 520/530 VPN; supplied annealed to 170/180 Brinell, for machining and subsequent heat treatment; or may be supplied heat-treated to 280/320 Brinell for direct machining and immediate use.

Wm. Jessop & Sons Ltd. : J. J. Saville & Co., Ltd.—"Jessop-Saville R.I.A.", oil- or air-hardening; carburizes to 60 Rockwell-C; supplied annealed.

Richard W. Carr & Co., Ltd.—"Pluteus 1000", air-hardening to 286/302 VPN; supplied annealed. "Pluteus 1001", oil-hardening to 454/627 VPN; supplied annealed for machining and subsequent heat treatment, or supplied heat-treated to 300 Brinell for direct cutting and immediate use.

English Steel Corporation Ltd.—"PLME/1", supplied annealed for machining and subsequent heat treatment, or heat-treated to a tensile of 60/65 tons for direct cutting and immediate use.

Hall and Pickles Ltd.—"Resource", oil-hardening to 53 Rockwell-C; may be carburized for higher surface hardness. Supplied annealed to 217/262 Brinell for machining and subsequent heat treatment, or, alternatively, supplied heat-treated to 269/321 Brinell for direct cutting and immediate use.

OTHER MOULD STEELS. Some other mould steels are employed to a lesser extent for specific applications or to suit mould-manufacture preference. These include carbon-manganese, chromium-vanadium and carbon-tungsten-chromium steels. In addition, mild steel is not infrequently employed for use in injection moulding, particularly with low-density polythene. The material is inexpensive and quite satisfactory for non-critical moulds.

Hobbing. The hobbing process is a method of producing cavities in which a hardened male form, or hob, is forced under great pressure into a cold block of steel. The hob is made to the final size of the moulding required plus the estimated shrinkage of the plastics material within the mould cavity. Upon extraction of the hob from the mould blank, a female mould is produced which is a faithful reproduction of the original master and whose surface requires only final polishing. The operation itself is carried out in a hobbing press which is of massive rigid construction, built with total loading capacities of 50 to 5,000 tons, and fitted with low and

Fig. 1.8. Sack and Kiesselbach Hydraulic Hobbing Press. [*Courtesy: Rockwell Machine Tool Co. Ltd.*]

high pressure control systems, and with the operating pressure clearly indicated on a dial. A micrometer gauge also is fitted to indicate the amount of penetration of the hob into the blank. A typical hobbing press is shown in Fig. 1.8.

The process of mould hobbing is indicated in Fig. 1.9. The blank *A*, retained in the hobbing ring *B*, is placed upon the bottom platen of the press with the hob *C* above it. As the press is closed, the hob *C* is steadily forced into the blank as indicated, the displaced metal being caused to flow up and around the hob. Copper sulphate is used as a lubricant, if deemed necessary.

An abnormal rise in pressure on the gauge as the hob is sunk indicates that the steel is work-hardening and, when this occurs, removal and annealing of the blank is necessary before proceeding further.

The hobs require steels of great strength and good wear resistance, and must be highly polished before use. As shown in Fig. 1.9, the hob is carried beyond the form length required in order to provide an overrun x,

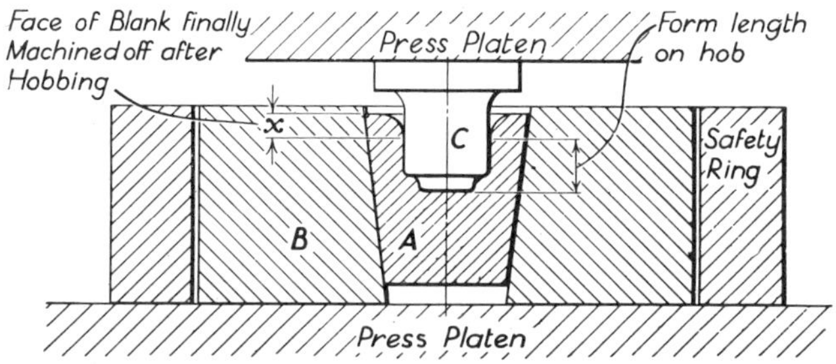

Fig. 1.9. Hobbing the Cavity in the Die Blank.

which is necessary to clear the build-up of steel from the blank as it flows up around the hob. The hob is then run out to the backing plug which is provided to be set against the press platen. It is also used to remove the hob after sinking. The hob must be free of sharp corners and designed for maximum strength.

The blank itself may be produced from steels of varying hardness as required for the particular usage, but must be capable of easy hobbing and receiving a high polish. Hard steels, for instance, are required in cavities having raised lettering or detail. In proportion, the diameter of the blank varies between two or three times the diameter of the hob, and the length is about twice the depth of the cavity. In form, the blank may be parallel or may be tapered—one reason for the latter being the ease with which the blank may be freed from the hobbing chase. The taper on the blank is normally 3 deg. The top of the blank should be well polished prior to hobbing, since any marks at points of low metal flow will not be ironed out on sinking, and may well be difficult to polish out subsequently.

The pressure required to force the hob into the blank generally varies between 100 and 200 tons per sq. in. of impression. In some cases, the back of the blank may be relieved to assist flow and reduce pressure. It may be noted that the hobbing process itself improves the grain of the steel on the hobbed cavity surface, and hence confers improved finish and longer wear resistance.

After hobbing is completed, the top of the blank is machined back by the amount x (Fig. 1.9) to the correct depth, and the outside of the blank is machined to fit the bolster. On occasions, hobbings in the harder types of steel such as nickel-chromium are used without final heat treatment, but otherwise they are generally hardened, usually by carburizing, and polished prior to final fitting in the tool.

The blank is retained within the hobbing ring or chase, of which the centre hole is parallel or tapered as required. Should the ring be considerably deeper than the blank in a parallel chase, a backing block may be used. If necessary, a mild-steel outer safety ring may also be placed round the hobbing chase.

The use of the hobbing process is advantageous in a number of circumstances. It is of particular advantage in multi-impression tools, since, once the master hob has been made, sinking of the cavities from the one hob is reasonably rapid, and certainly a great deal faster than machining out each impression. Close uniformity of each cavity is additionally ensured. Again, in many cases, it is far easier to machine and polish a male form on a hob than to die-sink and polish the corresponding cavity. This is clearly demonstrated in the common practice of hobbing radio-knob forms, of which an example is given in Fig. 1.10.

Hobbing is frequently used where recessed lettering or design from the female tool is required on the final moulding. This necessitates raised lettering on the female tool, a matter of considerable difficulty, especially if the final moulding is in transparent material. If the cavity is machined, such lettering can be machined out from the solid metal to leave the characters proud. However, this machining and especially the subsequent polishing of the mould surface at the root of the raised design are a matter of considerable difficulty. To overcome this, if hobbing is not practicable or available, separate characters can be let into the mould, provided these are of suitable design and sufficient size. If, however, the cavity is to be hobbed, it is a matter of comparative simplicity to engrave the hob and thus produce a raised impression in the mould during the hobbing operation.

The production of a typical radio knob by hobbing is indicated by Fig. 1.10. This shows:

(*X*) the component to be produced drawn in section and plan, with an enlarged view of the knurl;

(*Y*) a plan and side view of the master hob used to sink the blank; and

(Z) a side section of the hobbed and finished die insert produced by the hob.

It will be noted that the knob is produced in an opaque material with the lettering "VOLUME" and the direction arrow moulded and viewed on the front face of the component, the characters being recessed into the moulding. To achieve this, the characters are engraved into the hob the same way round as shown in the plan view. When this is hobbed into the mould blank, the lettering will be reversed and raised. On moulding, the lettering will again be reversed to read correctly, and recessed into the surface as required. If, however, a moulding is to be in a transparent material with the characters on the back and viewed from the front through the thickness of the material (e.g. three-dimensional badges), then the opposite applies. That is, the characters are engraved in reverse on the hob to produce the design the right way round in the die.

Also illustrated by Fig. 1.10 is the case where the position of knurling

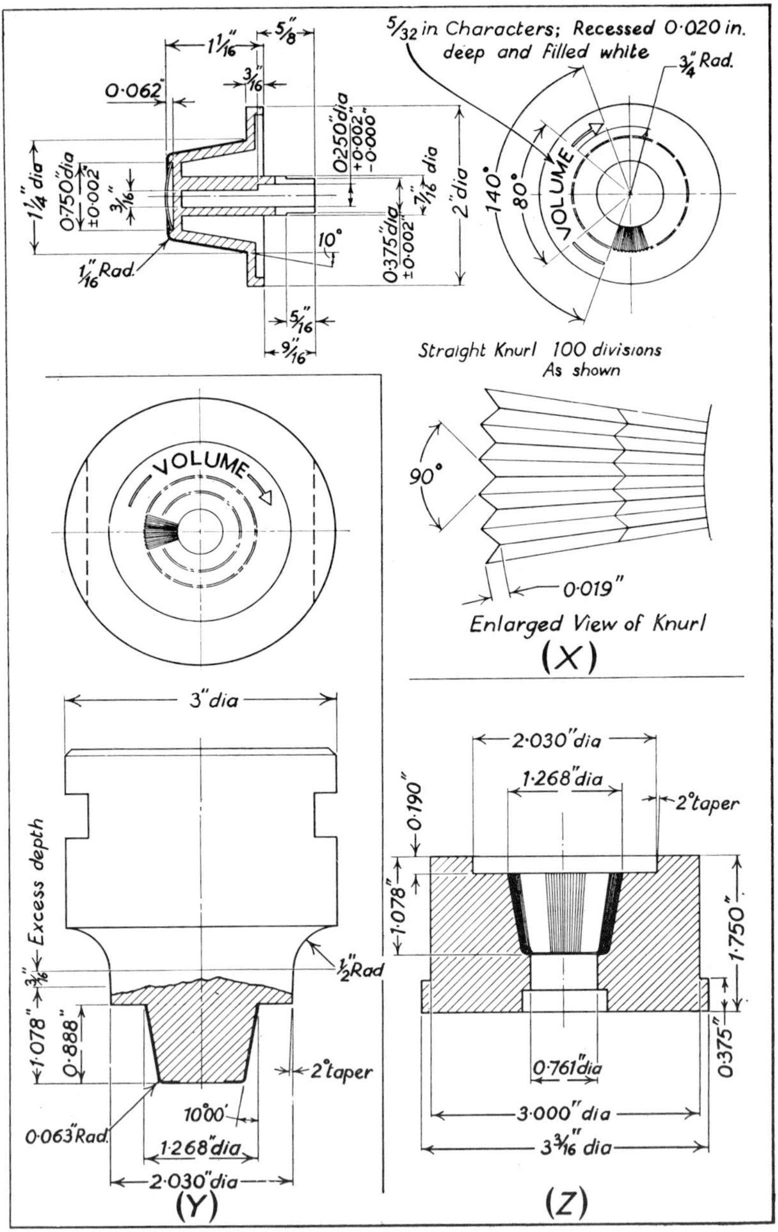

Fig. 1.10. (X) A Moulded Knob in Nylon with 0.015 per cent Shrinkage that is produced from a Hobbed Cavity; (Y) The Hob from which the Cavity was produced; (Z) Die Cavity dimensioned to allow for Shrinkage.

called for on the component would be almost impossible to machine in the female cavity, but which can be readily produced by hobbing.

The knob is intended to be moulded in nylon having a shrinkage of 0.015 in. per inch; hence the hob dimensions are made correspondingly larger. The $\frac{5}{32}$-inch characters in the final knob are 0.020 in. deep and filled in with white. The corresponding characters in the hob are engraved to the same depth.

Cavities intended for hobbing must generally be of balanced design, as an unbalanced form may easily result in a broken hob. It is, of course, possible to combine machining with hobbing, so that features such as open-sided cavities may be obtained by machining away one wall of the cavity after hobbing.

Alternatively, where raised lettering or design is required on the final moulding from the female mould, this may be achieved by engraving the cavity after hobbing. Draft must be provided on the hob, and the minimum values for this vary between 0.005 and 0.010 in. per inch per side. Where lettering or design is to be incorporated by engraving the hob, care should be taken that the depth is not too great in relation to the width, since difficulty may be encountered in getting the flow of steel of the blank to fill the recessed hob. While fluting can be hobbed up to angles of 30 deg. or more from the vertical, lettering or designs are generally difficult to hob on a sloping surface. If the lettering is to be raised in the cavity (and hence recessed in the hob) there is danger that the hobbed lettering will shear off under the continuing flow of the blank metal. Should the characters be raised on the hob, these may similarly be sheared off the master hob itself.

Any lettering that is included on an angled surface is generally limited to 45 deg. from the vertical and, it should be borne in mind, that the design should be such that no undercuts occur on the hob. It is sometimes convenient to extend portions of the recessed lettering or design to add strength to the raised hobbed form which otherwise might be sheared off. This excess metal is subsequently machined off the finished cavity. On occasions, such lettering or design may be added by means of a second hob after the main cavity has been sunk. Two typical hobbings and their respective hobs are shown in Figs. 1.11 and 1.12.

A commonly accepted limitation for straightforward work is that the depth of hobbing should not exceed twice the diameter, but considerably deeper hobbings are performed by specialists. If the ratio of depth to diameter is 1 : 1, it is normally necessary to sink the hob more than once, annealing the blank and repolishing the hob between sinkings. Should the design necessitate a hollow form on the front of the hob, it is advisable to ensure that this does not run out near the edge, since pressure of the metal in this recess during sinking may quite likely break the hob. Similarly, due to the high pressures involved, it is not possible to hob complex or delicate forms; features of this sort may sometimes, however, be added to the hobbed cavity by subsequent machining.

The hobbing process itself is of great value in the manufacture of plastics moulds, but, while the method is of comparative simplicity, great experience is necessary before success and reliability can be achieved.

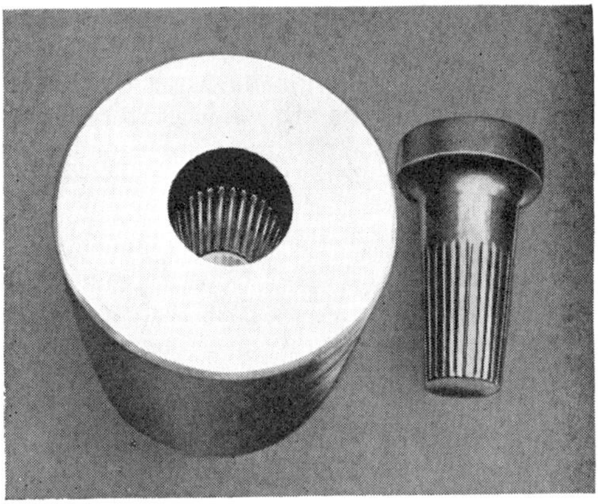

Fig. 1.11. Hobbed Die Insert and Master Hob. The Blank is 8-in. diameter and 8-in. high, with a $4\frac{11}{32}$ in. deep Cavity. The Blank is relieved at the Bottom and on the Cylindrical Face and hobbed in Four Stages without intermediate Annealing. The Blank is of PREMO Steel and the Hob of UHB SVERKER - 3 Steel. [*Courtesy: Uddeholm Ltd., Birmingham.*]

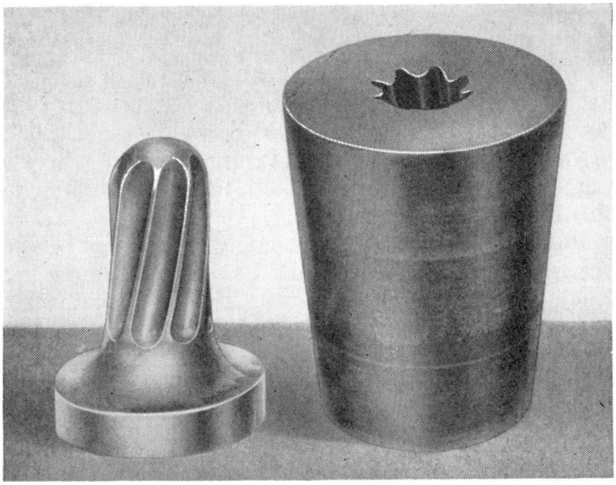

Fig. 1.12. A Large Die Insert after Hobbing and prior to Final Machining and Fitting to the Bolster, together with the Master Hob. The Blank is 7.6 in. diameter and 8.9 in. high, and the Splined Cavity is 2.75 in diameter by 5 in. deep. The Maximum Hobbing Pressure was 700 tons; approximately 110 tons/sq. in. on Projected Area. Hobbed in four Stages with Intermediate Annealing at 750 deg. C. Blank in PREMO steel and Hob in REGIN type Steel. [*Courtesy: Uddeholm Ltd., Birmingham.*]

An interesting demonstration of metal flow during hobbing is afforded by the illustration, Fig. 1.13. Here, the hobbing blank was prepared in two halves and ruled on the abutting faces as shown on the right. After hobbing, the distortion and flow of the metal were indicated by the distortion of the ruled lines as shown at the left.

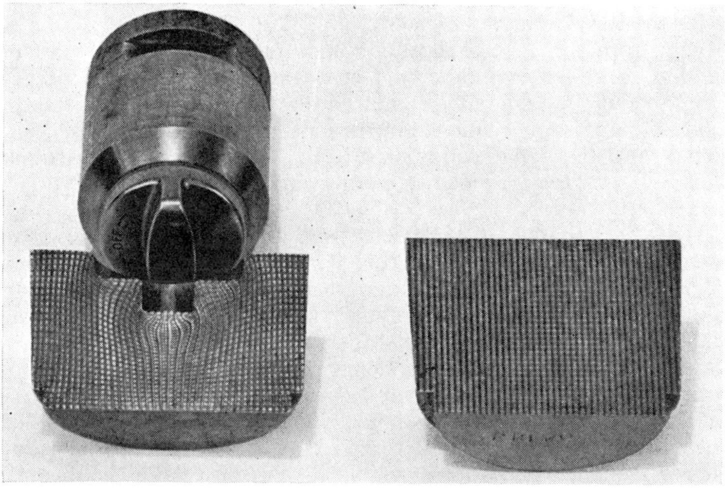

Fig. 1.13. Demonstration of Metal Flow under Hobbing Pressure.
[*Courtesy: Ekco Plastics Ltd.*]

Steels for Hobbing

STEELS FOR BLANKS. The prime requirements in steels for hobbed cavities are softness and ease of flow under the hob, ability to harden with a hard surface and a tough core, freedom from non-metallic inclusions, good reproduction of detail, and ability to take a high polish. In practice, all these ideals cannot be completely attained in one steel, and a compromise has to be made, according to the particular conditions, between the low-carbon and the alloy steels.

The former have excellent ease-of-hobbing properties without work-hardening, and such steels are normally supplied in the dead-soft condition. For this reason, they are extensively used, both for general work and for deep and slender cavities. Their principal drawback is that, although they may be case-hardened before use to a hardness of about 64 Rockwell-C, the core remains soft and may not be suitable for use under heavy pressures. Also, the heat treatment, generally with water-quench, is liable to introduce distortion and brittleness in thin sections or at sharp edges. Apart from these special low-carbon hobbing steels, general-purpose mild steel, case-hardened, is also used by some toolmakers for hobbing blanks.

The advantage of alloy hobbing steels is that, while providing a hard surface of some 61/65 Rockwell-C, they also possess a high core strength, and hence are more suitable for heavy moulding pressures. On the other hand, ease of hobbing is reduced because of the greater hardness of the blank, while work-hardening may also occur during sinking, causing the hobbing pressure to rise excessively. The blank must then be withdrawn again and annealed, before continuing. For these reasons, alloy steels are used for the less deep types of cavity.

Apart from the special-alloy hobbing steels, the normal oil- or air-hardening nickel-chromium mould steels are used to a limited extent, but due to their considerably higher hardness in the annealed state, they are suitable for only shallow impressions. They may be used in the hardened and tempered condition to give a surface hardness of 49/50 Rockwell-C, or case-hardened to yield a surface hardness of 59/60 Rockwell-C.

As has been stated, the choice of steel for a hobbing blank must be dictated by individual circumstances. The requirement of ease of flow is, of course, closely allied to the strength of the hob to be used, both in design and material, since it is clearly inadvisable to attempt to sink a delicate hob into a blank of high hardness.

Such care of the hob is of paramount importance because of its expense and, in many cases, it is preferable to use easily hobbed soft cavities and replace them when damaged in moulding, rather than hazard the hob by sinking it into a hard blank. This choice is, of course, dependent on the particular design, since alloy-steel blanks can be readily hobbed if the master is sufficiently robust.

Both the low-carbon and alloy steels are normally carburized after hobbing, although it is common practice to confine the hardened case on cavities for injection moulding to providing only a scratch-resistant surface.

Stainless or corrosion-resisting steels are available for blanks to be used with corrosive moulding materials.

A short list of typical hobbing steels is as follows:

Low-carbon Steels

Edgar Allen and Co., Ltd. ..	"Plastic Hobbing Steel" (Carburize or cyanide; case-hardening; oil- or water-quench) Supplied annealed to a hardness of 100 Brinell
Wm. Jessop and Sons, Ltd.	
J. J. Saville and Co., Ltd. ..	"A.00 Hobbing Steel" (Carburize and water-quench)
Hall and Pickles Ltd. ..	"Casona" (Carburize and water-quench) Supplied annealed to a hardness of 95 Brinell
Uddeholm Ltd.	"Uhb, Forma. 1" (Carburize and water-quench) Supplied annealed to 85 Brinell

Non-alloyed Mild Steels

Richard W. Carr and Co., Ltd...	"Plutonic 151" (Case-hardening and water-quench)

Alloy Hobbing Steels

English Steel Corporation Ltd.	"PLMA/2: 1-per-cent chromium case-hardening steel" (Carburize, oil- and water-quench) Supplied annealed to 120 Brinell "PLMA/5: 5-per-cent chromium-molybdenum-vanadium steel" (Carburize and air-harden) Supplied annealed to 125 Brinell
Wm. Jessop and Sons Ltd. J. J. Saville and Co., Ltd. ..	"H49 Alloy Hobbing Steel" (Carburize and air-harden)
Uddeholm Ltd.	"Premo"—A special low-carbon alloy steel for deep cavities (Case-hardening; cyanide or carburize; oil- or air-quench) Supplied annealed to 110 Brinell

Stainless or Corrosion-resistant Hobbing Steels

Edgar Allen and Co., Ltd. ..	"Imperial R.I.O. Stainless Hobbing Steel" (Case-hardening; cyanide or carburize; oil-quench) Supplied annealed to 135/145 Brinell
Wm. Jessop and Sons, Ltd. J. J. Saville and Co., Ltd. ...	"R.I.A. — Non-corroding Mould Steel" (Carburizing or oil-hardening; for shallow impressions)
Uddeholm Ltd.	"Uhb Stainless 31" (Oil- and air-hardening)

STEELS FOR MASTER HOBS. The steel used for the master hob must obviously be of great strength, both in compression and tension, in order to withstand the high crushing and distorting loads encountered in sinking the cavity. In the annealed state, the steel should allow ready machining, and should harden throughout with a dense structure, and without distortion or excessive scale. In the hardened state, it should be capable of receiving a high polish, and should possess maximum wear resistance.

To fulfil these requirements, the steel used most widely for master hobs is the high-chromium high-carbon type, with carbon from 1.5 to 2.5 per cent, and chromium from 12 to 13 per cent. This may be oil- or air-hardened up to 57/63 Rockwell-C, and in the annealed state has a hardness of 217/255 Brinell.

Among steels of this type are:

Hall and Pickles Ltd.	"Vital-Alloy Die Steel"
Edgar Allen and Co., Ltd. ..	"Double-Six Steel"
English Steel Corporation Ltd...	"Super C.12"

Wm. Jessop and Sons Ltd.	
J. J. Saville and Co., Ltd. ..	"CD"
	"Alloy C"
	"W.P.S."
Richard W. Carr and Co., Ltd...	"Motor 235"
Uddeholm Ltd.	"Uhb Sverker 3"

Some less expensive steels for master hobs are of the carbon-tungsten type. These are oil-hardening to give 60/64 Rockwell-C. Among steels of this type are:

Edgar Allen and Co., Ltd. ..	"Imperial Oil Hardening Steel"
Richard W. Carr and Co., Ltd. ..	"Motor 09B"
Uddeholm Ltd.	"Uhb Regin 3"

Finishing and Polishing of Moulds. After the completion of machining on the mould, the remaining operations comprise:

1. Bench fitting;
2. Finishing of the mould surfaces;
3. Hardening;
4. Final polishing.

Bench fitting includes the assembly and fitting together of mould blocks, plugs, blades and inserted pads in the case of built-up moulds; the fitting of ejectors, moving cores and pins, threaded pins or sleeves and insert carriers; and, finally, the mating of the flash faces. In addition, some extra hand working may be necessary on the machined form, such as, for example, the completion of square corners from the radius left by the cutter, and the manual blending of radii.

As left by nearly all machining operations, the surface is covered with cutter ridges, burrs and machine-tool marks. All these within the mould area must be removed and the surface brought to a high finish. A polished surface is essential in a plastics mould to ease ejection of the mouldings, to provide a high finish on the moulding (particularly important on mouldings in clear materials), to facilitate the flow of the plastics material within the mould, and to minimize the danger of local abrasive or chemical attack. Similarly, a high degree of polish is necessary on master hobs to facilitate the flow of metal and to ensure a high finish in the hobbed cavity.

The initial removal of ridges and machining marks is effected with scrapers, files, special grinding and deburring wheels, rotary files and burrs, and abrasive cones and discs. While some of this work is carried out manually, much is performed with rotary flexible-shaft equipment or with hand-held tools having a power-driven reciprocating motion and into which files or abrasive shoes, etc., can be fitted. Filing is performed with needle or precision files, or with riffler files. The former are straight files made in various qualities, cuts and lengths, and with round, knurled or square handles. They are made in a wide variety of cuts and cross-sectional shapes. Rifflers are shaped double-ended files, again available in various qualities, cuts and lengths, and with the cutting ends in a large range of shapes, profiles and angles. These enable filing to be carried out in any corner, recess, slot or radius. Standard files are used for soft steels, while diamond files are used for hardened steels in the repair and

modification of completed and hardened tools; diamond and sapphire discs, and similar tools, are used for the same purpose.

With circular moulds, finishing is considerably eased, since the initial finish is higher and the polishing operations can be facilitated by spinning the tool.

After the initial operations have been completed, honing and lapping follow. In honing, the tool used is itself abrasive, whereas, in lapping, an abrasive fluid or paste is introduced between a soft lap and the workpiece.

Synthetic honing stones of either carborundum or carbide are supplied in sticks of various lengths and sections, in a number of grits and in specific grades such as "deburring", "finishing" and "polishing". Other hones are of natural stone such as Washita or Arkansas. Synthetic stones produce the fastest cut, and it is common to commence with these and to finish with a Washita stone, and then finally with Arkansas.

Lapping is carried out with laps of cast iron or copper, and in the later and polishing stages, with felt or wood. As with honing, the operations may be performed manually, or with rotary tools into which hobs and cones can be fitted, or reciprocating hand-held power-driven tools into which pads and shoes are fitted. The lapping compound may be carborundum powder, silicon carbide (for fast cutting or hardened steels) or aluminium oxide (for slow cutting and use on soft steels, or for polishing). All are supplied in various grades in fluid or paste form, the vehicle being oil or water. In one range of a proprietary brand of lapping compound, it is

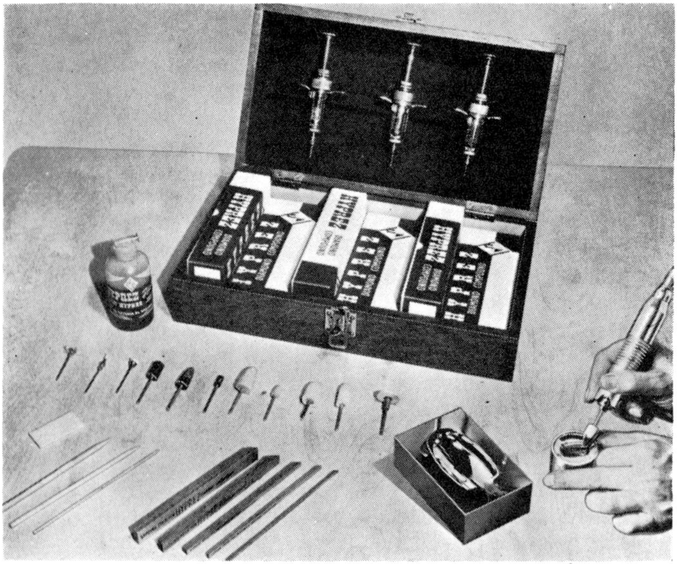

Fig. 1.14. Typical Diamond Lapping Kit consisting of various Lapping Compounds, Application Guns, Lapping Sticks, Separating Fluid, Rotary Brushes, Cones and Discs. [*Courtesy: Engis Ltd.*]

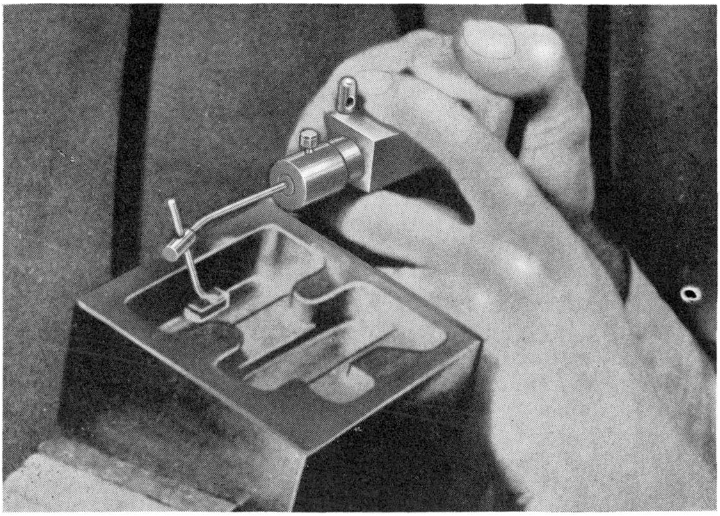

Fig. 1.15. Lapping the Mould Cavity with Diamond Lapping Compound and Reciprocating Hand Tool. The Shoe may be of Cast-iron, Copper, Brass, Wood or Felt, according to the Finish required and the Grade of Compound used. [*Courtesy: Engis Ltd.*]

arranged that the water paste is the fastest cutting compound, with water fluid, oil paste and oil fluid following in decreasing rates of cutting performance.

Polishing is strictly considered as a non-cutting operation, and, in this sense, is carried out by hand with polishing cloths or with rotary equipment using a felt mop with buffing compounds or polishing materials such as rouge or aluminium oxide. Often, however, it is more a final lapping operation in which a soft lap such as felt is used with a very fine lapping compound. The highest degree of polish is obtained on a hardened tool, and thus the final polishing is often carried out after heat treatment. However, it should be noted that scaling during hardening is minimized if the finish is high prior to heat treatment.

An alternative method to the honing, lapping and polishing procedure described is to employ diamond abrasives throughout, using coarse grits to commence with and progressively employing finer grades right up to and including final polishing. Initial lapping may commence with 80 grit, proceeding down to pastes of the order of 320, and using hand laps of densified wood or mechanically operated felt or wood laps. Final lapping is then carried out with diminishing grades and decreasing hardness of lap. The mould is cleaned at intervals. When eventually brought to a polish, the mould is hardened and then given a final polish with a polishing or super-polishing grade. Typical equipment is shown in Fig. 1.14 and a lapping operation in Fig. 1.15.

The advantage of diamond finishing is primarily the rapid cutting that is possible, often providing substantial reductions in completion times.

Additionally, the method provides a complete finishing process in itself by the employment of special scheduled grades and the lapping and polishing accessories provided.

After hardening and polishing, the mould may be completed by hard chromium plating, although this procedure is by no means universal. This type of plating is quite satisfactory on most nickel-chromium mould steels, giving a hardness of 900–1,000 VPN. A high finish on the mould prior to plating is essential if a highly polished surface is to be achieved afterwards. Plating is normally employed to improve abrasion resistance, or to confer chemical resistance against corrosive moulding materials such as P.V.C. In some cases it is used to retain the high finish on the mould against natural degradation. One disadvantage of the process lies in the fact that once the mould is plated, subsequent corrections or modifications to it are rendered more difficult, since the plating must be penetrated and then afterwards replaced.

Beryllium-copper Moulds. By the addition of beryllium to copper, alloys are produced which, after heat treatment, have properties in tensile strength and hardness approaching those of some moulding steels. Beryllium copper alloys possess a high thermal conductivity, and hence moulds of these materials can often achieve faster moulding cycles by the more rapid dissipation of heat. This can be beneficial with most moulding materials, especially the polyamides with which increased cycle rates of 20 to 30 per cent have been achieved. After heat treatment, the beryllium copper hardens uniformly throughout and can be readily brought to a high polish. Its corrosion resistance is high, and tools do not need greasing for storage; however, it is attacked by the vinyls unless electro-plated.

One of the alloys most generally used for moulds has the percentage composition: 2.75 beryllium, 0.5 cobalt, and 96.75 copper.

Typical properties after heat treatment (dependent on the alloy and specific treatment) are:

Ultimate tensile strength	50 to 90 tons per sq. in.
Elongation	1 to 6 per cent
Coefficient of thermal expansion ..	0.000017 per deg. C.
Thermal conductivity at 20 deg. C.	800 to 900 B.Th.U. per sq. ft. per in. per hour
Ultimate compressive strength ..	80 to 150 tons per sq. in.

Beryllium copper is available as sections for machining or as alloys for casting. As cast, the material has a hardness of 7/12 Rockwell-C., and, when directly hardened at 300–340 deg. C. for 2 to 3 hours and air-cooled gives a hardness of 20/25 Rockwell-C. This method does not, however, give optimum strength, hardness or conductivity. When the casting is solution-annealed at 800 deg. C. for 1–3 hours and oil- or water-quenched, subsequent hardening at 300–340 deg. C. for 3 hours will impart a hardness of up to 40/45 Rockwell-C.

Hobbing of beryllium copper can be carried out at lower pressures than those necessary for steel, and the hobbed cavities are usually set in a steel bolster. Since the material work-hardens, it is usually employed for shallow impressions, but if deeper cavities are required, the blank must be annealed between hobbing stages. On the other hand, work-hardening may be kept to a minimum by hobbing at an elevated temperature of about 800 deg. C.

Sand casting offers the most economic method of producing mould impressions in beryllium copper, but considerably more finishing work is generally required to produce a high surface finish in comparison with other mould-producing methods. Wood or plaster patterns can be employed, and tolerances of ± 0.015 in. can be achieved. Flash faces must, of course, be machined. Investment casting by conventional techniques offers good detail and finish with tolerances down to 0.002 in. per inch.

The most widely used method employs pressure casting using molten alloy and a steel hob. In this process, molten metal at 1,000–1,050 deg. C. is poured into a chase heated to 700–750 deg. C. The hardened steel hob is then applied under a pressure of 2–5 tons per sq. in. and left for 5–20 minutes during cooling. Since the master hob has to withstand thermal shock it is often made of a hot die steel. Removal is accomplished by a stripper plate fitted to the hob. This method produces a good finish which usually requires only final lapping and polishing. Excellent definition is achieved with greater detail, thinner sections and less breakage than is obtained with cold hobbing. This process has extended the applications of hobbed cavities considerably, allowing the method to be used on deeper, larger or delicate work. Beryllium copper hobbings are produced by specialist firms and are normally somewhat more expensive than steel hobbings. Additionally the 'as hobbed' finish is slightly lower than in steel and the surface hardness slightly less. For these reasons beryllium copper hobbings have not replaced steel but are valuable and popular in cases where steel would be impractical or difficult.

Cast-steel Moulds. The development of precision investment casting methods using refractory moulds has led to some use of this process for the production of cast moulds in alloy steels, including stainless and corrosion-resisting steels. These moulds can be used for full production work, since they are much harder and abrasion-resistant than zinc-alloy tools. They are not normally, however, produced by the mould maker himself but are cast by specialist firms. They are particularly applicable where there is a complication of shape since this would mean high costs in machining.

In cases, however, where machining of the mould would be simple and straightforward, there is normally little to be gained in steel casting unless the amount of metal to be removed is considerable. The process can be used for the production of injection moulds, and the alloys used can normally be heat-treated in the conventional manner. As cast, the mould has a satin-type finish which must be completed by riffler filing, followed by lapping and polishing. The castings are normally supplied in the annealed condition, to allow for such finishing work and any subsequent machining that may be necessary.

This includes the flash faces for which a $\frac{1}{16}$-in. allowance is usually made. Machining of the outside surfaces of the casting is also necessary, especially as the castings are frequently set in steel bolsters in the final tool, and a similar allowance is made for this purpose.

The accuracy that can be expected is generally not greater than 0.005-in. per inch of any length or section and, for this reason, investment cast moulds are more suitable for components where dimensional accuracy is not of major importance. This is due to the fact that with castings it is difficult to estimate closely the amount and direction of casting contraction. Where there are only a limited number of close-tolerance dimensions, it is general practice to provide a machining allowance on the casting and finally machine the cast mould to size. Cast steel has not been widely adopted for the manufacture of moulds, chiefly due to its limited accuracy and to problems of distortion and surface defects.

Cast moulds are also produced in Meehanite and a number of large moulds have been successfully produced by this method. Castings completely free of blowholes are essential and these are produced by specialist foundries. In general, the surface of the casting is not adequate for cavity or flash faces and these surfaces are invariably machined and polished—a $\frac{1}{8}$-in. machining allowance being provided for this purpose. Care must be exercised that the design allows adequate strength to resist internal moulding pressure and, if necessary, the cavity casting should be encased in a steel chase. Cast, thin or weak sections in the mould should also be avoided. Cast Meehanite moulds do reduce tool costs by obviating the need to machine cut large sections of metal, a point particularly relevant on larger moulds. However, the casting will not carry the high polish possible with a steel tool and hence is not so suitable for mouldings in transparent materials or materials where a high surface gloss is important.

Moulds in Zinc Alloy. While the steel mould is pre-eminent for length of life, accuracy and finish, occasions not infrequently arise when close accuracy and high finish are not essential; and when the production runs are limited and the cost of a steel mould is not warranted. In such instances, the use of zinc-alloy moulds is of considerable advantage, particularly for prototype moulds, short-run technical mouldings and for toys and fancy goods.

Several alloys are available, each having slightly different properties in the following range:

Melting point	358 to 390 deg. C.
Elongation on 2 in.	0 to 3 per cent
Ultimate tensile strength	9.5 to 17 ton/sq. in.
Charpy impact strength ($\frac{1}{4}$ by $\frac{1}{4}$ in.)	0.4 to 4 ft. lb.
Brinell hardness	110 to 143
Compressive strength	20 to 26 ton/sq. in.

The primary application of zinc alloys is in injection moulds which are normally produced by casting from a wood or metal pattern of which the dimensions must include both casting and moulding shrinkage allowances.

One method used to produce castings having the highest finish employs a polished metal pattern which is firmly secured to a mild-steel backplate by screws, the pattern being relieved on the back face to ensure tight seating against the backplate. Four mild-steel plates are then secured to

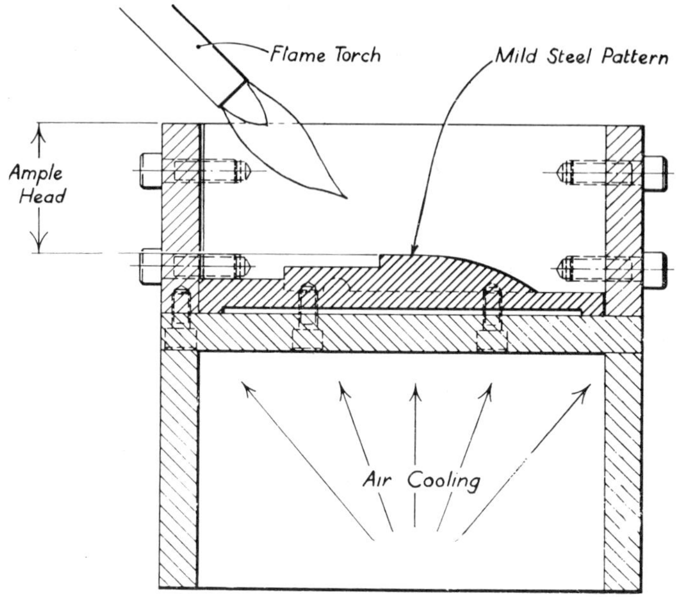

Fig. 1.16. Method of Casting a Zinc-alloy Mould around a Mild-steel Pattern.

the sides to form an open-topped box with the pattern plate at the bottom. The pattern is polished with blacklead and then dusted with french chalk. The whole assembly is then heated up to 400 deg. C. Next, the molten zinc alloy is poured into the open top of the box, a flame torch being directed on the top face of the metal so that fluidity of the metal is maintained for sufficient time to ensure even shrinkage. At the same time, a jet of air is directed against the bottom of the box to produce a temperature gradient through the pattern, thereby assisting the formation of a homogenous casting at the pattern face. The method is indicated in Fig. 1.16.

As soon as possible after the casting has solidified, it is removed from the box to allow shrinkage to take place free of the pattern. If it is allowed to cool on the pattern, a highly-stressed casting will result. The depth of the molten metal above the pattern must be sufficient to provide a pressure head and include sufficient metal for a machining allowance. Brass or copper must not be employed as patterns as they are liable to attack by the zinc alloy.

Other methods of casting employed are those based on conventional foundry practice. These employ for the mould suitable sands such as Southampton sand, or special plasters, in conjunction with wood, plastics or plaster patterns. Very large castings can be produced by these methods, although measures must be adopted to prevent shrinkage distortion with large moulds.

One of the advantages of zinc alloy is that, due to its moderate melting point of about 380 deg. C., a casting can easily be produced by the mould maker himself, without elaborate equipment. The alloy can also be remelted for re-use when the tool is no longer required.

Mouldings produced from patterns cooled by forced air have, in general, a good finish and require only a small amount of finishing and polishing. With sand-cast and plaster-cast moulds, the surface may not be as good, and, in particular, may be liable to local porosity or pitting. This must be rectified by preheating and welding the affected area, or by drilling and pinning, since the slightest defect in the mould surface would mark the final moulding and could cause difficulties on ejection.

To resist moulding-pressure deformation, zinc-alloy cast impressions are often set in a steel bolster.

Electro-deposition. This method, also known as electroforming, is a process of mould making by the electrolytic deposition of metal on to a master in a plating bath. It is primarily used for the manufacture of cavities, since it is in the manufacture of female forms that its economic advantage lies; male punches can, however, be produced by this process. It is essential that the metal deposited should have a high hardness and toughness and be free from internal stress, and special alloys and processing techniques have accordingly been developed.

The first requirement is to produce a master on which the metal will subsequently be deposited. This master can be metallic or non-metallic and, although the latter is preferred, brass and aluminium masters can be used if necessary for greater accuracy or rigidity. The ideal requirements in a material used for the master are ease of machining, ease of cementing (in cases where the master is built up from a number of pieces), inertness and dimensional stability in the plating vat, ready availability in convenient forms of rod and block, and, finally, ability to soften in hot water or air to facilitate withdrawal from the completed mould.

In practice, acrylic materials such as Perspex or Plexiglass are widely used. Existing components such as plastics mouldings, metal components and sculptors' models can be utilized, but these of course do not carry the final moulding shrinkage allowance normally added to masters and hence will mould slightly smaller than the originals. Wood and plaster are not suitable materials for such masters, but it is the practice to take from the model a cast in P.V.C. and subsequently to cast an acrylic resin master into the P.V.C. mould so obtained.

The master is the reverse of the form required, e.g. a male form if a cavity is to be produced, but besides the form (including the shrinkage allowance) it also carries such of the die face as will be exposed in the bolster, together with any runners (in reverse form) that are to be included. The finish on the electrodeposited cavity will be a reproduction of the standard of finish on the master. Normally the finish on the master is smooth and matt, and the resulting matt finish on the cavity is brought to a final polish by manual work. For some components, however, such as reflectors, the master itself is first polished, and the polish is, in turn, reproduced on the mould.

After the master has been prepared and checked, if of non-metallic material it is coated with a conductive covering of chemically deposited silver; if of metallic material, with a separating coating which is too thin to affect the accuracy of the mould, the master is then immersed in the plating vat under controlled conditions, and hard nickel-cobalt is deposited to give a final thickness of between $\frac{1}{8}$ and $\frac{3}{16}$-in. This nickel shell will have a hardness of between 40 and 45 Rockwell-C., although hardness up to 50 Rockwell-C. can be obtained if required. After completion of the nickel deposit, the shell is electrolytically backed with copper, with a hardness of 24 Rockwell-C., to a thickness that is sufficient to permit final machining.

When the master has been removed, the electrodeposited mould is turned, milled or ground to a shape suitable to allow fitting to the steel bolster. It may be fitted directly into the recess of the bolster, or fitted into a separate steel sleeve which in turn is mounted in the bolster—see Fig. 1.17.

As indicated at (*Y*), the electrodeposited form is normally secured to the bolster by means of bolts passing from the back of the bolster into holes drilled and tapped in the copper-nickel shell. Alternatively, a flange can be electroformed on the shell as shown at (*X*), thus allowing the shell

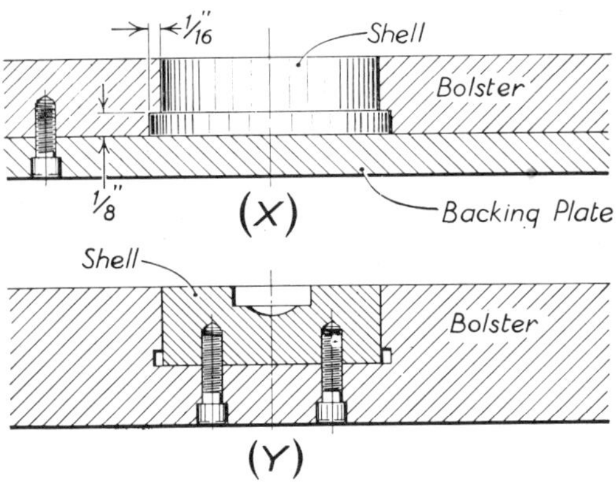

Fig. 1.17. (*X*)—Securing the Electro-deposited Shell into the Bolster by employing a Retaining Flange and Back-plate; (*Y*)—Securing the Electro-deposited Shell into the Bolster by employing Retaining Screws tapped into the Shell.

to be inserted in the bolster from the back, the flange locating in a counterbore in the steel. The shell is secured into position by mounting a retaining plate across the back of the bolster. Whichever method is adopted, a close fit between the copper-nickel shell and the steel mounting is desirable. The moulding faces are treated to remove the initial deposit of silver, and finally polished. No hardening of the mould is required, and one of the advantages of this process is the avoidance of hardening distortion problems.

Moulds in which cavities are required in both halves, and possibly also with a parting line at various levels, can be readily produced from one master. In such cases, one half of the master is first embedded to the parting line in dental wax and the exposed section electrodeposited. By then removing the wax, the remaining section may be similarly electro-

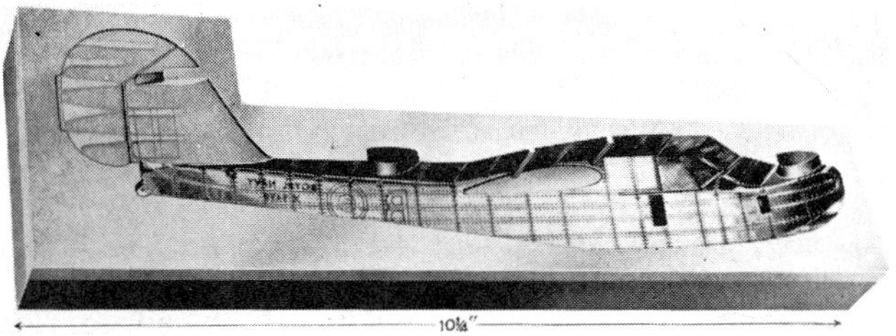

Fig. 1.18. Electro-formed Cavity for a Model-aircraft Kit showing the Large Amount of Fine Detail that can be achieved economically by the Process. [*Courtesy: London & Scandinavian Metallurgical Co. Ltd.*]

formed. On separating the two shells and removing the master, opposite cavities are obtained with an accurate mating flash face. This method particularly applies to small solid mouldings such as miniature electric plugs and toy-boat propellers.

Electro-deposited moulds are chiefly used in injection moulding, although a limited number have been used for intricate compression and transfer moulding. One of the advantages is similar to that of hobbing—namely, the far greater ease of making a male form than the corresponding female form. However, whereas the male master hob has to be made from steel and carefully hardened, the non-metallic master for electrodeposition is obviously easier to machine or even make-up from individual constituent parts. The electroforming master is clearly more economical than hobbing if a single-impression cavity is required. While multi-impression tools are made by electroforming, the economic advantage is not so great compared with single impressions, but clearly depends on individual circumstances—some designs, for instance, are too complex or delicate for hobbing.

Examples of two electroformed cavities are illustrated in Figs. 1.18 and 1.19.

Since a large part of the processing time required for an electro-deposited shell is spent in the plating vat, the direct labour time may be small compared with that required for a machined cavity. If, however, the

Fig. 1.19. A 2-impression Electro-formed Cavity for a P.V.C. Radio Component, showing the Hard Nickel-cobalt Alloy Shell and the Hard Copper Backing. [*Courtesy: London & Scandinavian Metallurgical Co. Ltd.*]

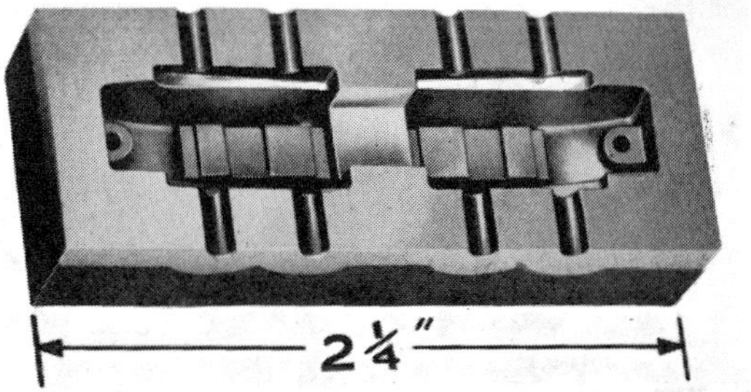

mould required is in fact of simple profile, the cost saving may not be significant and it may be quicker to machine the mould from steel. The electrodeposited shell is, of course, in one piece, and this provides freedom from dangers of flash lines that can occur with built-up moulds—a feature of particular use in the moulding of Nylon. In this connection also, the method is of particular value for moulding gears. All types may be produced—pinion, bevel, helical and worm. It is with the last type that the process is particularly useful, since it is very difficult to machine the negative of a worm, which in many cases cannot be hobbed. Usually for smaller gears, shrinkage is allowed on the diameter but not on the tooth form—thus allowing standard cutters to be used for producing the master.

Another useful application is for the production of deep slender cavities such as may be used to mould fountain-pen caps and barrels.

The accuracy obtained in the electroformed mould will be determined by the accuracy of the master, and in this respect, tolerances of ±0.001 to ±0.002 in. can be expected. The life of an electrodeposited mould is claimed to be not less than that of a steel mould, and if extra abrasion resistance be required, the mould may be chromium plated. It might be noted too, that the nickel-cobalt is inert to all plastic materials, including P.V.C.

The limitations on size that may be produced by this process are largely controlled by the necessity to maintain stress-free conditions in the deposited form. As the size becomes larger, so these difficulties increase, and the requirements of flatness, etc., are more difficult to achieve. At present, common practice is limited to projected areas of about 50 sq. in., or below, although the actual feasible size is dependent on each individual case. It is hoped that this limit on size may be increased in due course.

The chief applications of this process lie particularly in the fields of small intricate mouldings, deep slender moulds, and especially in the moulding of highly decorated fancy goods, three-dimensional name-plates and badges. In this latter field, the process avoids the difficulties of machining and polishing the form and background of raised lettering or decoration in a female cavity. Typical mouldings produced in electro-deposited moulds include nylon gears, pen caps and barrels, toys, ball-race cages, buttons, brooches, buckles, dials, key tips, reflectors, radio knobs, and motor-car horn buttons.

Electro-erosion. This process—also known as spark-erosion or spark machining—is one of the electro-chemico-mechanical methods of machining cavities into hardened metals. Other processes include electrolytic erosion, arc-erosion and ultra-sonic abrasive erosion. It is the electro-erosion method, however, which is primarily of potential interest in mould manufacture.

This system comprises placing an electrode of the desired shape above the block of metal forming the workpiece, the two being separated by a minute gap and immersed in a liquid dielectric. An electric potential applied to electrode and workpiece causes the dielectric between them to break down at whichever point the electrode and workpiece are nearest each other. A spark results and its intense heat produces localised melting and partial evaporation of the metal. The process is illustrated schematically in Fig. 1.20.

"The spark will pass between electrode and workpiece at that particular point where the electric field strength in the interspace happens to be the greatest. Upon removal of material from workpiece and electrode at this point by the spark, and the subsequent decrease in electric field strength owing to the now greater gap at this particular point, there will exist another point where the electric field strength is now greatest, so that a spark discharge causing erosion at this other point will result. Thus successive spark discharges wander over the entire surface and cause its gradual removal."*

The electrode, which is generally of brass, is made to the form required and this, by erosion, will reproduce its profile as a cavity in the workpiece.

The operating principle of this process may be gathered from examina-

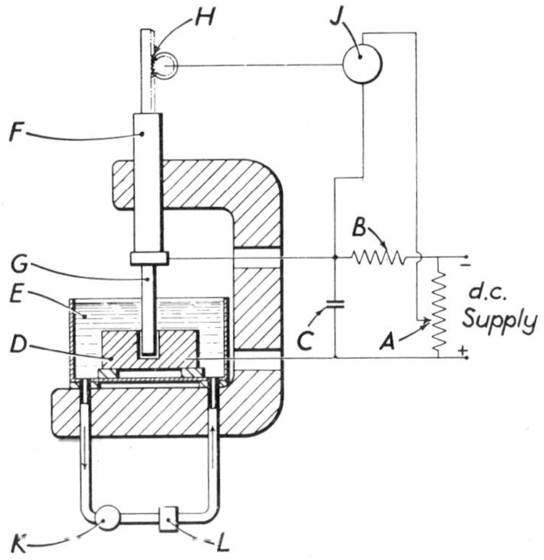

Fig. 1.20. Diagrammatic Arrangement of the Spark-machining Process.

tion of Fig. 1.20 which indicates a typical basic arrangement and circuit.

The block of metal *D* comprising the workpiece is immersed in a tank of liquid dielectric *E*, generally paraffin oil, mounted on the base of the machine. Mounted above the tank is the feed spindle *F* carrying the electrode *G* in the spindle chuck—however, except in the case of round-hole erosion, the electrode does not rotate.

The electrode is brought down to the workpiece by means of the rack and pinion *H* which in turn are operated by the geared servo-motor *J*. Electrical leads are brought to the electrode and the workpiece while the

*"Principles and Applications of Spark Machining," D. W. Rudorff. INSTITUTION OF MECHANICAL ENGINEERS.

dielectric is constantly circulated by means of the pump *K* and the sludge formed by the eroded particles removed by the sludge filter *L*.

The spark itself is produced by discharging a condenser *C*, and the duration of the spark is controlled by the capacity of the condenser itself. By thus limiting the duration of the spark instead of providing a continuous arc, overheating affects are avoided and the finish improved. The condenser is recharged after every spark, from a direct-current source through a ballast resistance *B*. This resistance controls the rate of recharging and hence the sparking rate necessary to allow the dielectric to become deionised after every spark, and to allow the eroded metal particles to be washed away. The sequence is thus that the condenser is charged at a controlled rate until its potential is sufficient to break down the dielectric in the gap separating electrode and workpiece. A spark then occurs of controlled duration as the energy of the condenser is given up. The condenser is then recharged, and this cycle is repeated at a rapid rate.

It is essential that the gap between the electrode and workpiece is maintained to a close value, and this requires that the electrode be fed downwards as the erosion proceeds and the electrode penetrates the workpiece: also that it should retract if short-circuiting occurs through direct contact of electrode and workpiece, or by lodged sludge. In a simplified form, the control works in the following manner:

"The armature of a servomotor is connected across a bridge network, two adjacent arms of which consist of a potential divider *A* connected across the direct-current supply, whilst the other adjacent arms consist of the ballast resistance *B* and the condenser *C* of the charging circuit, the latter arms also being connected across the supply. The two terminals of the condenser are also connected to the workpiece and electrode respectively.

"This control gear works as follows. Assume the electrode initially to be widely spaced from the workpiece and the current supply switched on to the condenser. This will cause charging of the condenser through the ballast resistance and, after a relatively short time, voltage across the condenser will have risen to approximately the supply voltage. The supply voltage will therefore prevail across one lower arm of the bridge. The voltage across the other arm of the bridge will depend upon the potentiometer setting and, if this setting is midway, then the voltage across the bridge (that is, difference between the voltages across these two lower limbs) will be equal to half the supply voltage. This voltage tends to rotate the motor causing the electrode to close the gap.

"When the electrode reaches the correct position, sparking takes place and the condenser rapidly charges and discharges, so that a saw-toothed voltage waveform across its terminals is produced. The electrode will cease to move when the average value of this voltage equals that prevailing across the lower limb of the potentiometer. Under this condition the bridge will be balanced and no armature current will flow. Should the electrode overshoot the demanded position and establish a gap width smaller than required the average condenser voltage will drop, since now the condenser will no longer be able to charge up to the specific voltage.

Since the condenser voltage is now less than that prevailing across the lower limb of the potentiometer the bridge becomes unbalanced again but with a reversed polarity. As a result the motor will reverse its motion and widen the gap until the correct electrode position is attained. Should the electrode touch the work the condenser will be shorted, causing all the supply volts to appear across the ballast resistance. A similar action takes place when the gap is bridged by an arc or by swarf particles. In these cases the electrode will be caused to rise until the gap is cleared and the control gear will again act to re-establish the correct spark gap."*

The process is completely automatic and machines of various capacities are available. The process is pre-eminent for the machining of very hard metals and carbides and, in this respect, is largely used for the piercing of drawing and extrusion dies of all shapes, and for the diesinking of forging and moulding cavities. It will produce holes and recesses of any form as well as threaded holes and effect the removal of broken drills and taps. Its interest in the field of plastic mould manufacture lies in the ability to machine cavities after hardening—either as a finishing cut, to correct distortion, to sink small delicate cavities or to carry out later design modifications.

It is also used for the sinking of submarine gates and also to cut thin sections after hardening (where hardening after machining might well cause cracking of the section).

---

*"Principles and Applications of Spark Machining", D. W. Rudorff. INSTITUTION OF MECHANICAL ENGINEERS.

CHAPTER II

ASPECTS OF INJECTION PRESSES IN RELATION TO TOOLS

The plastics mould is, in itself, a complete entity, capable of producing mouldings to the designed form when fitted to the injection press. It is essential, however, that the mould is so designed that it can be accommodated and secured to a press of sufficient capacity, that it can utilize the cooling system provided, and that the ejector system on the tool is suitable for the ejector-operating gear on the press. Hence it is clearly necessary to design the tool with the press requirements and capacity in mind. These press features are considered in detail in the following sections.

Platen Mounting. The two halves of the tool must be mounted in alignment to the two platens of the moulding press and secured by means of direct bolts or bolted clips. On injection-press platens, a pattern of threaded holes is provided, and the bolt holes or clipping slots on the tool must be positioned to suit these. Similarly, the tool design must take into account the distance between the tie bars, on presses of this type, to enable the tool to be passed through them for fitting. For satisfactory mounting, operation of the tool and removal of the component, account must be taken of the maximum and minimum die thickness possible with the press, and its die-closing stroke.

Thus, in order that the tool may be properly designed to suit the particular press on which it is intended to operate, certain basic fitting data from the press are required, and this information is commonly obtained from the press-manufacturer's machine handbook, a typical example for a tie-bar press being shown in Fig. 2.1. Platen designs vary between different presses, and a number of these are shown in Fig. 2.2.*

In the main, the design of horizontal injection machines can be placed in two categories. The first are those based upon tie-bar construction and in which the distribution of stresses can be readily calculated; this design is the most common. The second type is the side-frame design, in which the loading stresses are taken by the side frames of the machine. The stress distribution in this design is more complex, but the machines are very robust and rigid. Many of these presses employ flat faces as the sliding surfaces, giving even wear and low stress values. In the tie-bar press, the bearing between the tie-bar and the moving-platen bush can

*Attempts are being made to standardize bolt holes on injection presses and the British Plastics Federation has issued a "Specification for Mould Fixing Centres on Injection Moulding Machines".

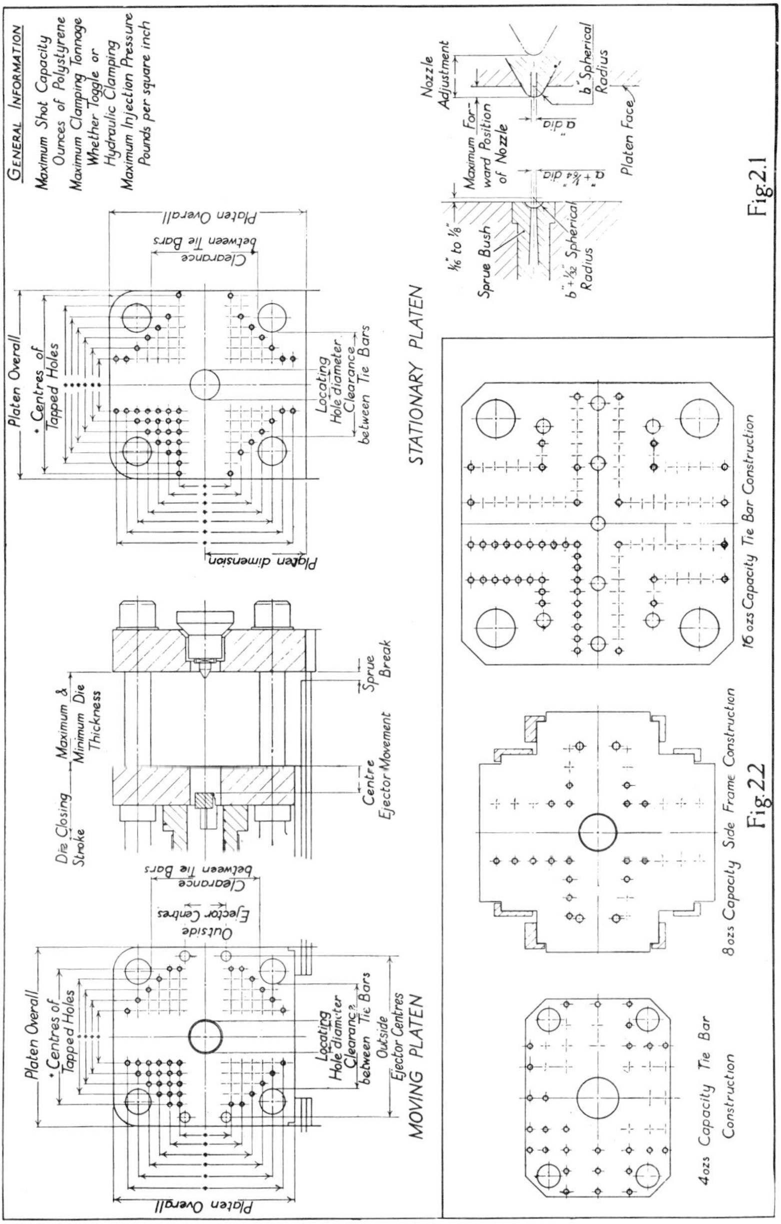

Fig. 2.1. Typical Tie-bar Press Data that is required when a Mould Design is commenced. Such Data is usually obtainable from the Press Manufacturer's Handbook. Fig. 2.2. Various Layouts of Press Platen.

result in eventual wear. Thus, after some period in service, wearing of the bushes may allow the moving platen to drop out of line with the stationary side. The purpose of the location holes in the platen is to position the moulds by means of a spigot secured to each tool half and intended to enter the location hole. However, because of the possible drop due to bush wear, the location hole provided in the moving platen should not be used to line up the two tool halves, unless the alignment of both locating holes has been checked beforehand. If the holes are misaligned or unchecked, the spigot on the moving half of the tool is omitted and other means such as guide pins are employed to set the two

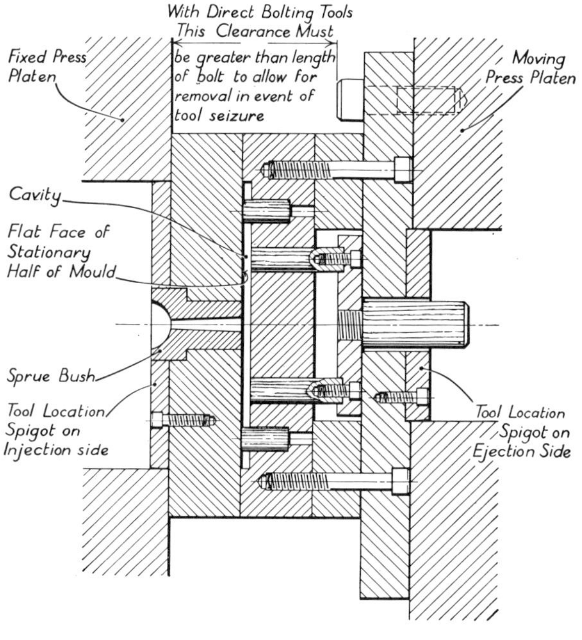

Fig. 2.3. Simple Mould with all the Form on the Moving-platen Side.

halves in correct relationship. An exception to this precaution lies in the case of a mould in which one side has no form on it. In Fig. 2.3 is shown a simple mould with all the form on the moving platen side; therefore, providing sufficient overlap of polish has been allowed on the flat face of the mould on the stationary side, the relative positions of the two halves is of little importance. Thus, this is an instance where spigots provided in each half to enter the two location holes will give adequate alignment; guide pins would be superfluous.

A common tolerance provided by the press makers on the location-hole diameter is plus 0.001 in. minus 0.003 in., but due to the many tool changes that the normal machine experiences, the holes become damaged and the tolerances have only nominal values. The purpose of the locating spigot on the injection side is to provide alignment of the injection-cylinder nozzle with the sprue bush in the tool, but since the aperture in the sprue bush is usually made $\frac{1}{32}$-in. larger in diameter than the nozzle orifice, exceptional accuracy in alignment is not necessary. Fig. 2.4 shows the relationship between the locating spigot or ring dimensions and the corresponding dimensions of tool and platen. Should a closer fit than that indicated be attempted, it may be found that difficulty will be experienced

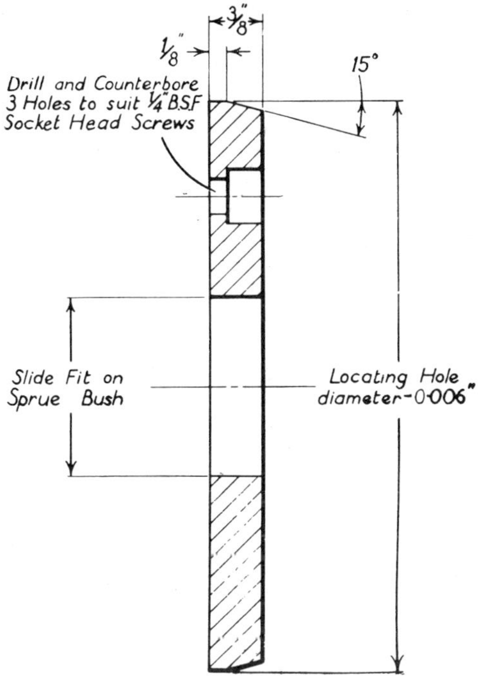

Fig. 2.4. Tool-location Spigot.

when assembling the tool on the press. This is especially true for larger moulds, where the lack of sensitivity of the lifting tackle hampers precision positioning of the tool.

Tool Clamping. Two methods of securing the tool to the press platens are in general practice, namely, direct bolting and clipping. The former employs bolts which pass directly through holes in the backplate of the tool and into the threaded holes in the press platen. Clipping makes use of a metal dog which is pulled down by a bolt, the latter engaging a conveniently positioned threaded hole in the platen, with one end of the dog resting on packing against the platen and the other end on the edge of the tool backplate, or in a recess provided in the tool dieplates.

Direct Bolting. This method has the advantage that the maximum possible load that can be transmitted by the bolt is available for clamping. The fact that, on the horizontal injection press, the mould has to be secured to a vertical platen requires that the clamping is completely positive and secure, since any slipping of the tool can cause extensive damage as the press is closed, or, in the extreme case, may fall completely off the press. Direct bolting limits this slipping to the clearance around the screw, which is usually sufficient to prevent excessive damage. The size of the bolt for fastening a tool to a given press is determined by the size of the threaded holes in the platen. The number of bolts required is influenced by several factors. Some of these are as follows:

(a) The size of the tool limits the number of threaded holes in the platen that are available for bolting.

(b) In some cases, high drawback pressures are required, due to the resistance to the die opening offered by the completed moulding within the tool. In extreme cases, this may result in complete seizure. The mould

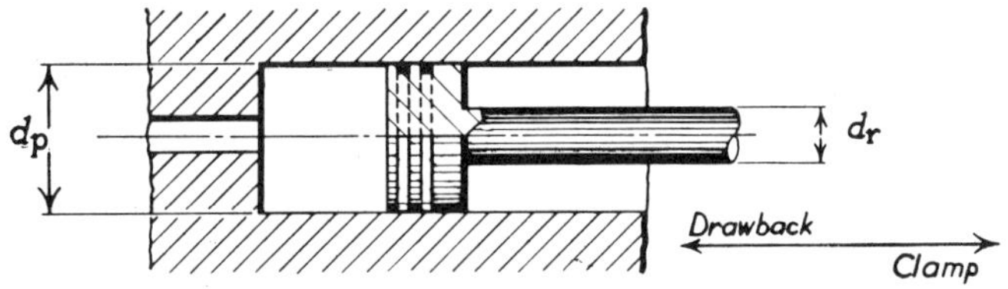

Fig. 2.5. Showing how the Difference between the Diameter of Ram and Piston affects the Difference between Clamping and Drawback Pressures.

bolting, therefore, must be capable of resisting the drawback pressure.

The maximum drawback pressure on an injection press of the direct-hydraulic clamp type can be calculated from the following:
Referring to Fig. 2.5,

Drawback pressure (lb.) $=(A_p-A_r)\times$(maximum hydraulic line pressure in lb./sq. in.)

$=\frac{1}{4}\pi(d_p^2-d_r^2)\times$(maximum hydraulic line pressure in lb./sq. in.)

Where: A_p = area of piston (sq. in.); d_p = diameter of piston (in.);
A_r = area of ram (sq. in.); d_r = diameter of ram (in.).

In presses employing a toggle-closing mechanism, the result is expressed as a proportion of the clamping tonnage, viz.,

Drawback tonnage $=[(d_p^2-d_r^2)/d_p^2]\times$[clamping tonnage].

(c) Bolts must be so spaced as to avoid long spans of unclamped tool, since the tool may deflect on opening, due to component resistance.

(d) The sliding of heavy tools under the bolt heads must be prevented. This is a function of the coefficient of friction and the maximum load which can be transmitted by the bolt on tightening. Thus,
total friction force = weight of tools $=\mu\times n\times p$, where μ = the coefficient

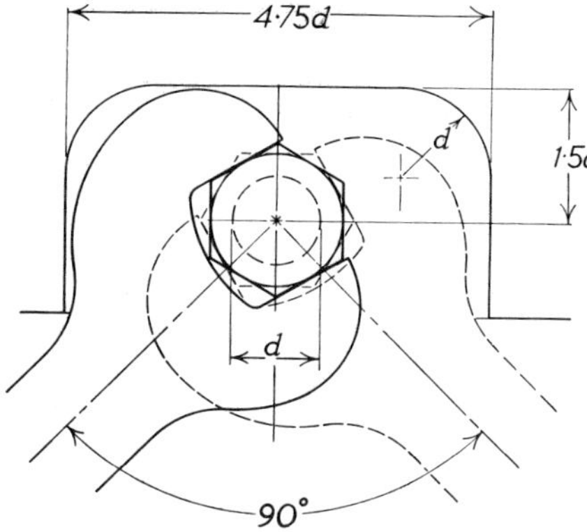

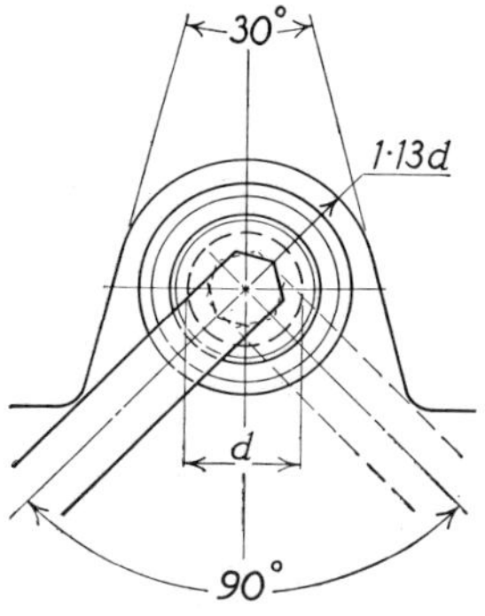

Fig. 2.6. [*Left*] Minimum Wrench Clearance for Hexagon-headed Screws. Fig. 2.7. [*Right*] Minimum Wrench Clearance for Socket-headed Cap Screws.

of friction, which may be taken as 0.2; n=number of bolts; and p= permissible load transmitted by each bolt.

It is necessary that consideration is given to the length of the bolts. If this is excessive, then the bolt will undergo bending due to the downward weight of the tool, and thus give rise to a relaxation of the clamping load. The bolts must thus be kept to a minimum length, although the thickness of the tool backing plate clamped under the bolt heads must still be ample to withstand the stresses on moulding. Adequate clearance must be provided in the tool design to allow free operation of spanners or keys to tighten the bolts, and the application of this clearance to bolt-access slots is shown in Figs. 2.6 and 2.7.

Safe loads on Standard Whitworth screws are:

B.S.W. (in.)	Safe load (tons)
$\frac{1}{2}$	1.21
$\frac{9}{16}$	1.63
$\frac{5}{8}$	2.04
$\frac{11}{16}$	2.56
$\frac{3}{4}$	3.04
$\frac{7}{8}$	4.22
1	5.54

Since these safe loads can easily be exceeded on tightening, the use of a

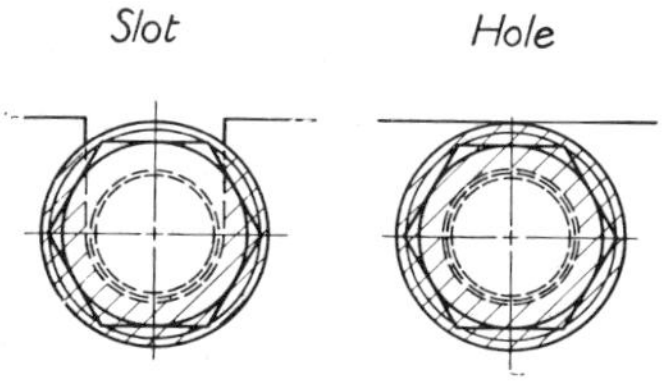

Fig. 2.8. [*Left*] Clamping at a Slot; [*Right*] Clamping at a Bolt Hole. The Shaded Areas represent the Clamping Area under the Washer.

preset torque spanner is strongly recommended. It may also be noted that the use of bolt holes is to be preferred to slots, Fig. 2.8, since in the latter case the clamping load is applied to a reduced area under the bolt head because of the slot, and this, under the same load, causes an increased stress which may, in some instances, exceed the yield stress and cause deformation of the clamped plate, or hobbing of the bolt washer into the plate.

Consideration must sometimes be given to the possibility of the tool seizing and opening being thereby prevented. In such a situation, it may be necessary to remove the tool bodily from the press. If the mould has been secured by clipping, this may be readily accomplished by loosening the clips, but if the mould has been direct-bolted, it will be necessary to remove all the screws completely. It is thus essential to ensure that there is sufficient space above the bolt head *with the tool closed* to allow the bolts to be extracted: see Fig. 2.3.

Clamping by Clipping. It must be understood from the outset that on horizontal moulding presses the hazards involved by clipping are great enough to deprecate the use of this method unless absolutely unavoidable. With direct bolting, movement of the tool is generally limited to the clearance of the bolt holes, but with clipping any relaxation or slipping of the

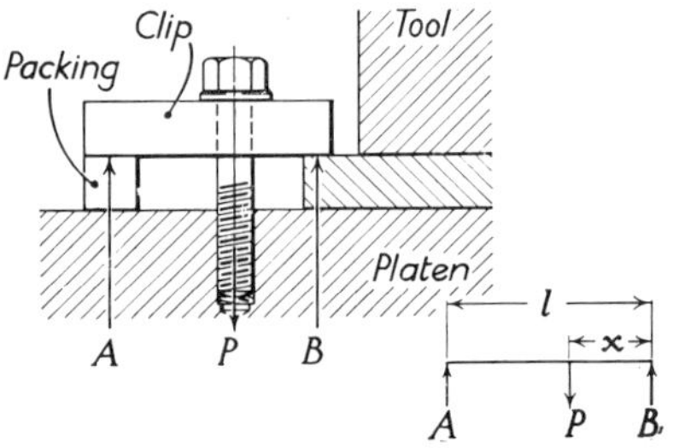

Fig. 2.9. Showing the Effect of the Distance x on the Clamping Force and the Maximum Bending Moment.

clips may be disastrous. There are, however, occasions when the use of clips cannot be avoided; notably where the bolt holes in the backplate will not meet the tapped holes in the platen and where the plate cannot be increased in size without fouling the tie bars.

If clips are essential, then certain points must be emphasized. The seating face on the tool should be at least $\frac{1}{2}$-in. wide, flat and smooth, and the packing the same height on the clamping face. To reduce deflection, the clip should be very robust, and the span from clamping face to packing kept down to the minimum. The bolt position must be as close as possible to the tool, in order that the greatest proportion of the clamping load is applied to the seating face of the tool itself.

Referring to Fig. 2.9, it will be seen that the clamping force B would be equal to the load P on the bolt if the distance x were zero. This would represent the optimum direct-bolting conditions, and thus the smaller that the distance x can he made, the nearer will the clamping force B

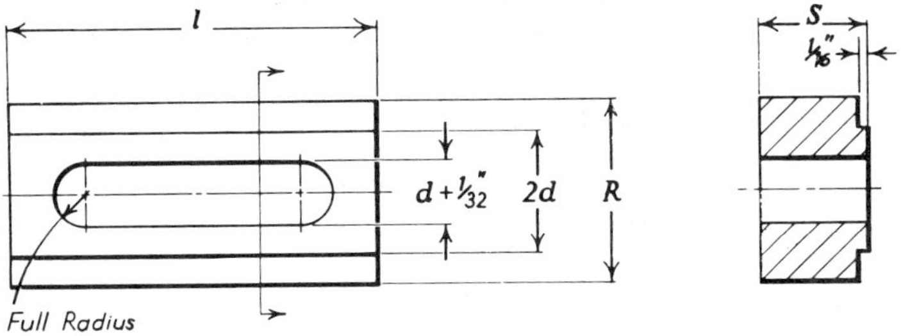

Fig. 2.10. Proportions of Clip.

approach this ideal condition. The greater the span l, the more robust must be the clip. By moments it is shown that

$$B = P[(l-x)/l].$$

Proportions for Clips. Referring to Figs. 2.9 and 2.10, let P = load applied to bolt (tons); l = effective span (in.); d = diameter of bolt (in.); R = width of clip (in.); S = thickness of clip (in.); Z = section modulus of clip (in.^{3}); and, f = maximum permissible stress (ton/sq. in.).

The maximum bending moment will occur at the point of application of the bolt, and this bending moment will be $M = P(x - x^2/l)$.

But, $M = fZ = f[R-(d+\frac{1}{32})]\ [S^2/6]$.

$$\text{Hence } S = \left[\frac{6P(x-x^2/l)}{f\{R-(d+\frac{1}{32})\}}\right]^{\frac{1}{2}} \quad \ldots \quad \ldots \quad \ldots \quad \ldots \quad \ldots \quad (1)$$

Example : Let $R = 3$ in.; $d = 1$ in.; $P = 5.54$ tons; $f = 8$ tons/sq. in.; $x = 2$ in.; and $l = 5$ in.

Then from (1),

$$S = \left[\frac{6 \times 5.54(2-4/5)}{8[3-(1+\frac{1}{32})]}\right]^{\frac{1}{2}} = 1.59 \text{ in.} = 1\tfrac{5}{8} \text{ in., say.}$$

With regard to clamping by clipping, several of the considerations noted under direct bolting also apply—notably the sufficiency of clips to resist deflection of the tool on opening, and the prevention of the sliding of the tool under the clamps due to its own weight.

Taper on Guide Pins to correct Tool Slip. It has been mentioned that with direct bolting, if slipping of the tool takes place, the clearance in the bolt holes will be taken up. Any tendency towards damage that might result can often be avoided or minimized by providing that the guide pins will align the tool on closing. This is accomplished by providing on the end of the pin a taper which will reset the tool on the pin engaging its mating bush. This is indicated in Fig. 6.36. It is essential that the guide pin is made robust enough, either to accomplish this re-alignment or else to stall the clamping motion.

Press Capacity. Before tool design is commenced, it is necessary to determine the press capacity that will be required for successful operation, and thence the particular press to be employed. When the actual press is known, the information necessary in respect of design data for mounting, platen area, tie-bar clearance, etc., can be obtained as previously indicated.

The essential considerations are:

(1) Shot capacity; (2) Plasticising rate;
(3) Clamping force; and (4) Injection pressure.

Injection presses are normally rated by the maximum weight of moulding material that may be injected per shot. Originally, this was given as the weight of cellulose acetate, but present practice is to express the weight in ounces of polystyrene. If the moulding material to be used differs from that upon which the machine rating is based, the latter must be corrected for any difference in specific gravity and bulk factor. Thus to determine the shot capacity of a machine with any other material B from a machine specification that is based on material A,

Shot capacity (oz.) in material B =

$$\left[\begin{array}{c}\text{Shot capacity (oz.)}\\ \text{in Material A}\end{array}\right] \times \left[\frac{\text{Spec. Gravity of B}}{\text{Spec. Gravity of A}}\right] \times \left[\frac{\text{Bulk factor of A}}{\text{Bulk factor of B}}\right]$$

The press selected, therefore, must be capable of injecting sufficient material at each shot to meet the estimated weight of the moulding, including the runners and sprue. This would comprise the weight of all mouldings in the case of multiple-impression tools.

PROPERTIES OF THERMOPLASTIC MATERIALS
IN RELATION TO PRESS CAPACITY

Material	*Bulk Factor*	*Specific Gravity*	*Specific Heat*
Cellulose acetate	2.4	1.24–1.34	0.3–0.42
Cellulose acetate butyrate	2.2	1.15–1.22	0.3–0.4
Nylon	2.0–2.1	1.09–1.14	0.4
P.V.C.—Rigid	2.3	1.35–1.45	0.2–0.28
,, —Flexible	2.3	1.16–1.35	0.3–0.5
Methyl methacrylate	1.8–2.0	1.17–1.20	0.35
Polystyrene	1.9–2.15	1.04–1.06	0.32
Acrylonitrile butadiene styrene (A.B.S.)	1.8–2.0	1.0–1.1	0.35–0.4
Styrene acrylonitrile	1.9–2.15	1.1	0.33
Polyethylene—Low-density	1.84–2.3	0.91–0.94	0.55
,, —High-density	1.725–1.9	0.94–0.965	0.55
Polypropylene	1.92–1.96	0.90–0.91	0.46
Polycarbonate	1.75	1.2	0.30
Acetal	1.8–2.0	1.4	0.35

The plasticising rate is expressed as the number of pounds of material that the machine can bring per hour to moulding temperature. Polystyrene is often the standard material on which the plasticising rate is based, but the actual value will depend upon the particular material to be moulded. The plasticising rate is a function of the heating capacity of the machine, and hence the amount of material that can be brought to moulding condition in a given time is dependent upon the moulding temperature necessary and the specific heat of the material.

The plasticising rate of a machine with any material B can be determined approximately from the machine specification plasticising rate based on material A (usually cellulose acetate or polystyrene) from the formula:

Plasticising rate with material B =

$$\begin{bmatrix}\text{Plasticising rate}\\ \text{with material A}\end{bmatrix} \times \begin{bmatrix}\dfrac{\text{Specific heat of A}}{\text{Specific heat of B}}\end{bmatrix} \times \begin{bmatrix}\dfrac{\text{Moulding temp. of A}}{\text{Moulding temp. of B}}\end{bmatrix}$$

If the total heat content per pound of the plasticised material is known and is equal to Q (see page 59), the plasticising rate may also be determined from:

$$\begin{bmatrix}\text{Plasticising rate}\\ \text{with material B}\end{bmatrix} = \begin{bmatrix}\text{Plasticising rate}\\ \text{with material A}\end{bmatrix} \times \frac{Q_a}{Q_b}.$$

It is necessary, therefore that the press selected should be capable of plasticising sufficient material to maintain the moulding cycle expected with the tools, and this may be determined from:

$$\begin{bmatrix}\text{Plasticising rate}\\ \text{(lb./hour)}\end{bmatrix} = \begin{bmatrix}\text{Weight of mould-}\\ \text{ing (lb.)}\end{bmatrix} \times \begin{bmatrix}\text{Number of mould-}\\ \text{ings per hour}\end{bmatrix}$$

It is good practice to select a machine having a greater plasticising rate than actually necessary, to provide a margin to take care of an eventual improvement in the moulding cycle, and to ensure that the plastics material is in an even plasticised condition.

For maximum efficiency, it is generally considered that an injection press should not operate at above 80 per cent of its rated performance on either shot weight or plasticising capacity.

The clamping tonnage of the press controls the maximum projected area of moulding that can be produced. The injection pressure exerts within the mould cavity a force which tends to open the mould. This force is proportional to the projected area of the moulding and runners, and must be opposed by the clamping force. Only a proportion of the pressure produced by the injection cylinder is transmitted to the cavity, various losses occurring in the heating cylinder, nozzle and gate. The actual cavity pressure to be counteracted by the clamping force is, therefore, only a fraction of the injection gauge pressure, and it is usual to take this fraction as between one-half and one-third. Thus:

$$\begin{bmatrix}\text{Clamping force}\\ \text{necessary (tons)}\end{bmatrix} = \begin{bmatrix}\text{Projected area of}\\ \text{mouldings* (sq. in.)}\end{bmatrix} \times \begin{bmatrix}\tfrac{1}{2}\text{ to }\tfrac{1}{3}\text{ of injection}\\ \text{pressure (ton/sq. in.)}\end{bmatrix}$$

The exact proportion of injection pressure to be adopted is dependent upon the thickness of the moulding section and the ease of flow of the moulding material used. Thin sections require a high injection pressure to fill, and thus need more clamping force. Relatively easily flowing materials such as high-melt index polythene and polystyrene fill more readily and hence require a lower clamping force.

The injection pressure may be found from the following:

$$\begin{bmatrix}\text{Injection pressure}\\ \text{(lb./sq. in.)}\end{bmatrix} = \begin{bmatrix}\text{Injection hydraulic line}\\ \text{gauge pressure (lb./sq. in.)}\end{bmatrix} \times \frac{d_i^2}{d_p^2}$$

---

*Including runners. For multiple-impression tools, the area of all cavities must be included. See also page 77, Chapter III.

where d_i = diameter of injection cylinder (in.); and
d_p = of heating-cylinder ram (in.).

On many plunger injection presses, $d_i^2/d_p^2 = 20$.

As has been stated, the injection pressure necessary on the machine is dependent upon the section thickness, length and size of runners and gates, and the material to be employed. Pressure requirements for various materials are given in manufacturers' data books, while the influence of gates and runners is considered in another section.

Most modern hydraulic presses are capable of delivering injection pressures which are sufficient for most normal requirements and generally fall into the range of 20,000 lb./sq. in. and below. It is quite frequent practice, however, to fit where necessary a machine with a special high-pressure injection cylinder. This is done by reducing the diameter of the ram and the bore of the heating cylinder, thus increasing the specific pressure although reducing the shot capacity.

In the case of direct screw injection or transfer preplasticising machines, somewhat lower injection pressures are used depending on the type of machine employed and the injection cylinder fitted. Pressure losses in the cylinder are less with these machines although lower nozzle pressures may be used to fill the cavity. Maximum injection pressures may be obtained from the press manufacturer's specification sheet and for clamping purposes a value between $\frac{2}{3}$ and $\frac{1}{2}$ should be taken.

Mould Cooling. With the moulding of thermoplastic materials it is necessary to reduce the temperature of the hot plastic material injected into the mould cavity to a point when the material freezes into a sufficiently rigid state to allow the moulding to be extracted. Thus the temperature of the tool must be maintained sufficiently low to cause the hot material to give up both its sensible and latent heat of fusion to the tool surfaces. Cooling time increases with the square of the wall thickness of the moulding. If the heat input to the tool from the hot material is greater than can normally be dissipated by the mould through conduction, etc., then other means have to be introduced to remove this excess heat. Such additional heat removal is usually accomplished by water cooling, although high-pressure air cooling is employed on occasions.

Water Cooling. By far the greatest number of tools are water-cooled, the water being directed through passages provided in the tool.

Most injection presses incorporate valves for mains water, and these are connected to the inlet and outlet of the tool by means of flexible pipes, the valves being adjusted to control the flow of water to maintain optimum tool temperature. The water from the outlet side is either run to waste or recirculated through a cooling tank or tower. Water cooling allows an increase in the production rate, especially when the shot weight/tool weight ratio is high. The water passages may either be drilled through the tool or be formed from copper tubing embedded in the tool with low-melting-point alloy. Examples of tools thus equipped are shown in Figs. 2.11 and 2.12. The following particular points should be observed by the tool designer.

(i) The cooling medium must not be allowed too close to the mould surface, since this results in cold spots and may lead to moulding faults such as weld lines. Care, too, must be taken to prevent the cooling holes from being so close to the mould surface that there is danger that the cavity pressure of the plastics material could cause localized collapse of the tool. At the same time, however, the cooling should not be so distant from the heated surface that cooling is ineffective.

(ii) The amount of cooling provided must be sufficient to maintain the temperature of the tool when the latter is running at maximum production

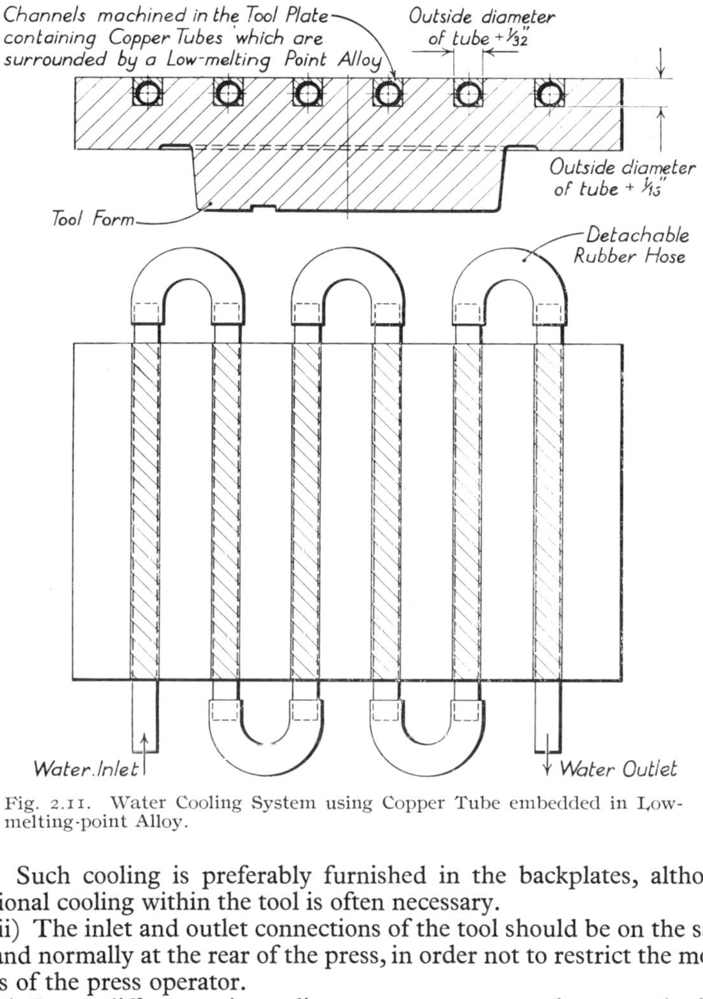

Fig. 2.11. Water Cooling System using Copper Tube embedded in Low-melting-point Alloy.

rate. Such cooling is preferably furnished in the backplates, although additional cooling within the tool is often necessary.

(iii) The inlet and outlet connections of the tool should be on the same side and normally at the rear of the press, in order not to restrict the movements of the press operator.

(iv) Large differences in cooling-water temperature between the inlet

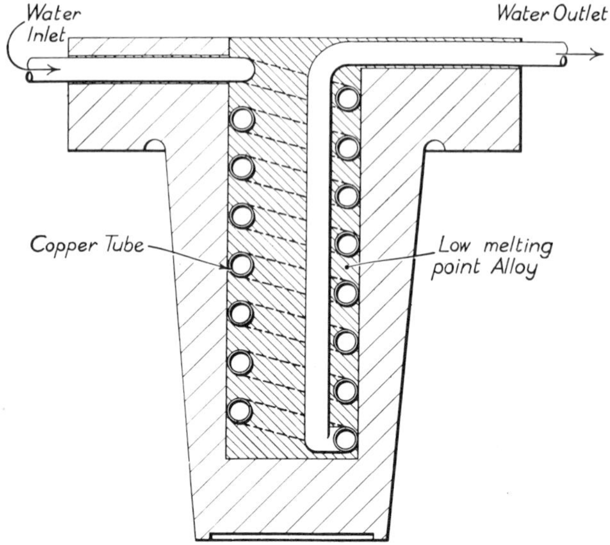

Fig. 2.12. Cooling-water Method using a Coiled Copper Tube positioned in a Pocket and surrounded by Low-melting-point Alloy.

and outlet, or across subsidiary water circuits, should be avoided, since these lead to differing temperatures in various parts of the tool, and may possibly cause moulding difficulties.

(Other detail features on cooling are considered in Chapter VII.)

To calculate the quantity of heat Q_c (B.Th.U.) to be extracted from the tool per hour, let

m = mass (lb.) of plastics material injected per hour into the tool;
C_p = specific heat of material;
T_1 = injection temperature of material (deg. F.);
T_2 = tool temperature (deg. F.); and
L = latent heat of fusion of material (B.Th.U./lb.).

Then, $Q_c = m[C_p(T_1 - T_2) + L]$ .. .. (2)

Alternatively, if the total heat per pound of the plasticised material is known, then

$Q_c = m \times a$ .. .. (3)

where m = mass of plastics material injected into the tool in lb./hour; and, a = total heat content in B.Th.U. per lb. of the plasticised material. Typical values at normal moulding temperatures are given in the table on page 59.

The weight of water m_2 to be passed per hour to dissipate the heat is derived from:

$H = Km_2(T_3 - T_4)$ .. .. (4)

where m_2 = weight of water passed (lb./hour);
T_3 = outlet temperature of water (deg. F.);
T_4 = inlet temperature of water (deg. F.);

K = constant to allow for heat-transfer efficiency; and
H = quantity of heat in B.Th.U. extracted per hour.

Values of the constant K may be taken as follows:

	K
Cooling channels bored in cavity plate or male cores	0.64
Cooling channels bored in backplates	0.50
Cooling channels using copper pipes	0.10

Now, as H is equal to Q_c as defined by formula (2),

$$m_2 = \frac{m[C_p(T_1 - T_2) + L]}{K(T_3 - T_4)} \quad \ldots \quad (5)$$

TYPICAL VALUES OF TOTAL HEAT CONTENT OF PLASTICISED MATERIAL

Material	*B.Th.U. per lb.*
Cellulose acetate	124
Cellulose acetate butyrate	111
Nylon	300–350
P.V.C.	90
Methyl methacrylate	123
Polystyrene	120–150
Acrylonitrile butadiene styrene (A.B.S.)	140–170
Styrene acrylonitrile	120–150
Polyethylene—Low-density	250–300
,, —High-density	300–350
Polypropylene	250–300
Acetal	180

Air Cooling. There are occasions when water cooling is difficult to apply to a particular mould piece, or where water would cause too rapid cooling. In such cases, air cooling at standard air-line pressure may be employed. Examples of air-cooled core pins are shown in Figs. 2.13 and 2.14.

Injection-mould Heating. A number of thermoplastic materials, e.g. acrylic, will not mould satisfactorily in cold moulds, and hence the

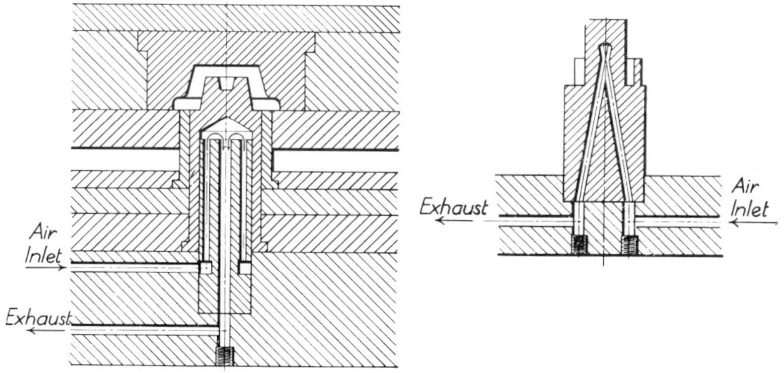

Fig. 2.13. [*Left*] Air Cooling of a Male Core Pin, with Directional Inserts. Fig. 2.14. [*Right*] Air Cooling of a Male Core Pin, with Cross-drilled Holes.

tool temperature must be raised before sound mouldings can be achieved. This can be done by continuing to inject material until the tool temperature rises, but this is slow and wasteful, hence external methods of heating the tool to operating temperature are adopted.

Another critical material is nylon. This has a narrow moulding-temperature range and readily freezes in the cavity unless the tool temperature is maintained at a sufficiently high value. Heating may be effected electrically by employing strip or cartridge heaters similar to those used in compression-mould heating, and operated at mains voltage.

Fig. 2.15. A Portable Mould Heater and Temperature Regulator. [*Courtesy: Weston Evans and Co. Ltd.*]

Alternately, there can be used portable mould-temperature controllers which not only raise the tool to the required temperature but also hold this temperature to the close limits now demanded by high-class moulding techniques. The controller normally comprises a unit in which the heating liquid (generally oil or water) is heated and pumped to the tool through flexible connecting pipes. After leaving the tool, the liquid is returned to a tank in the unit for recirculation. The temperature of the liquid is controlled by a thermostat and, in some units, means of cooling the heating liquid also are provided, to assist in faster and closer temperature control. The unit often includes multiple outlets and returns to enable parallel circuits to be used for maintaining each side of the tool at identical or different temperatures. An example of a mould-temperature regulator is shown in Fig. 2.15.

Hot-runner Moulding. With the increase in runnerless mouldings, the hot runner technique has been developed in which very difficult heat conditions are present. The runner or direct sprue must be maintained hot enough to pass the material, but the immediately adjacent tool portion must be at a substantially lower temperature. As the necessary temperature control also presents a tooling problem, it is considered in more detail in a later section.

Press Ejector Systems. The actual ejection of the moulding is performed by the ejector pins that form part of the moulding tool. These ejector pins are secured to, and operated by, an ejector plate or grid within the tool, this grid being brought into operation by the ejector system of the press. On considering any proposed tool for use on a specific press, it is essential to ensure that the ejection stroke available is adequate.

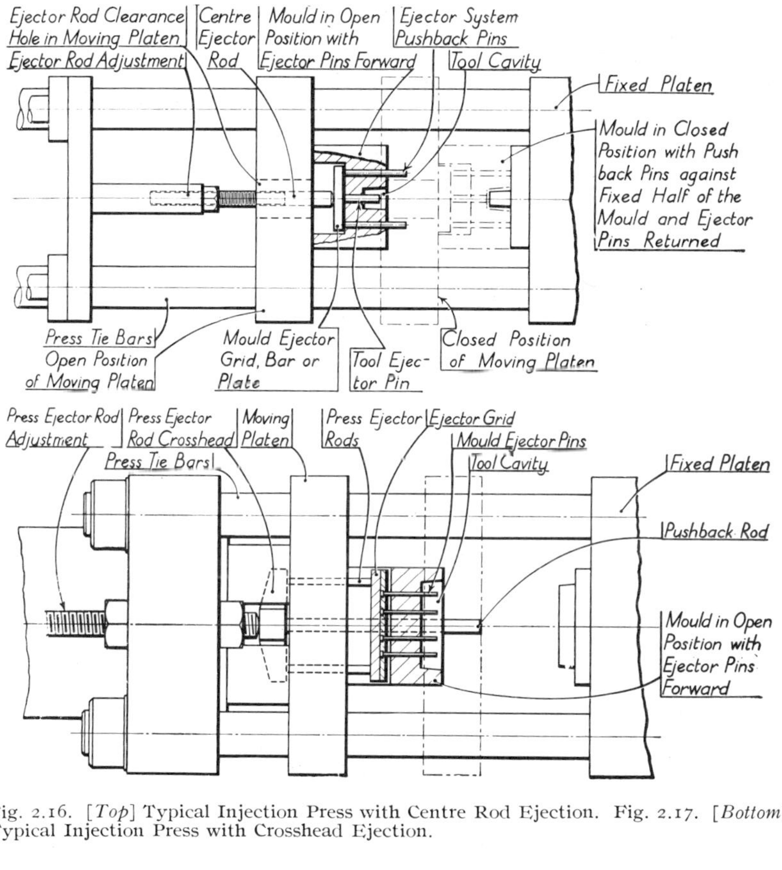

Fig. 2.16. [*Top*] Typical Injection Press with Centre Rod Ejection. Fig. 2.17. [*Bottom*] Typical Injection Press with Crosshead Ejection.

On injection presses, the ejector system is frequently of either the centre-pin or cross-head type as depicted in Figs. 2.16 and 2.17, or a chain type of ejector may be used. Here, the ejector grid in either half of the tool is connected to the opposite platen by means of balanced chains. As the press opens, the chains first become taut, and further movement causes the ejector pins to move forward relative to the tool. Care must be taken that the chains do not obstruct the operator's access to the tool.

In the centre-pin or crosshead type, the rearward movement of the platen first brings the ejector grid or plate in the moving half of the tool into contact with the stationary press ejector rod or rods. Further backward movement of the mould ejector plate is thus prevented, thereby causing the ejector pins in the tool to move forwards relative to the tool. The position of the press ejector rod is adjustable to suit different tools and to provide varying lengths of ejection stroke. The ejector system within the mould is returned by the closing movement of the press, when the ejector push-back pins contact the stationary half of the tool.

Tool Handling. All moulds but the very smallest must be provided with means of slinging and hoisting in order to facilitate handling during press setting, tool examination and repair. Normally, the tool is provided with a number of tapped holes in which forged eye bolts can be screwed. These must be positioned in the solid portion of the tool, and be of ample proportions for the weight to be carried. They should be provided in both halves of the tool, so that each can be handled separately. On injection tools, the position of the eye bolt should be such that, when suspended, the tool hangs reasonably square, to facilitate setting.

On occasion, some injection tools are necessarily too large to pass through the tie bars of the press, and in certain instances the press is fitted with a removable tie bar which allows the tool to be mounted on the platen. In other cases, the tool can be swung sideways on the hoist and passed through the tie bars at an angle, being returned to the square once it is through. Needless to say, the exact method of setting such an oversize tool must be determined before finalizing the tool design.

Auxiliary Cylinders. Due to the increasing complexity of plastics mouldings, it is necessary in a great number of cases, for certain sections of the tool to be withdrawn prior to ejection of the moulding, these tool sections forming apertures or moulded features out of the normal line of draw. Such moving parts are, for the most part, actuated as the press opens, by means of angle pins, cam faces, etc., incorporated in the tool, and these methods are discussed in a later section. However, such actuation is not always possible, particularly in designs where withdrawal is essential before any tool-opening movement takes place, and for these, auxiliary cylinders are employed. The operation of these cylinders is normally automatically initiated by sequence controls fitted to the press, and the cylinders may be either hydraulic or pneumatic. Many hydraulic cylinders are built into the tool as an integral part, but, while the principle is the same, the present discussion is primarily concerned with auxiliary cylinders fitted externally to the tool.

The choice between hydraulic and pneumatic operation is determined partly by the power required, but chiefly on whether or not the tool component being moved is exposed to plastics material pressure in the cavity and thus requires locking in position to prevent its being forced back. Although mechanical locking can sometimes be incorporated, hydraulic operation, which provides positive pressure, can effect this locking, but air pressure, being an elastic medium, can easily be recompressed. Should locking not be essential, pneumatic cylinders can readily be employed if a suitable press is available. Normally they tend to be less expensive than hydraulic cylinders.

Hydraulic Cylinders. Since hydraulic cylinders are generally operated at high line pressures, they provide a very convenient means of moving core blocks, etc., with comparatively small rams. As has been mentioned, they are especially useful if the design of the moving part results in high resistance to motion, while, additionally, the part can be locked hydraulically to resist the material pressure on the core block when a mechanical locking device is not easily incorporated.

The cylinders can be operated very conveniently by tapping into the

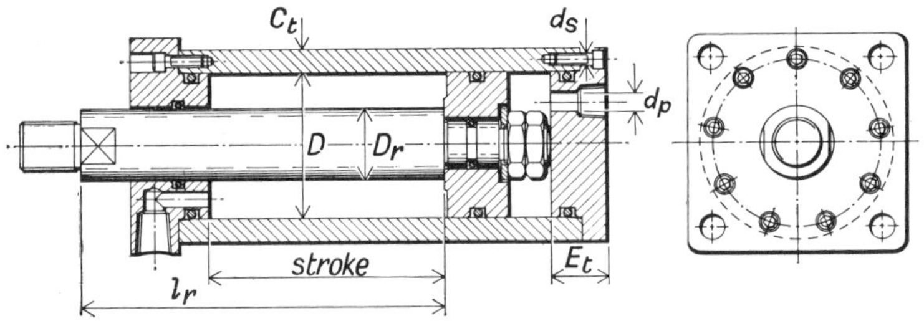

Fig. 2.18. Flange mounting 3-in diameter Hydraulic Cylinder.

normal hydraulic circuit of the machine, usually without decreasing the speed of operation of the press.

Care should be taken, however, to ascertain that full hydraulic pressure is available at all times during the period of operation. On many machines, there is a fall in pressure during the commencement of the injection stroke, and no pressure during "mould open" (i.e. gate open) periods. The danger of fall in pressure can in many instances be avoided by the use of efficient non-return valves, or better still, by the use of a separate hydraulic unit.

Many firms specialise in the supply of standard hydraulic cylinders, and these can often be used with advantage. However, due to the great variety and particular requirements of plastics moulds, it is often necessary that they are individually designed and manufactured by the actual tool maker.

Almost invariably, the cylinder required is of the double-acting type in which hydraulic pressure can be applied to either side of the piston to provide positive power to the tool part on both insertion and withdrawal. A non-commercial hydraulic cylinder of this type is shown in Fig. 2.18.

Stroke and Load on the Ram. The stroke of the cylinder is clearly determined by the requirements of the tool design and the withdrawal distance necessary on the moving part. The power required from the cylinder, expressed as load on the ram, may be determined from the force required to operate the moving part, and whether or not it is intended to use hydraulic pressure to lock the core against the pressure of the material. If the tool part being operated by the ram is subject to cavity pressure, the load necessary to oppose this pressure is given by:

$$\text{Load} = \begin{bmatrix}\text{Projected area of} \\ \text{tool part subject} \\ \text{to material pressure}\end{bmatrix} \times \begin{bmatrix}\text{cavity pressure} \\ \text{(say, 9,000* lb./sq. in.)}\end{bmatrix}$$

Bore of Cylinder. It is always desirable to arrange that the maximum area of the piston is open to the hydraulic line pressure in order to provide maximum outward pressure on the cylinder ram to oppose the cavity pressure. To this end, the cylinder should be designed so that, on load, the oil is fed to that end of the cylinder through which the piston rod does not pass.

The cylinder bore diameter can be found from

$$D = 1.13\sqrt{(L/p)} \qquad (6)$$

where L = load on the ram (lb.);
p = hydraulic line pressure (lb./sq. in.); and
D = diameter of cylinder bore (in.).

Having been ascertained by equation (6), the value of D must be adjusted so that standard sealing rings can be accommodated.

Sealing Rings. The hydraulic operating pressures of most injection machines in the range of up to 16 oz. is 1,000 to 1,500 lb./sq. in. Where this is the case, all sealing can be effected by standard O-rings, both for reciprocating members (e.g. pistons and rams) and for the static members of the cylinders such as end-plates.† The dimensions of recesses, leads and fits to house these seals are important and can be obtained either from the suppliers' handbooks or from the appropriate British Standard (BS 1.806)

Although it is not generally recommended that more than one O-ring be fitted on any one shaft, it is in fact quite usual to do so in this type of work, but there is a tendency for pressure to build up in the space between the rings.

If the line pressure is greater than 1,500 lb./sq. in., then it is essential to incorporate a back-up washer. To obtain optimum conditions of

*This value for the cavity pressure is adequate for most applications, but cases can arise where a higher pressure should be adopted. Some of the larger injection presses, for instance, operate at a higher injection pressure than normal—i.e. above 20,000 lb./sq. in.—while others have been fitted with special injection cylinders giving higher pressures than standard. Freak conditions can also arise, as in the instance of a press restarting after idling, when the material in the cylinder will have been brought to a more fluid state than in normal operation, a condition more likely in presses working at maximum pressure on materials such as acrylic. Other cases have been known, with materials such as low-density polythene, where the cavity pressure has risen to 80 per cent of the injection pressure. A final consideration will be the cavity section—higher pressures will be encountered in thick sections rather than in thin.

†Special high-pressure 'fluid seals' are also manufactured for this purpose.

service from O-seals, the bore should have a honed surface with a roughness not exceeding 15 micro-inches.

Speed of Operation. On occasions, the withdrawal or insertion of the ram must take place in a given time interval. To achieve this, the output from the pump in cu. in. per minute must be known, or a flow-control valve fitted into the auxiliary circuit, to give the correct quantity of oil required.

The velocity of the oil through the piping is usually kept to 180 in. per second, or below, as the pressure drop in the pipe becomes excessive above this velocity.

The bore of the pipe required to feed the cylinder can be found from

$$d_p = D\sqrt{(L/Vt)} \quad \ldots \quad (7)$$

where d_p = bore of the pipe (in.);
D = diameter of cylinder (in.);
L = length of stroke (in.);
V = velocity of oil through the pipe (in./sec.); and
t = time required to operate (sec.).

If the diameter of the bore is decided, then the velocity of oil may be determined from

$$V = \frac{\text{Quantity of oil supplied (cu. in./sec.)}}{\text{Area of bore of pipe (sq. in.)}} \quad \ldots \quad (8)$$

Example : Let $D=3$ in.; $L=5$ in.; $V=180$ in./sec.; and $t=2$ secs.
Then, $d_p = 3\times\sqrt{[5/(180\times 2)]} = 3/8.5 = 0.375$ in. internal diameter.

The maximum speed of rubbing for O-rings should not be greater than 8 in. per sec.

Nearly all machines are fitted with constant-volume pumps, but there are exceptions.

Thickness of Cylinder Wall. The Lamé equation (9) is normally used for calculating the thickness of cylinder walls, but this formula gives very thin walls on small cylinders and yields a bursting pressure somewhat lower than that obtained experimentally. To accommodate this discrepancy and to provide sufficient thickness in which to set the end-cap retaining screws, the equation is used with a low maximum permissible stress value. Where, however, end-plate fixing screws are used, the wall-thickness is practically determined by equation (10). A practical rule is to determine the thickness C_t by both (9) and (10) and to adopt whichever is the larger.

The Lamé equation may be expressed as

$$C_t = r\left[\left(\frac{f_t+p}{f_t-p}\right)^{\frac{1}{2}} - 1\right] \quad \ldots \quad (9)$$

or, giving an equation dependent on the diameter of fixing screws

$$C_t = 2d_s \quad \ldots \quad (10)$$

where C_t = thickness of cylinder wall (in.);
f_t = maximum permissible stress (say 5,000 lb./sq. in. for mild steel);
d_s = diameter of end-plate retaining screws (in.);
p = maximum hydraulic line pressure (lb./sq. in.); and
r = radius of bore of cylinder (in.).

The usual practice is to make the cylinder of mild steel and the ram of heat-treated nickel-chromium steel. The increase in thickness gained by taking such a low permissible stress as 5,000 lb./sq. in. compensates to some degree for the lack of end reinforcement normally required to counteract the bending moment on the cylinder walls due to the axial pressure on the ends of the cylinder.

Example: Let $r=1.5$ in.; $p=1{,}000$ lb./sq. in.; $f_t=5{,}000$ lb./sq. in.; then by (9)

$$C_t=1.5\left[\left(\frac{6000}{4000}\right)^{\frac{1}{2}}-1\right]=0.337\text{-in.}$$

If the end-plate retaining screws are assumed to be 0.25-in. diameter, this value is less than $2d_s=0.5$ in., hence the value to be taken for the cylinder thickness is that given by (10), i.e. $C_t=0.5$ in.

Piston-rod Diameter. The diameter of the piston rod may be determined from Rankine's formula, based on free-end fixing to provide a conservative estimate. This formula is

$$L_b=\frac{f_cA}{1+c(l_r/k)^2} \qquad (11)$$

where: L_b = buckling load (tons) = maximum load on ram multiplied by a factor of safety of, say, 4;
f_c = compressive yield stress (ton/sq. in.);
c = a constant = 1/7500 for steel with both ends free;
k = least radius of gyration (in.);
l_r = length of piston rod (in.);
A = cross-sectional area of piston rod (sq. in.);
D_r = piston-rod diameter (in.).

Equation (11) may be put in the form

$$L_b+(L_bcl_r^2)/k^2=f_cA\ ;$$

i.e.

$$L_bcl_r^2=f_cAk^2-k^2L_b \qquad (11a)$$

and since $k^2=D_r^2/16$, and $A=\pi D_r^2/4$, equation (11a) becomes

$$L_bcl_r^2=\pi f_cD_r^4/64-L_bD_r^2/16 \qquad (11b)$$

This is a quadratic in D_r^2 with the solution of

$$D_r^2=[2L_b+8\sqrt{(L_b^2/16+\pi cf_cL_bl_r^2)}]/\pi f_c.$$

Putting $c=1/7500$,

$$D_r=[\{2L_b+8\sqrt{(L_b^2/16+\pi f_cL_bl_r^2/7500)}\}/\pi f_c]^{\frac{1}{2}} \qquad (12)$$

Example: Find the diameter D_r of the piston rod when the maximum load is 3.13 tons and the length l of the piston rod is 7 in., taking the factor of safety as 4, and $f_c=20$ ton/sq. in. The buckling load is

$L_b = 3.13 \times 4 = 12.52$ tons, and by (12)

$$D_r = [\{(2 \times 12.52) + 8\sqrt{(12.52^2/16) + (\pi \times 20 \times 12.52 \times 49)/7500\}}/(20\pi)]^{\frac{1}{2}}$$

$$= \left[\frac{25.04 + 8\sqrt{9.797 + 5.140}}{62.832}\right]^{\frac{1}{2}} = \sqrt{0.8905} = 0.9437 = 1.0 \text{ in., say.}$$

Thickness of Cylinder End-plates. This can be determined from consideration of the internal pressure, using an equation of Reuleaux, which may be stated as

$$E_t = 0.81\, r(p/f_t)^{\frac{1}{2}} \quad \ldots \quad (13)$$

where: p = internal fluid pressure;
f_r = maximum permissible stress;
r = radius of cylinder bore (in.);
E_t = thickness of end-plates (in.).

Example : $2r = 3$ in.; $p = 1{,}000$ lb./sq. in.; $f_t = 5{,}000$ lb./sq. in.;
For these values,

$$E_t = 0.8 \times 1.5 \times (1000/5000)^{\frac{1}{2}} = 0.539 \text{ in., say } \tfrac{5}{8} \text{ in.}$$

An alternative method is based on the consideration that the greatest stress condition may arise on the return stroke of the ram when the head of the piston rod is brought to a stop against the end plate. In this case, the return load is transmitted directly to the centre of the end-plate. This may be determined by Grashof's equation

$$E_t = \left[\frac{P}{\pi f_t}\left(2 \log_e \frac{D}{d_1} + \frac{3}{2}\right)\right]^{\frac{1}{2}} \quad \ldots \quad (14)$$

where P = total load on centre of inside of end plate

$= p.\ \frac{1}{4}\pi(D^2 - D_r^2)$;

d_1 = diameter over which P acts (i.e. diameter of piston rod in contact with end plate);

$$\text{Then } E_t = \left[\frac{p(D^2 - D_r^2)}{4f_t}\left(2 \log_e \frac{D}{d_1} + \frac{3}{2}\right)\right]^{\frac{1}{2}} \quad \ldots \quad (14a)$$

Example : Let $D = 3$ in.; $D_r - 1$ in.; $d_1 = 0.75$ in.; $p = 1{,}000$ lb./sq. in.; and $f_t = 5{,}000$ lb./sq. in.
then,

$$E_t = \left[\frac{1000(3^2 - 1^2)}{4 \times 5000}\left(2 \log_e 4 + \frac{3}{2}\right)\right]^{\frac{1}{2}}$$

$$= \left[\frac{8}{20}\left(2 \log_e 4 + \frac{3}{2}\right)\right]^{\frac{1}{2}}$$

$= 1.31$ in., say $1\frac{5}{16}$ in.

The end-plate thickness should thus be determined by the equation giving the greater value.

Number of Retaining Screws in End-plate. This may be determined from

$$N = pD^2F/d_s^2 f_r \quad \ldots \quad (15)$$

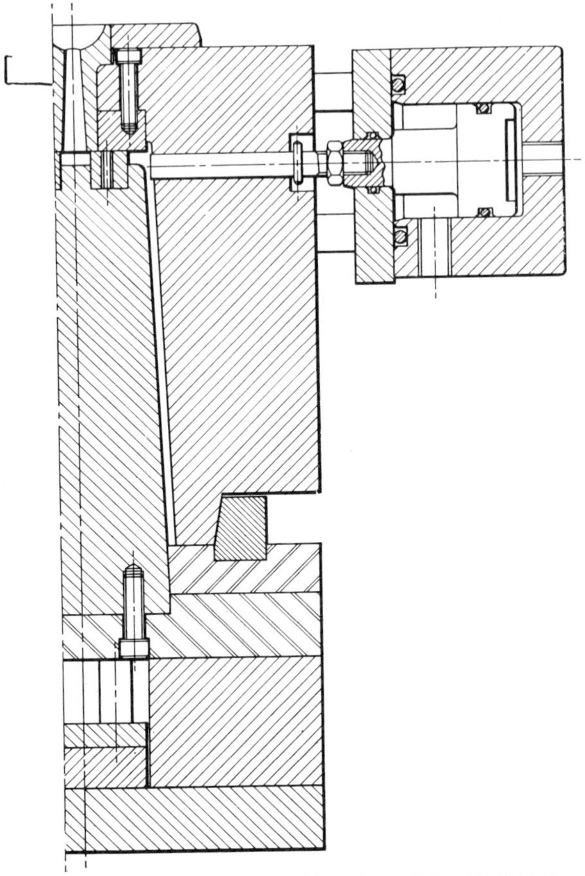

Fig. 2.19. Short-stroke Double-acting Hydraulic Cylinder.

where F = factor of safety, say 4;

d_s = diameter of retaining screws (in.); and

f_r = ultimate tensile strength of retaining screws (lb./sq. in.), say 100,000 lb./sq. in.

The core diameters and areas of B.S.F. screws are given in the table on page 69.

Using the example of the 3 in. diameter cylinder already given, and assuming the use of 0.25-in. diameter screws (0.2-in. core diameter), the number of screws required to secure each end plate is

$$N = (1000 \times 3^2 \times 4)/(0.2^2 \times 100000) = 9.$$

Cores employed as shut-off faces are not normally exposed to full cavity pressure tending to force them back and, in the case of small cores, may be capable of pneumatic operation. For large blocks, however, where high operational frictional forces may be encountered, hydraulic operation should be adopted. Also to be considered is the effect of incomplete shut-off on the face of the block through damage, wear or inaccurate tool-making. In such instances, the moulding will flash, and the core will be exposed to partial or total cavity pressure. An example of a hydraulic cylinder operating a mould pin is shown in Fig. 2.19. The cylinder operates a ram which carries an embossed design at its extremity within the mould cavity. Retraction is necessary to allow extraction of the moulding. It will be noted that the rotation of the core is prevented by a pin that passes through the ram and operates in a rectangular recess.

BRITISH STANDARD FINE SCREW THREADS

Nominal Size (in.)	*Threads per Inch*	*Core Diameter (in.)*	*Area at Bottom of Thread (sq. in.)*
$\frac{1}{4}$	26	0.2007	0.0316
$\frac{5}{16}$	22	0.2834	0.0508
$\frac{3}{8}$	20	0.3110	0.0760
$\frac{1}{2}$	16	0.4200	0.1385
$\frac{5}{8}$	14	0.5335	0.2235
$\frac{3}{4}$	12	0.6433	0.3250
1	10	0.8719	0.5971

Air Cylinders. These are manufactured by a number of concerns, in a range of standard and special sizes. For most mould work, however, cylinders of 1.5 or 3.0 in. are employed. Due to their availability, these cylinders are invariably purchased rather than made by the mould maker. Various forms of mounting are available, together with a wide range of fittings, and full details may be obtained from the manufacturers' handbooks. Two typical cylinders are shown in Figs. 2.20 and 2.21.

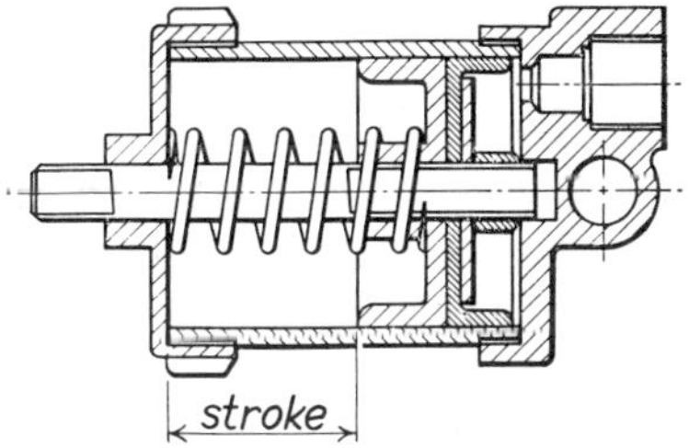

Fig. 2.20. Single-acting Pneumatic Cylinder.

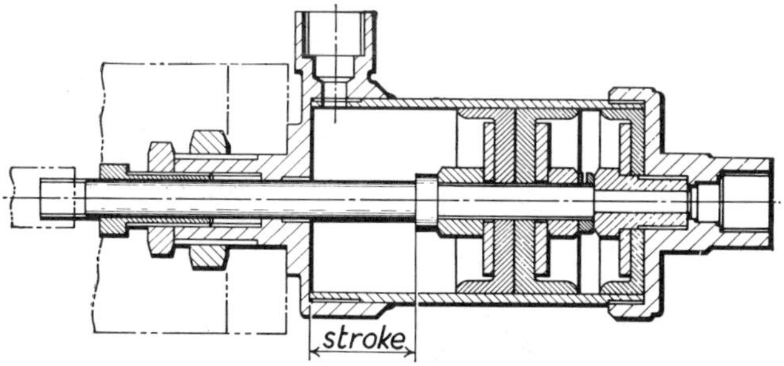

Fig. 2.21. Double-acting Pneumatic Cylinder.

Most standard cylinders are designed for operation on a 100-lb./sq. in. supply. In considering operation of the moving cores with air cylinders, there must be borne in mind the possible damage that may result if the cylinder fails to operate during the moulding cycle. This is perhaps most likely on the return stroke, either by shrinkage or by flashing of the moulding causing the core to seize. For this reason, double-acting cylinders are preferable to spring-loaded single-acting cylinders for most applications.

Operation of Auxiliary Cylinders

HYDRAULIC CYLINDERS. The operating circuit for auxiliary hydraulic cylinders may be built into the press pilot circuit which operates the valves in the main machine hydraulic circuit. For this, the pilot circuit is tapped by a connection into which a solenoid-actuated four-way valve is fitted. This is connected to the auxiliary double-acting cylinder while the operating solenoid is controlled from the press-cycle time circuit or limit switches. This method will allow nearly any of the usual sequences of operation. Normally, it is required to withdraw the ram from the tool before the

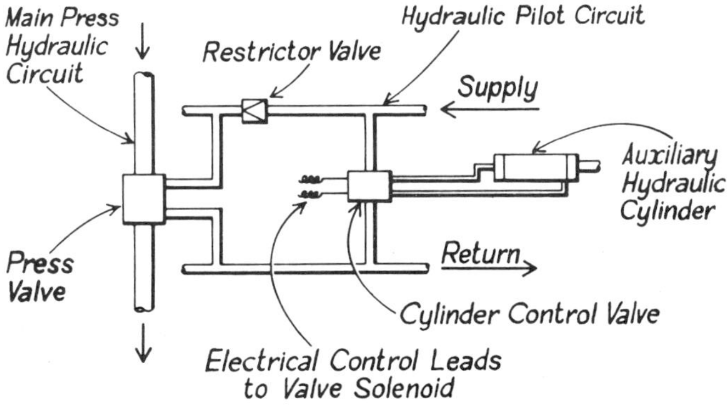

Fig. 2.22. Method of Connecting the Auxiliary Cylinder to the Main Circuit of the Press.

latter opens, and to enter the ram after the tool has closed. To ensure that the auxiliary cylinder will operate before the main hydraulic valves of the machine, a restrictor valve is sometimes fitted beyond the cylinder valve, in the pilot circuit leading to the main hydraulic valves. This ensures a build-up of pressure in the auxiliary cylinder before the pilot circuit operates the main valves. This arrangement is indicated in Fig. 2.22.

AIR CYLINDERS. The sequencing of an air cylinder is achieved by utilizing some convenient moving part of the press to actuate air-control valves or micro-switches connected to solenoid-operated air valves. The latter method is useful if the movement of the machine is small or the operation must take place at a precise moment. With direct-hydraulic-clamp machines, the moving platen can usually be utilized to operate the valves. On a toggle-closing mechanism, the toggles themselves can be used. In Fig. 2.23 is shown diagrammatically a set-up of this type.

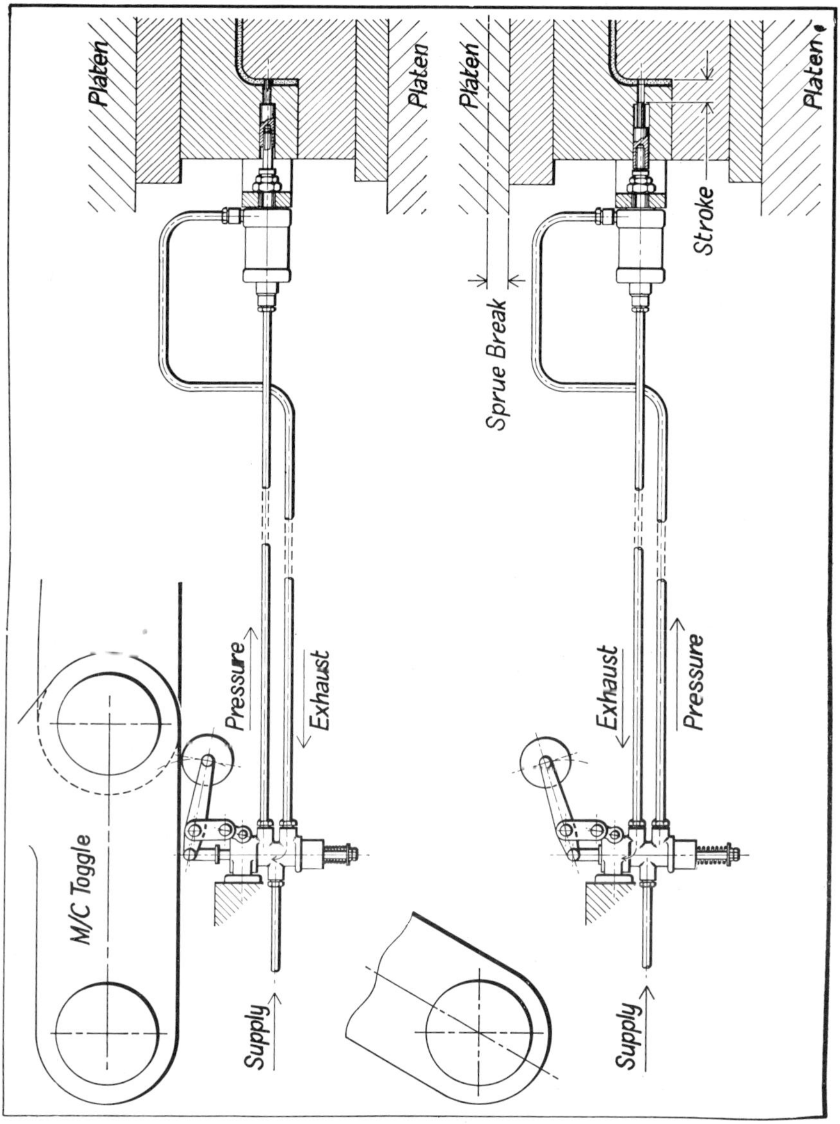

Fig. 2.23. Controlling the Operation of Air Cylinders from the Toggle-closing Mechanism.

CHAPTER III

INJECTION MOULDS—INITIAL CONSIDERATIONS

Before actual tool design can be commenced, a number of preliminary factors must be considered, and a decision made on the method of approach to be adopted in the design. These initial considerations include:

Determination of shot weight and clamping necessary; estimation of the moulding cycle.

Number of impressions to be included in the tool.

Disposition of tool flash faces and the line of draw.

Position and type of feed.

Press Requirements. The first consideration is to determine the weight of the proposed moulding, and hence the shot capacity required from the press. Thus,

$$\text{Shot capacity} = \begin{bmatrix}\text{Volume of}\\ \text{moulding in}\\ \text{cub. in.}\end{bmatrix} \times \begin{bmatrix}\text{Weight of}\\ \text{plastics material}\\ \text{per cu. in.}\end{bmatrix} \times \begin{bmatrix}\text{Number of}\\ \text{impressions}\end{bmatrix}$$

The volume of the moulding is determined by the normal methods of mensuration, while the weight of the moulding material can be obtained using the information from the material manufacturers' data sheets and the conversion relationships:

Weight of material in grams/cu. in. $= 16.4 \times$ Specific Gravity; or, Weight of material in oz./cu. in. $= 0.578 \times$ Specific Gravity. The weight per cu. in. for a number of typical materials is given in the following table.

WEIGHT OF MOULDED PLASTICS PER CUBIC INCH

	oz.	*grams*
Cellulose acetate	0.775	22.0
Cellulose acetate butyrate	0.705	20.0
Nylon	0.658	18.7
P.V.C.	0.838	23.8
Methyl methacrylate	0.687	19.5
Polystyrene	0.600	17.0
Toughened polystyrene	0.600	17.0
Polyethylene—Low-density	0.532	15.1
,, —High-density	0.555	15.8
Polypropylene	0.525	14.8
Acrylonitrile butadiene styrene (A.B.S.)	0.636	18.0
Styrene acrylonitrile	0.636	18.0
Acetal	0.81	23.0
Polycarbonate	0.7	20.0

Calculation of the number of mouldings likely to be produced per hour is very difficult and complex: in practice it is estimated by experience. There are many factors influencing the length of the actual moulding cycle, but the principal item is generally that of the cooling period necessary before the moulding can be extracted. Factors controlling the moulding cycle are:

Thermal characteristics of moulding material;
Weight of moulding;
Thickness of section;
Surface area of moulding;
Weight of tool and efficiency of cooling;
Efficiency of ejector system; and
Handling of inserts, loose blocks, etc.

Cooling Period of Cycle. One method of estimating the cooling period of the cycle is to use the formula,

$$\theta = \frac{-t^2}{2\pi\alpha} \log_e \left[\frac{\pi}{4} \left(\frac{T_x - T_m}{T_c - T_m} \right) \right]^{\star}$$

Where, θ = minimum moulding set-up time in seconds;
t = thickness of moulding in inches;
α = thermal diffusivity of material in sq. in. per sec.;
T_x = heat distortion temperature of material in deg. C.;
T_m = temperature of mould in deg. C.; and
T_c = cylinder temperature in deg. C.

Example : Let $t = 0.05$ in.; $\alpha = 1.34 \times 10^{-4}$; $T_x = 90$ deg. C.; $T_m = 70$ deg. C.; and $T_c = 250$ deg. C. Then

$$\theta = \left[\frac{-0.05^2}{2\pi \times 1.34 \times 10^{-4}} \right] \log_e \left[\frac{\pi}{4} \times \frac{90-70}{250-70} \right] = 7.2 \text{ secs.}$$

It must be remembered that this is the minimum theoretical time to set up, and a number of the assumptions on which the theory is based are not likely to be always satisfied in practice—hence a somewhat longer period may be required. Again, this cooling period can be extended considerably if the shape of the part would result in distortion at high ejection temperatures (e.g. through internal stress). Similarly, local thick sections may require a longer cooling period.

The three graphs A, B and C show this theoretical set-up time in seconds against moulding thickness in inches, for a number of common materials, based on average moulding temperatures, and provide a convenient approximate guide to minimum cooling periods.

Some useful thermal properties of a range of materials are given in the table on page 76.

When the estimated cycle is set against the output performance of the proposed press, two principal items have to be considered: viz., the plasticising capacity of the press, and the maximum operating speed of the press.

It is generally the first of these which is the controlling factor, in

*R. L. Ballman and Tevis Shusman, Modern Plastics, Vol. 37, No. 3; E. C. Bernhardt, "Processing of Thermoplastic Materials".

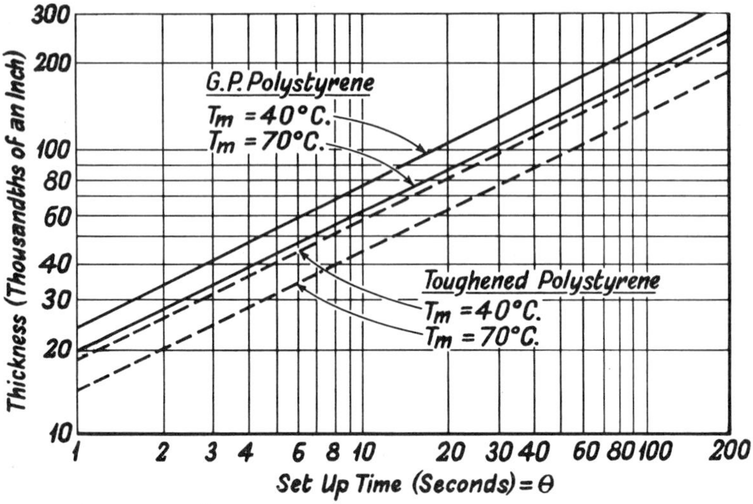

Graph A—Set-up Times for General-purpose and Toughened Polystyrene.

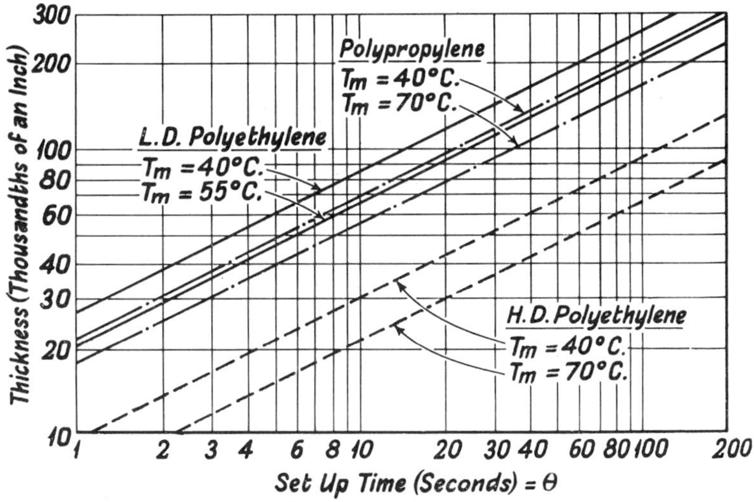

Graph B—Set-up Times for Low- and High-density Polyethylene, and Polypropylene.

considering whether a proposed tool can operate on a particular press (assuming that other factors of shot weight, clamping force, platen area, etc., can be satisfied).

The determination of plasticising capacity has been considered previously on page 54.

The maximum operating speed of the press may be important in some cases—notably where automatic high-speed tools are employed, or where

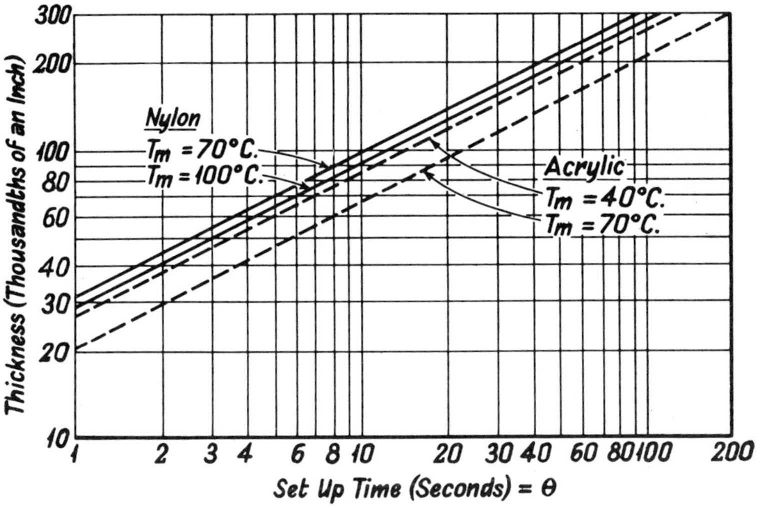

Graph C—Set-up Times for Nylon and Acrylic.

a moulding of small weight but large surface area is being produced on a press of excessive shot and plasticising capacity in order to utilise the high clamping tonnage. The speed is usually expressed as the "maximum dry-running speed", which may be defined as the maximum speed of operation of the press as limited by its own mechanical construction, i.e. without moulding material. This value is slightly more than the fastest cycle that can be achieved, and may be determined from the manufacturer's handbook or specification, or from:

$$\text{Maximum number of cycles per minute} = \frac{\text{Output of machine hydraulic pump (gals./min.)}}{\text{Oil requirements of press (= injection hydraulic cylinder consumption + clamping hydraulic cylinder consumption, etc.) in gallons per cycle}}$$

Based on plasticising capacity the determination of maximum possible cycle on a specific press is given by:

$$\left[\begin{array}{c}\text{Theoretical maximum number}\\ \text{of cycles per minute}\end{array}\right] = \frac{\left[\begin{array}{c}\text{Effective plasticising rate of press}\\ \text{in lbs. per hour}\end{array}\right]}{\text{Weight of moulding in lbs.} \times 60}$$

Example: Weight of moulding = 6 oz.; Press shot capacity = 8 oz.; Press plasticising rate = 60 lb. per hour.

Assuming that 60 per cent of the plasticising capacity represents the maximum operating efficiency, then the capacity is equal to 36 lb. per hour. Hence, the maximum number of cycles per minute = $(36 \times 16)/(6 \times 60) = 1.6$.

Assume that the maximum dry running speed = 6 cycles per minute.

The number of mouldings produced per hour will, therefore, be controlled by the plasticising capacity of the press. For maximum

TYPICAL THERMAL PROPERTIES OF SOME MOULDING MATERIALS

	Thermal Conductivity cal./cm./cm.^{2}/sec./deg. C. /10^{-4} k	Density gr./cm.^{3} $\dot{\rho}$	Specific Heat cal./deg. C./gr. C_p	Thermal(2) Diffusivity cm.^{2}/sec./10^{-4} α	Thermal(2) Diffusivity in.^{2}/sec./10^{-4} α	Distortion Temperature (deg. C.) T_x
General-purpose polystyrene	3.0	1.06	0.32	8.86	1.34	90
Toughened polystyrene ..	2.0	1.06	0.33	5.72	0.89	80
Polyethylene—low density ..	8.0	0.92	0.55	15.8	2.45	60
,, —high density ..	12.0	0.96	0.55	22.5	3.65	90
Polypropylene	3.3	0.91	0.46	7.9	1.23	90
Acrylic..	5.0	1.18	0.35	12.1	1.88	85
Nylon	5.5	1.1	0.4	12.5	1.96	160
Cellulose acetate(1)	6.0	1.3	0.35	13.2	2.05	70
Cellulose acetate butyrate(1)..	6.0	1.18	0.35	14.5	2.28	70
P.V.C.(1)	3.5	1.3	0.35	7.7	1.2	—
Acrylonitrile styrene	2.9	1.08	0.35	8.15	1.26	90
A.B.S.(1)	5.0	1.03	0.38	12.8	2.0	98

(1) Properties particularly dependent on formulation.
(2) Approximate value based on $\alpha = k/\dot{\rho}C_p$.

efficiency, it is generally considered that an injection press should not operate at above 80 per cent of its rated performance on either shot weight or plasticising capacity.*

The final major consideration is that of clamping capacity. The general requirements have been covered previously (see page 55) but the clamp tonnage required on a given moulding is given by

$$\begin{bmatrix}\text{Clamping force} \\ \text{necessary on} \\ \text{press (tons)}\end{bmatrix} = \begin{bmatrix}\text{Projected area} \\ \text{of moulding} \\ \text{(sq. in.)}\end{bmatrix} \times \begin{bmatrix}\text{Clamping force neces-} \\ \text{sary to resist cavity} \\ \text{pressure (ton/sq. in.)}\end{bmatrix}$$

The projected area is the total area of the moulding as seen when viewed in the direction that it will be placed on the press in the plane normal to the press opening. View (A) in Fig. 3.1 illustrates a side elevation section of a component and the view (B) a plan view of the same

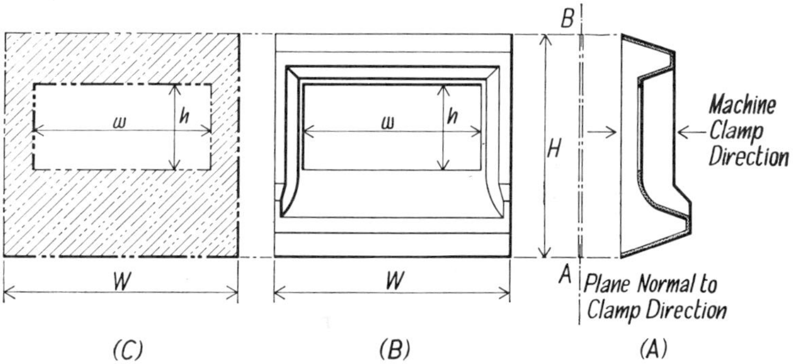

Fig. 3.1. Projected Area.

component. The moulding has a central cutout opening of dimensions $w \times h$. Thus the area of the cavity which sets up pressure opposing the clamp is the projected area $W \times H$ less the area of the opening $w \times h$. The total projected area is shown shaded in the view (C), and this is equal to $(W \times H)-(w \times h)$.

A further case exists with split tools, where not only has the press clamping force to resist the opening force described previously, but the tool itself has to resist the side pressure that tends to open the splits and arises from the projected area of the moulding in a plane normal to the split. View (C) of Fig. 3.2 shows the total projected area of component (A) and (B) which is opposed by the machine clamp. This area comprises the area of the large flange less the area of the hollow core. Since this component is to be produced in a split tool, with the opening line at right angles to the machine clamp direction, an opening force will also be set up in this direction by the cavity pressure which is proportional to the

*Many machines are supplied with alternative injection cylinders giving a range of maximum shot weights with corresponding changes in maximum injection pressure. Care must be taken, therefore, that the cylinder selected on the basis of shot capacity will also give adequate injection pressure for the particular tool and component.

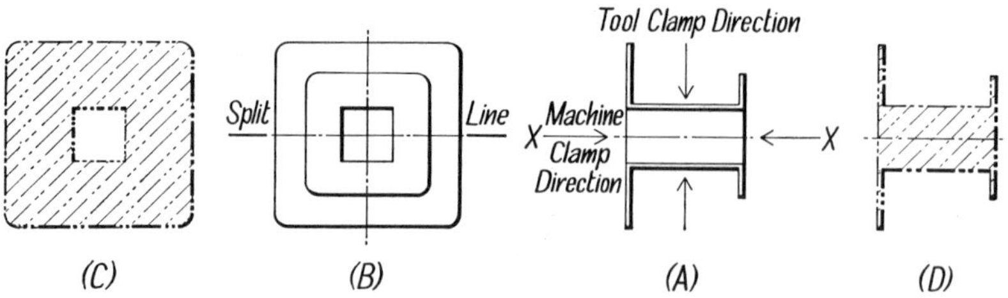

Fig. 3.2. Projected Areas on Split Tool.

projected area in this plane. This projected area is shown in view (*D*).

The clamping tonnage is often the major controlling feature with many mouldings, and often results in tools to produce mouldings of relatively large surface being placed on a press of excessive shot capacity in order to obtain the clamping capacity of the larger machine. In such cases it is useful, during initial tool consideration, to examine whether the moulding could be disposed otherwise to the flash line, in order to reduce both the projected area and clamp requirements. The ability to do so, however, is limited to dispositions which do not result in making the tool design and manufacture too difficult, do not interfere with the line of draw, nor present a flash line in an undesirable visible position on the moulding. An example of such a rearrangement is shown diagrammatically in Fig. 3.3.

At (*A*), the moulding is positioned in the normal manner. In this position, however, the maximum surface area of the moulding ($W \times L$ sq. in.) is opposed to the clamping force. If the moulding is disposed as shown at (*B*), the smaller surface area ($d \times L$ sq. in.) is normal to the press action, and hence, with the same cavity pressure, the clamping force

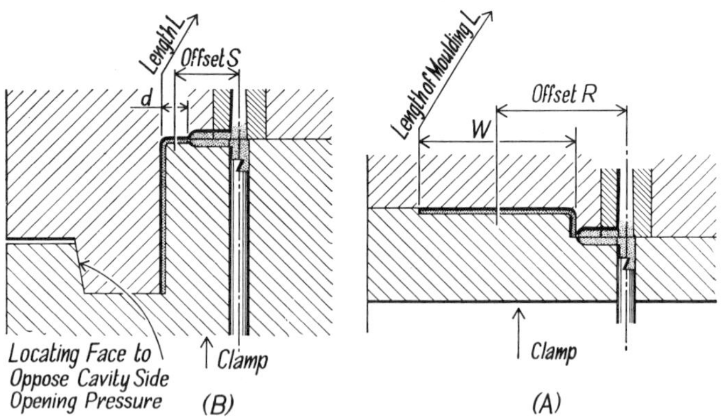

Fig. 3.3. Rearrangement of Disposition of Moulding to reduce Clamping Force.

required to keep the tool closed will be considerably less. If the mould is a single-impression tool, another advantage is demonstrated in this illustration. Since the cavity is offset to the press centre line, the cavity pressure will set up an eccentric load between the platens, giving an opening moment equal to the total cavity pressure times the distance of the centre of the load from the press centre line. In the case of (*B*), both the degree of offset is less (distance *S* being less than *R*) and the total cavity pressure is less ($d \times L$ being less than $W \times L$). The stresses set up in the press from the offset tool are therefore less. With the layout shown in (*B*) however, care must be taken that the high side-opening force generated by the cavity pressure is adequately resisted to prevent sideways deflection of the tool. This is done by providing adequate locking faces and mould thickness.

In the case of those tools whose cavity pressure centre is offset to the centre line of the platens, an adjustment must be made in the clamping force necessary. Here, a packing block is set up between the platens of thickness equal to the closed height of the tool, to balance the clamping force. However, the opening force of the tool set up by the internal cavity pressure then sets up a moment about the packing block, the value of the moment depending upon the degree of offset of tool and packing block. To oppose this moment and keep the tool closed, an increase in clamping pressure is necessary. In all but extreme cases, an increase in clamp of 50 per cent should be sufficient.

Another technique to reduce clamping requirements, in special applications, is that of the double-stacked tool. This latter is a multi-plate tool producing two or more components (usually of a flat nature) in separate cavities, all being closed by a common clamp.

Since the projected area is that normal to the line of press opening, the area of the second moulding lying below the first is held within the same clamp and is not included in the calculation of the total projected area. The double-stacked tool is confined to special cases, since the tool complexity is increased. This technique is illustrated in Fig. 3.4 which shows a three-plate tool for producing a number of saucer mouldings between the fixed

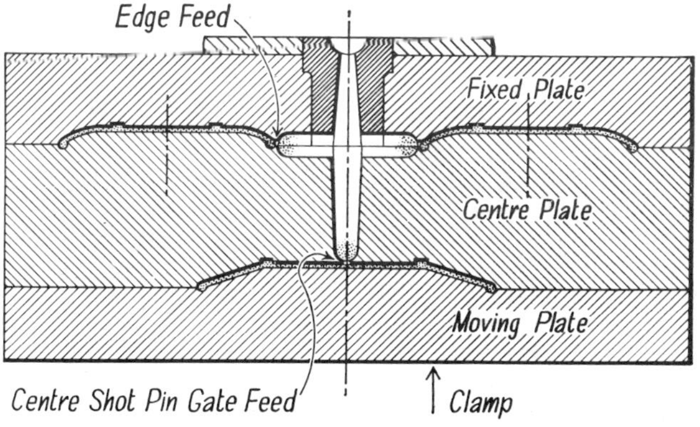

Fig. 3.4. Double-stacked Mould.

and centre plate, and a single moulding between the centre plate and the moving plate. The projected area of the latter which is shaded by the top saucer mouldings does not have to be included in the calculation of total projected area.

Another case is that of two dissimilar mouldings where one is a frame and the other flat—a toilet seat and cover is one example. These are set one above the other, the seat on the injection side and the cover directly in front of it. The feed is taken through the opening of the seat, branching off through runners to feed it on the inside rim, while the main feed continues straight through and gates directly into the cover. Here again only the overall projected area is relevant to the clamp requirements —generally that of the cover only.

Number of Impressions. The number of impressions to be laid down for a given injection tool is primarily determined by:

(1) the piece part cost required; and

(2) the output required in a given period.

Often the question of multiple impressions does not arise, since the total number of components required is insufficient, the moulding is too large, or the tool cost has to be kept down to a minimum. It is generally in mouldings of small or medium size which are required in fair numbers that multiple impressions are considered. Apart, however, from several impressions of the same component, it is quite frequent practice to design a multiple impression tool producing different components in each cavity—generally components comprising a common final assembly. By this means, the cost of producing individual components is decreased, although it does presuppose that equal numbers of each component are required and that they will be required in the same colour and material.

The use of a multi-impression tool will reduce the direct labour cost per moulding and increase output, but against this must be balanced the fact that the use of a greater number of impressions in a larger tool may increase the size of press necessary and hence involve higher press depreciation charges in the cost.

Further, should a larger press be necessary, the increase in output from the extra impressions may be partly offset by the slower rate of operation of the larger machine (i.e. fewer lifts per hour). Finally, the larger tool will obviously be more costly, although not necessarily in proportion.

The effect of varying the number of impressions on cost and output is illustrated by the following example in which two alternative quantities are considered. This method of determining the effect on cost from multiple impressions is similar for any other case, although the details may vary. It should be emphasised that this example is illustrative only. The underlying principles shown are true for nearly all cases but costing systems used in practice are generally more complex. In the example, the same moulding is considered produced from one, two, and four impressions, it being assumed that the press size necessary is increased in each case.

A—SINGLE-IMPRESSION TOOL ON A 2-OZ. PRESS

Lifts per hour	100
Tool cost	£200
Press rate charge	5/– per hour
Cost of material per moulding	4d.
Cost of labour and overheads per moulding ..	4d.

1. *Based on a quantity of* 10,000 *mouldings :*

Cost of material per moulding	4d.
Cost of labour and overheads per moulding ..	4d.
Press rate per moulding	0.6d.
Amortisation of tool per moulding	4.8d.
TOTAL	13.4d. each

This will give an output of 800 per 8-hour day, and the order will require 12.5 working days to complete.

2. *Based on a quantity of* 100,000 *mouldings :*

Cost of material per moulding	4d.
Cost of labour and overheads per moulding ..	4d.
Press rate per moulding	0.6d.
Amortisation of tool per moulding	0.48d.
TOTAL	9.08d. each

800 mouldings per 8-hour day requires 125 working days to complete order.

B—TWO-IMPRESSION TOOL ON A 4-OZ. PRESS

Lifts per hour	90 (i.e. 180 mouldings per hour)
Tool cost	£350
Press rate charge	8/– per hour
Cost of material per moulding	4d.
Cost of labour and overheads per moulding ..	4d. × (100/180)

1. *Based on a quantity of* 10,000 *mouldings :*

Cost of material per moulding	4d.
Cost of labour and overheads per moulding ..	2.2d.
Press rate per moulding	0.53d.
Amortisation of tool per moulding	8.4d.
TOTAL	15.13d. each

This will give an output of 1,440 mouldings per 8-hour day, and the order will require just under 7 working days to complete.

2. *Based on a quantity of* 100,000 *mouldings :*

Cost of material per moulding	4d.
Cost of labour and overheads per moulding ..	2.2d.
Press rating per moulding	0.53d.
Amortisation of tool per moulding	0.84d.
TOTAL	7.57d. each

1,440 mouldings per 8-hour day requires 69.5 working days to complete.

C—FOUR-IMPRESSION TOOL ON AN 8-OZ. PRESS

Lifts per hour	80 (i.e. 320 mouldings per hour)
Tool cost	£600
Press rate charge	12/- per hour
Cost of material per moulding	4d.
Cost of labour and overheads per moulding ..	4d. × (100/320)

1. *Based on a Quantity of* 10,000 *mouldings :*

Cost of material per moulding	4d.
Cost of labour and overheads per moulding ..	1.25d.
Press rate per moulding	0.45d.
Amortisation of tool per moulding	14.4d.
TOTAL	20.1d. each

This will give an output of 2,560 mouldings per 8-hour day, and the order will require just under 4 working days to complete.

2. *Based on a Quantity of* 100,000 *mouldings :*

Cost of material per moulding	4d.
Cost of labour and overheads per moulding ..	1.25d.
Press rate per moulding	0.45d.
Amortisation of tool per moulding	1.44d.
TOTAL	7.14d. each

Thus, 2,560 mouldings per 8-hour day requires just over 39 days to complete the order.

Apart, however, from economic considerations, certain other factors have to be taken into account in determining the number of impressions to be laid down.

Firstly, cognisance must be taken of the machine capacity position—an increase in the number of impressions may require a size of press that is already overloaded with other work. Here, it must be determined whether the extra advantage of the greater number of impressions must be forgone.

On the design side it must be realised that, although the impressions are nominally identical, in practice this perfection is not reached and the mouldings from each cavity will be slightly different. In most cases, these variations can be tolerated, but if exceptional accuracy is needed a single impression tool is preferable.

Finally, overall financial risks increase as the number of impressions increases: the greater the commitment if in fact the order is cancelled, the greater the losses involved if tool or component design faults become apparent on moulding and require rectification, and the greater the capital outlay on the initial tool.

Initial Tool Considerations

(1) FLASH FACE. Invariably, the first point to be determined in approaching tool design is the position of the flash line relative to the

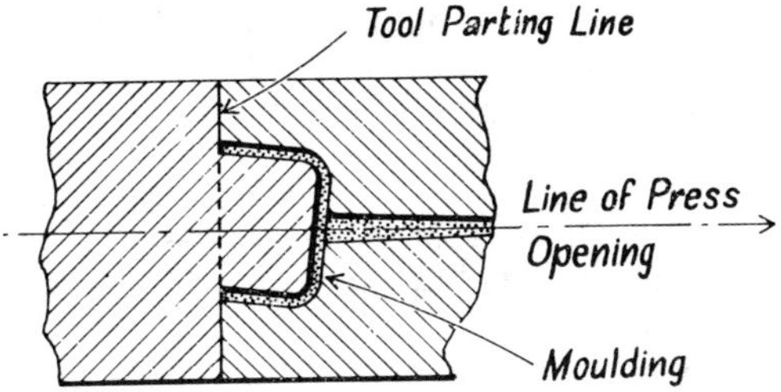

Fig. 3.5. Impossible Split-line Position; Male Tool forming Undercut.

component—i.e. the parting or split line between the two halves of the tool.

Here, various major considerations are involved:

(a) The split line normally leaves a witness line on the moulding where the moulding material creeps between the two faces of the tool. In bad cases, this creep may extend and a line of flash is formed which may have to be removed on finishing by filing or polishing. Thus, it is important that the flash line occurs at a point on the component which is visually and functionally acceptable.

(b) The relationship of the flash face to the line of press opening must be such that the moulding can be extracted without undercuts. Obvious undercuts are seldom overlooked (an example is shown in Fig. 3.5), but obscure slight undercuts commonly arise in blending geometrical profiles, and these require careful detection. This problem is also considered in a later section in conjunction with the line of draw.

(c) The split line should so divide the tool that the separate portions are capable of straightforward machining. Some choices of flash line give

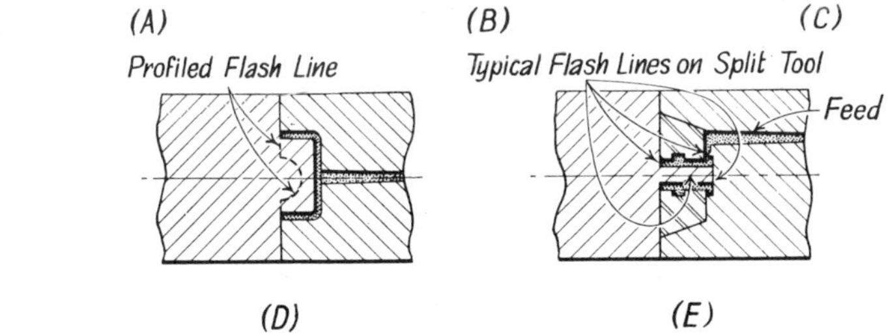

Fig. 3.6. Various Types of Flash Face.

rise to considerable difficulties in sinking the cavity or machining the punch, and these are best avoided.

(d) The flash line itself should be kept as simple as possible. A great number of tools have flat plain flash faces such as at (*A*) in Fig. 3.6, and these can be readily produced and give minimum trouble in service. If the flash line is stepped as at (*B*), or profiled as at (*D*), the difficulties in machining the split line are increased. The problem of bedding down evenly the various flash surfaces to produce a good shutout face is increased and the liability to flashing considerably raised.

(e) The choice of split line should not interfere with efficient ejection of the moulding, and the opened tool should present the finished moulding easily accessible for removal by the operator: see Fig. 3.7.

(f) Attention must be paid to any close tolerances that are called for on the moulding. It is undesirable that dimensions calling for high accuracy should be divided by the flash line, since any tendency of the tool to close incompletely in operation will affect this dimension.

The main flash line may take a number of forms—plain flat faces, stepped faces on two or more levels, angular and conical faces, and profile flash lines. In the case of split tools, flash lines occur on the dividing line of the splits and, in some cases, at one or both ends of the cavity, as shown at (*E*) in Fig. 3.6.

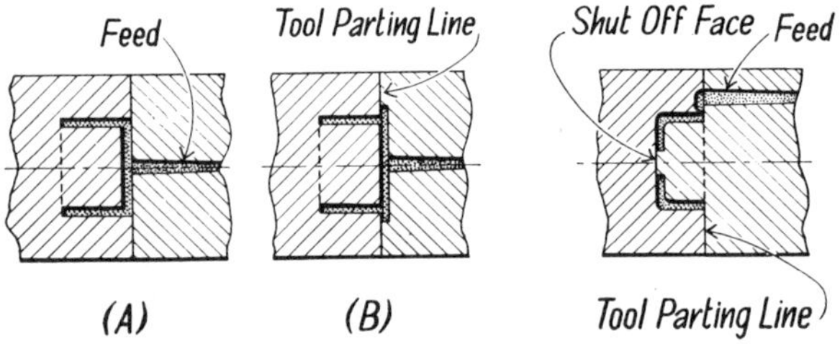

Fig. 3.7. [*Left*] (*A*) Undesirable Split-line Position with Moulding Difficult to extract; (*B*) Instance where the Previous Position cannot be avoided on Simple Tool. Fig. 3.8. [*Right*] Flashing will occur at Core Shut-off Face on Tool.

Apart, however, from the main-tool opening split line which is the initial consideration, subsidiary flash lines will, of course, occur at other shut-off faces—particularly those of fixed and moving cores, and the effect of these must be taken into account on the same lines. As can be seen from Fig. 3.8, any opening of the tool at the main split line will cause similar flashing at the core shut-off face.

(2) Line of Draw. The two parts of the tool must separate as the press opens, without interference, and, when open, the moulding must be capable of extraction from the tool without restraint. This obvious requirement must be considered in conjunction with the choice of split line.

This controlling feature of the line of draw is determined from the line of press action, and all tools must therefore open in this direction.

In a large proportion of tools, including those shown at (*A*) to (*D*) in Fig. 3.6, the moulding is also extracted in the line of draw.

With many components, however, this is not possible and a split type of tool is employed. Such cases arise where a feature on the moulding prevents straightforward ejection as at (*E*) in Fig. 3.6. On occasion, the undercut feature may be sprung from the tool. The ability to do this is chiefly determined by experience, being dependent upon the particular shape of the component, the proportions of the undercut and the material employed. If this method is doubtful or impossible, a split tool must be used. The tool is then made in two or more sections with the split line at an angle to the line of the press action (frequently at right angles), the split being so chosen that, on opening, the moulding is free of the two halves. These tool splits are then set in the bolster, the design often providing that the seating faces are tapered so that the two halves of the splits are tightly held together when the press is closed. Some form of mechanism is provided to open and close the splits, and frequently this is done by angle pins set in one half of the main tool and engaging in bushes

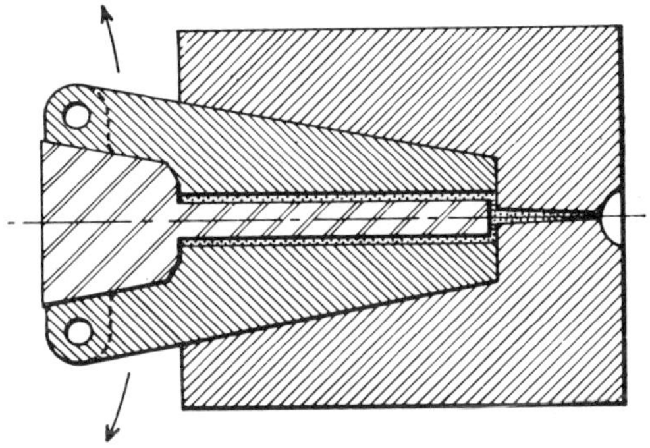

Fig. 3.9. Hinged Split Tool for producing a Moulding with Long Slender Core.

in the splits mounted on the opposite half. When the press is opened, the two main portions of the tool part while the angle pins cause the splits also to slide apart on the face of the bolster at an angle to the press action. By this means, the moulding is then clear in the line of the press opening to be extracted.

The flash line on this type of tool is, of course, that of the parting line of the splits, but an additional flash line at the normal flash face of the bolster and stationary portion of the main tool may also be involved if the tool cavity is not contained fully within the splits but is open to this face.

In Fig. 3.9 is illustrated a special application of splits to the moulding of a long, slender tubular component. This could have been moulded on its side in a normal plain tool, but this would mean that the ideal of feeding down the length of the tube would have been difficult to achieve, and the long core pins would be at right angles to the line of draw, and would have to be extracted by means of a hydraulic cylinder. By placing the component in splits in the manner shown, the feed can be led directly from the sprue bush to the end of the moulding, and the component can be ejected directly off the stationary core pin; e.g. by means of a sleeve ejector (not shown). Had the component been moulded in the position shown in a non-split tool, ejection could have been difficult.

The application is warranted here by the special circumstances, particularly the length and slenderness of the core; but usually split tools are more complex and expensive than the normal type. Generally, most injection moulding tools are based on one of these two types, or combinations of both, but either may, of course, have additional stationary or sliding cores, threaded pins, inserts, etc.

The actual position of the flash line may require consideration from the aspect of moulding accuracy as well as that of appearance. This is illustrated in Fig. 3.10 which depicts at (*A*) the component, an electrical switch plate, in which the diameters *E* and *F* must be concentric to 0.001 in. This requirement can lead to difficulties since the diameters will be

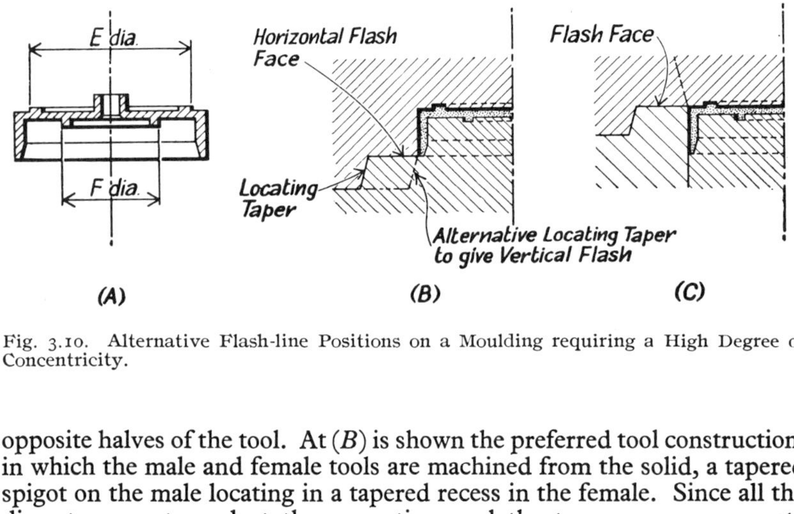

Fig. 3.10. Alternative Flash-line Positions on a Moulding requiring a High Degree of Concentricity.

opposite halves of the tool. At (*B*) is shown the preferred tool construction, in which the male and female tools are machined from the solid, a tapered spigot on the male locating in a tapered recess in the female. Since all the diameters are turned at the same time and the taper ensures accurate location, the concentricity of the diameters will be maintained. This design gives horizontal flash, but if vertical flash is required, the taper is positioned as indicated in chain-dotted lines.

If the flash is positioned as shown at (*C*), two principal drawbacks occur:

(i) The moulding is trapped between the two steel faces forming the inside and outside diameters, and hence eventual ejection of the moulding may be difficult.

(ii) The male tool would be very difficult to machine from the solid, because of the depth of the skirt. It would, therefore, have to be produced as a separate pad, let into the body, as shown. This would involve an increase in machining and fitting time when compared with the male tool of (*B*), although the female tool is, in fact, somewhat easier than the female of (*B*), due to the decreased depth of cavity to be machined in (*C*).

At the commencement of tool consideration, the component is considered in various approaches to possible tools, particularly the angle at which the moulding is placed to the line of draw. On this basis, the effect of line of draw and ejection are considered, together with the position of flash line, method and position of feed, and ease of tool construction. As a result of these considerations, the approach producing the most satisfactory compromise is adopted. On many components, the design of moulding can include features which can lead to undercuts in some tool designs, and so hamper ejection. Such undercuts can be dealt with by means of sliding cores, or by the adoption of a split tool. Alternatively, however, sometimes the component can be removed by placing the component at an angle so that it can be drawn off the undercut in the line of draw, or the component itself can be ejected at the angle of undercut. This is illustrated in Fig. 3.11, and demonstrates that consideration of the angle of orientation of the component within the tool may avoid difficulties of this nature.

In Fig. 3.11, (*A*) illustrates a component incorporating a handle recess. If this moulding is placed parallel to the flash face in a plain tool, the solid core forming the back angle of the handle will clearly form an undercut, and it will be impossible to extract the moulding.

The moulding can be positioned in this manner, however, if the tool is constructed as shown at (*B*), in which the whole of the undercut face is made as a separate tool component to form a combined moving core and ejector which can be moved forward parallel to the line of the press action. This core is in its downward position during moulding to form the undercut, but, when the tool is opened, this core is moved forward by the ejector gear, carrying the moulding clear of the moulding face. The moulding is subsequently lifted off the core in the direction of the undercut. It should be noted, however, that under certain circumstances this angular removal of the moulding off the core can present difficulties (due to space limitations, etc.).

If the moulding is repositioned at an angle relative to the mould face so that the undercut is parallel to the line of ejection, as shown at (*C*), the need for the moving core is avoided and ejection is straightforward. In adopting this technique, however, care must be exercised that placing the component at an angle does not increase the difficulty of machining the mould unduly, and that any side pressures set up by the cavity pressure on the tilted faces are secured by adequate counteracting locking faces.

(3) Feed. The method of feeding the components, that is the method of leading the incoming plastics material from the nozzle of the press cylinder through the sprue bush to the runners and thence to the cavity via the gate, must also form an equal part of the initial tool consideration.

The sprue bush is of course in a fixed position, but the gate feeding the moulding must be positioned to provide the best conditions of flow for the material in filling the cavity. The mark left when the gate is finally severed from the moulding must be visually acceptable or subsequently polished out. Finally, the solidified plastics material in the sprue and complete

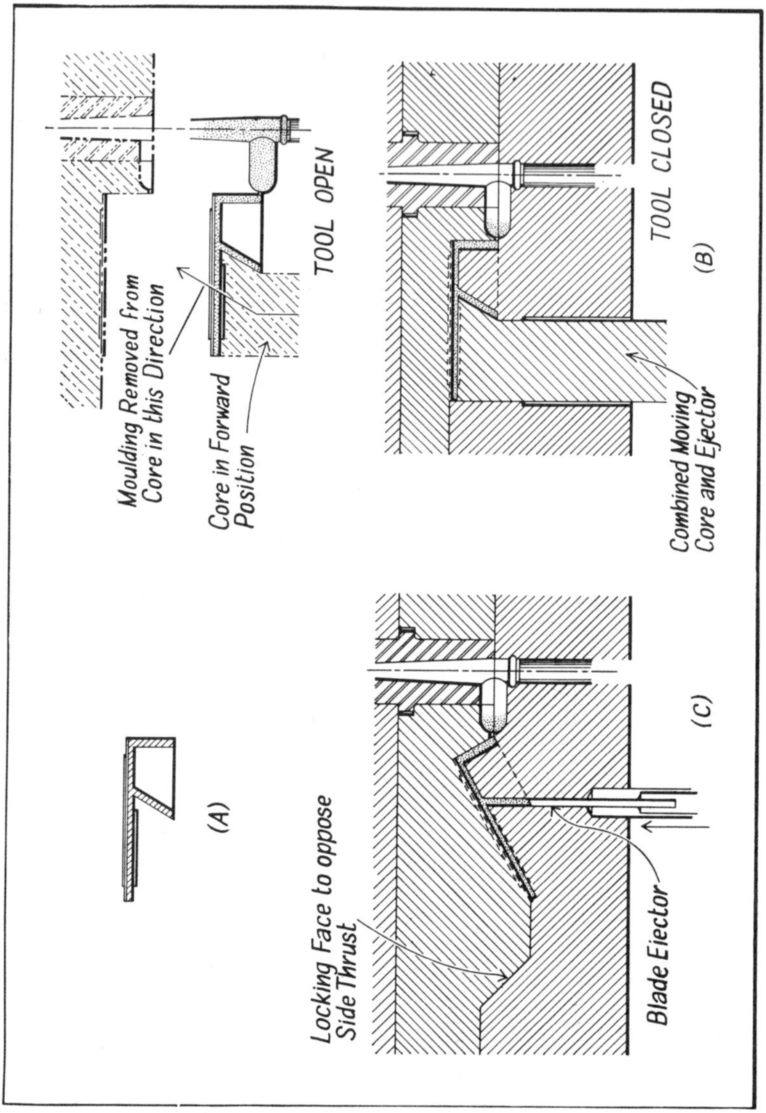

Fig. 3.11. Moulding of Undercut Faces.

runner system must be capable of extraction from the tool on opening (except hot-runner moulds). In the normal two-plate tool, the moulding is extracted with the sprue and runner still attached—either shot in the centre as at (*A*) in Fig. 3.12, or edge-fed as at (*B*)—the sprue being subsequently cut off. It is frequent, but not invariable, practice to ensure that, when the tool opens, the moulding goes away in the moving half of the tool therefore drawing out the sprue or sprue and runner from the fixed half of the tool at the same time. This is done because the ejector gear of the press operates on the moving half of the tool, and can thus be used to eject the moulding. If the moulding is to remain in the fixed half of the tool, it can be ejected only by employing such means as chain ejectors. These, however, although quite frequently used, tend to obscure the operator's access

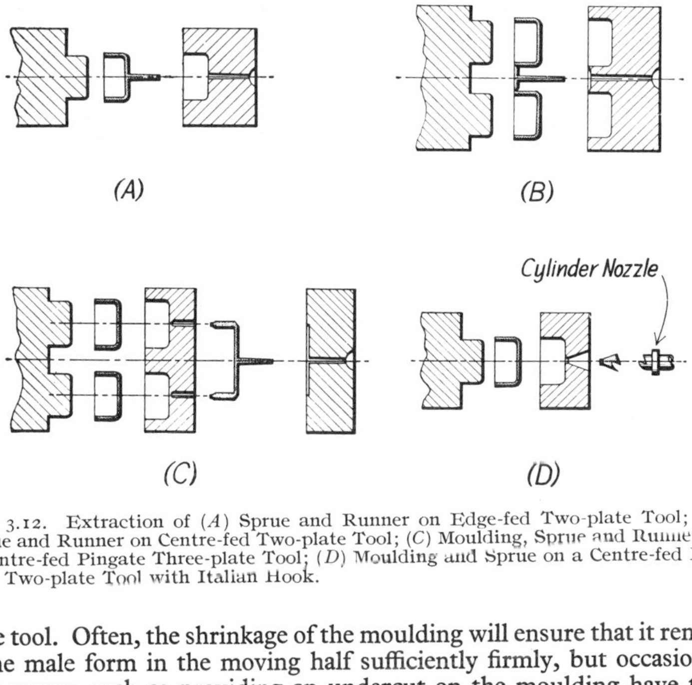

Fig. 3.12. Extraction of (*A*) Sprue and Runner on Edge-fed Two-plate Tool; (*B*) Sprue and Runner on Centre-fed Two-plate Tool; (*C*) Moulding, Sprue and Runner on a Centre-fed Pingate Three-plate Tool; (*D*) Moulding and Sprue on a Centre-fed Pingate Two-plate Tool with Italian Hook.

to the tool. Often, the shrinkage of the moulding will ensure that it remains on the male form in the moving half sufficiently firmly, but occasionally other means such as providing an undercut on the moulding have to be employed.

In the case of a three-plate tool, the mouldings are removed from the space between the moving half of the tool and the centre plate, whereas the sprue and runner are extracted from the space between the fixed half of the tool and the centre plate. This is illustrated at (*C*) in Fig. 3.12. Three-plate tools are obviously more expensive than two-plate tools, but among their advantages they allow centre-feed-pin gating on single or multiple impressions. With this type of gate, the sprue stalk terminates in a small orifice forming the gate to the moulding. This minute gate parts on opening of the tool, and thus no gate has to be cut off the moulding after removal and only a small mark is left on the component.

An attempt to achieve a similar result with a two-plate tool is shown at (*D*). Here, the sprue is pulled from the fixed half of the tool, it being arranged that the sprue remains partly attached to the specially designed cylindrical nozzle (often known as an Italian hook). On the retraction of the nozzle during the sprue break, the sprue is thus withdrawn from the tool, and is subsequently removed manually from the nozzle hook.*

The point of gating is generally either in the centre of the moulding (centre shot) or at the edge of the moulding. The first can be used with a single-impression, two- or three-plate tool, with a multi-impression three-plate tool or Italian-hook tool. Edge feeding is generally used with single- or multi-impression two-plate tools.

Precisely where the projected moulding is to be fed is a matter of great importance, not only in deciding the type of tool to be employed but also in having the greatest influence on the finish, strength and accuracy of the final moulding. Aspects of the position (centre shot, edge feed, etc.) and the type of gate (full, restricted, tab, etc.) are dealt with specifically in a later section on this subject, as well as the effects of gate position. Apart from the visual aspects of the gate already mentioned, considerable thought must be given to the path which the material will follow as it enters from the gate to fill the cavity.

Flow conditions will vary with every individual moulding due to particular circumstances of flow-path length and section thickness, and again will be dependent upon the thermoplastic material used, since these materials have different viscosities at melt temperatures and varying degrees of flow stiffness as the temperature of the material drops on entering the tool. There are, however, certain well-known points which apply to flow, including the following:

(a) The gate should not lead directly into an open cavity, otherwise "jetting" or "worming" will result, the incoming plastics material spurting uninterruptedly into the cavity in form of a jet or spiral and tending to produce surface marks on the final moulding. The gate should direct the material against some obstruction to break the flow, e.g. into a section or against a core or pin. A gate arrangement liable to cause jetting is shown at (*A*) in Fig. 3.13A, while an improved gate to avoid this is shown at (*B*). An example of jetting is shown in Fig. 3.13B.

(b) The unfilled mould is full of air and this has to be displaced by the incoming plastics material, the air being vented through the flash faces, ejectors and moving cores. If the air is not fully vented but is entrapped by the plastics material, this may lead to short mouldings, actual holes in the completed moulding, or to black burn marks caused by the com-

---

**Italian Hook :* a later development of this method does not employ the hook on the nozzle but allows the plastic material to remain in the 'pre-chamber'—the nozzle retracting or remaining in position as required.

The material in the pre-chamber acts as a thermal insulator, the centre remaining hot and allowing subsequent injections of plastic to be forced through. Should the press be stopped, of course, this insulator will solidify and require removal before restarting.

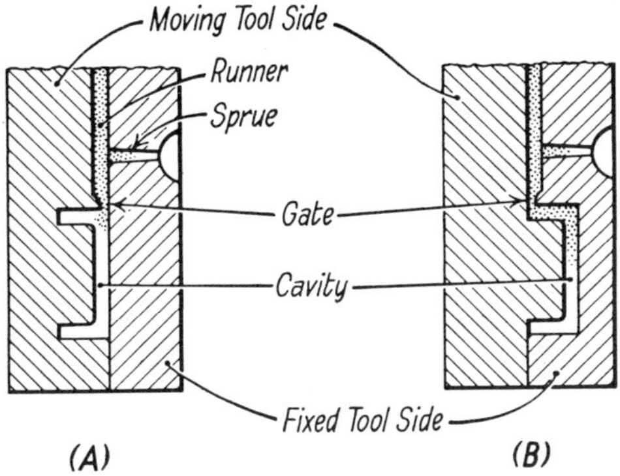

Fig. 3.13A. The Gate at (*A*) may lead to Jetting; That at (*B*) is designed to eliminate Jetting.

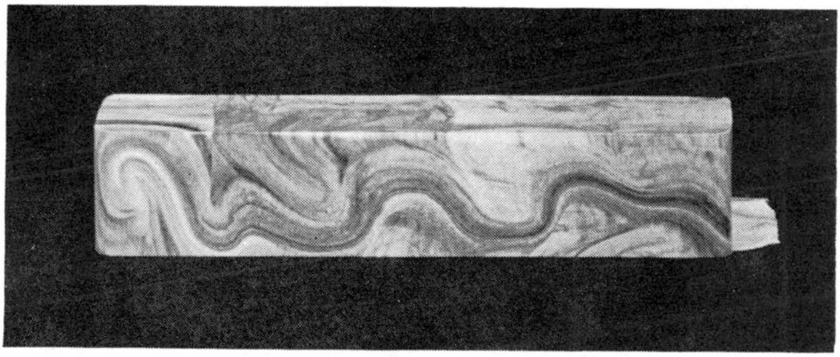

Fig. 3.13B. Example of Jetting from Gate into Open Cavity illustrated by Moulding in Multi-coloured Material. [*Courtesy: Ekco Plastics Ltd.*]

pressed air becoming superheated. Thus, it is most important that the expected flow path from any proposed gate position be considered in this respect. The incoming material must be arranged to expel the air from one of the vents mentioned, i.e. between the fitting faces of portions of the tool communicating with the cavity. One of these faces should be moved relatively to the other during some portion of the cycle, to ensure that the vent cannot become blocked by flash; hence, for this reason, venting between fixed blocks making up the tool construction cannot be considered as always satisfactory. In operation, it may be found necessary to file a minute path across the flash face or along the ejector or core pins, to improve the venting.

In considering the flow, due attention must be paid to the effect of section thickness, since the material will flow more easily in a thick section than in a thin section. Thus, in a moulding having thin and thick sections, the latter will act as runners and feed material ahead of that passing through the thin sections. Additionally, the material in the thin section will cool more rapidly and, in extreme cases, while the rest of the mould is filling, the material entering the reduced section may freeze and give a short moulding.

An example is indicated in Fig. 3.14. This is a circular moulding with the feed into the thick rim. Here, the material will rapidly flow right round the rim and then start flowing up the sides before the top of the moulding has filled. The air is thus trapped by the material coming up from the sides and may cause a hole or a burn. In such an instance, it is often necessary to increase the side-wall section immediately adjacent to the gate, in order to expedite the flow of material across the top and thus prevent air becoming entrapped. Another example is shown in Fig. 3.15, this type being similar to the gas-traps which arise from long bosses on many mouldings. Variation of wall thickness and provision of runners are common practice in order to overcome such difficulties.

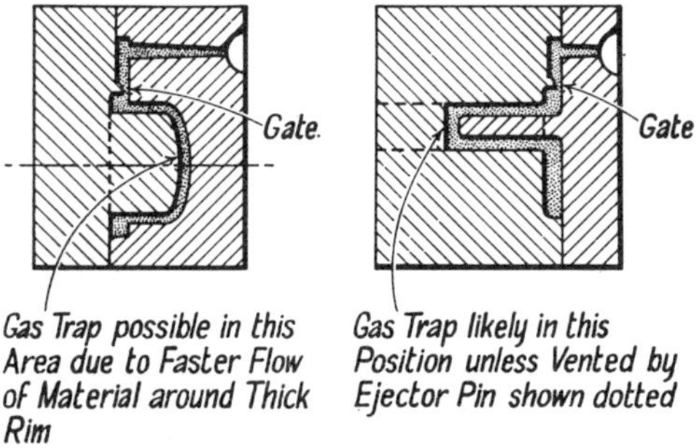

Fig. 3.14 [*Left*] and Fig. 3.15 [*Right*] Types of Gas Trap.

When a mould is designed, the most advantageous flow pattern available is adopted, but detail modification such as locally thickening the section, the addition of decorative ribs, runners, etc., is best left until actual sampling. Then, by moulding a succession of "short" shots as shown in Figs. 3.16 and 3.17, the actual flow can be analysed, and remedial action proposed.

An example of entrapped air can be seen in Fig. 3.16 of a styrene container. The other example shown in Fig. 3.17 is a knob moulding in cellulose acetate. At the extreme right hand is shown the complete moulding, and it will be seen that the hole caused by the entrapped air has been completely closed; but the actual moulding shows an unacceptable flow line and surface burning.

(c) It is general practice to lead the gate into a thick section of the moulding and allow the material to flow from there into the thinner sections. In a thin section, the material cools off much more rapidly, and hence retards the flow of material to other parts of the cavity. Very occasionally, however, it is only possible to feed into a thin section, and any resulting flow difficulties have to be overcome by press technique.

Another aspect of this problem is the control of sinking or voids in heavy sections. These are normally overcome by increasing the injection ram forward time (i.e. packing), but if the section leading to the area

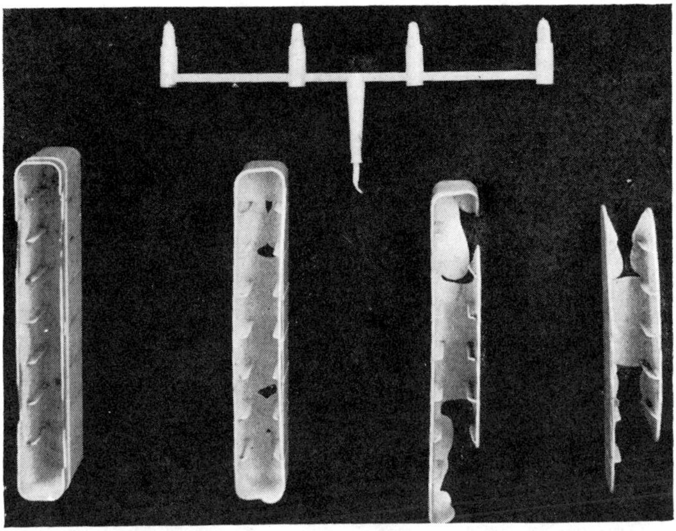

Fig. 3.16. Container Moulding in Polystyrene. The Illustration shows the Progressive Filling of One Impression in a Four-impression Mould, the Component being fed in the Centre. The Effect of the Side Ribs acting as Additional Runners can be clearly seen, the Ribs feeding Material ahead of the Flow from the Main Gate, resulting in the Formation of Gas Traps. There is also a Variation in the Thickness of the Two Long Walls as can be seen from the Uneven Flow. [*Courtesy: Ekco Plastics Ltd.*]

that is sinking is thin, this will tend to freeze off and make the packing ineffective. Thus an adequate thickness of section, or the addition of feeder ribs, etc., to ensure continuation of flow from gate to thick sections is advisable.

(d) The flow of incoming material may divide as it enters the tool, either by meeting core pins or inserts, by meeting projections on the male tool, or by the material flowing faster along certain sections of the tool due to thicker sections, or proximity to the gate. When these streams of material eventually rejoin, the joint forms a weld or knit line. The effect of these knit lines has to be carefully considered because:

(i) They may form a line of weakness, i.e. the strength of the weld may

Fig. 3.17. Radio Knob in Cellulose Acetate—a Series of Mouldings showing Progressive Filling of Cavity. The Material Flow Pattern is clearly seen leading to the Formation of a Gas Trap in the Side Wall. The Final Moulding carries a Marked Flow Line in this Area.

Fig. 3.18. The Injected Material divides on entering the Gate to flow round the Cavity and meets across the Slot. At the Line of Junction it forms a Knit Line (marked on the Moulding for Clarity). This Knit Line may be a Source of Weakness. [*Courtesy: Ekco Plastics Ltd.*]

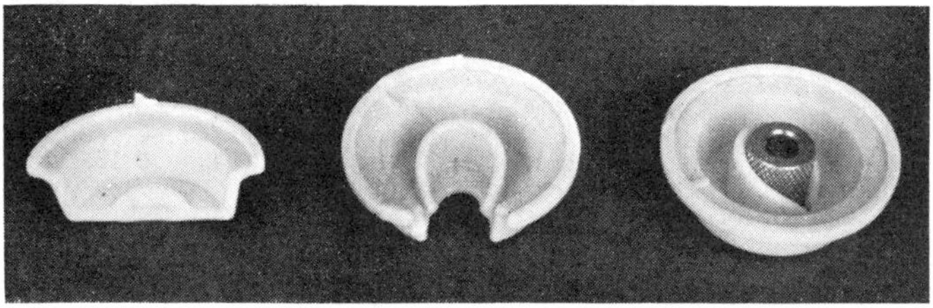

Fig. 3.19. Radio Knob in Polystyrene Copolymer—the Chilling Effect of the Large Cold Insert can be clearly seen.

be less than that of the parent material, especially if the weld also contains entrapped air. This is illustrated in Fig. 3.18. Here, the material has divided on entry at the gate and flowed into two paths around the moulding, to join again as a line across the front slot. On the actual moulding, this weld line may be scarcely visible, but has been marked in on the illustration, and if the moulding were bent, it would snap across at this line. This point is very important on mouldings subject to stress, and the probable knit-line position must be carefully estimated before the gate position is finalised.

Occasionally, the effect is minimised by adding a slug cavity adjacent to the weld line, this being connected to the moulding by a gate as shown in Fig. 3.20. By this means, the slightly cooler material at the head of each path of flow is diverted into the slug, the weld line in the component being effected by the hotter following material, and thus having increased strength.

Another related case is that of mouldings similar to, say, the typewriter spacing bar, usually consisting of a cored-out bar with fixed bosses at either end and produced on a two-plate tool. If it is edge-fed at one end, no weld line occurs, but it is difficult to prevent the sinking of the boss at the end remote from the gate. If it is gated at the centre, the condition is only slightly improved and occurs at both ends, while the component may also bow. The prevention of sinking often requires that the component is fed from gates at both ends adjacent to the bosses, the weld line that occurs at the centre being tolerated.

Multiple feeds often have to be resorted to, either to fill a long cavity (e.g. Fig. 5.10A), or to control sinking. Again, multiple feeds to a single moulding may be used on three-plate or hot-runner tools to prevent distortion. In all these cases, weld lines will be formed.

(ii) Under some circumstances, the weld line may form an unsightly mark on the moulding. Generally, on opaque mouldings, this may be overcome by correct press technique in all but mouldings of exceptional flow difficulties. In mouldings produced in plastics which reveal the flow

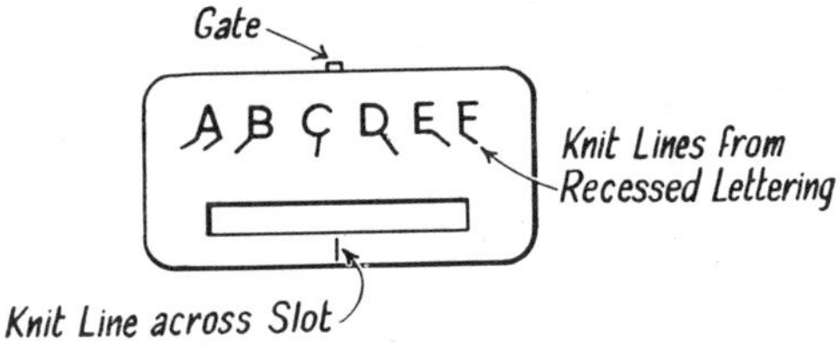

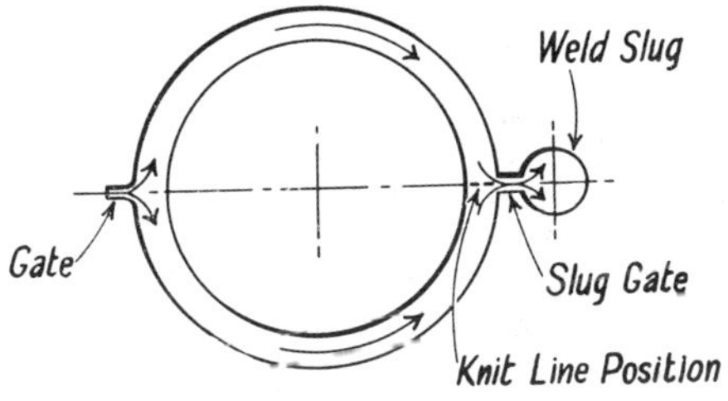

Fig. 3.20. [*Bottom*] Weld Slug to improve Knit-line Strength. Fig. 3.21. [*Top*] Knit Lines in Clear Mouldings.

pattern, such as mottles and pearlescent materials, the weld line will, of course, be apparent but need not be unsightly. Similar conditions can arise with matt-finish mouldings, or opaque materials subject to colour separation. In clear mouldings, the weld line will form a faint line internally across the moulding, and may not be acceptable if it has any optical use. To avoid these internal knit lines, the holes in items such as radio scales and motor-car speedometer dials are often drilled after moulding, because the knit line is caused by the flow of material around the core pin which forms the hole. A similar case arises with lettering recessed into a clear moulding. Here the material flows round the raised characters on the tool and rejoins on the opposite side, thus leaving a clearly visible knit line projecting from the letter. This is illustrated in Fig. 3.21, and for the reason stated, the depth of recessed characters relative to the thickness of the moulding is generally limited to 1 : 3, although with specialists in this type of work, greater depths can be achieved.

(4) Side of Feed. An early consideration is the decision of the position of feed, that is determining which point or which side the gate is to enter the moulding. This will depend to some extent on the visual aspect of the finished moulding, since on some portions the mark left by the gate or sprue may not be acceptable. However, on some occasions facility of tool design may be the overriding factor. In this respect, as explained previously, it is generally desirable to arrange that the moulding comes away on the moving half of the tool, since the press ejector is on this side. This requires that the moulding tends naturally to stay on this half as the tool opens. This is achieved by:

(i) Relying on the shrinkage of the moulding to retain it in the tool (e.g. in a box mould the moulding will shrink on to the male core and away from the cavity walls).

(ii) Utilizing any sliding cores on the moving half of the tool, by arranging that they are not withdrawn until the moulding is free of the stationary tool.

(iii) Adding artificial undercuts on the moving half to retain the moulding, these being subsequently forced free by the ejectors.

(iv) Employing push-off pins in the fixed half to keep the moulding in the opposite half; these generally require chains to operate, however.

(v) Utilizing any threaded core pins or insert carriers in the moving half of the mould to pull the moulding out of the fixed half of the tool and retain it in the moving tool. This requires that the core pins or carriers are locked in the moving tool and then freed prior to ejection of the moulding.

Consideration of alternative cavity disposition and side of feed are illustrated diagrammatically in Figs. 3.22, 3.23, and 3.24.

In Fig. 3.22 are shown three different methods of feeding a box-type component. At (*A*) is shown the normal method for this kind of component in which the male form of the tool is on the moving half, the moulding remaining on this half when the tool opens, due to its natural shrinkage on the core. The sprue is led centrally from the sprue bush into the centre of the base of the moulding. The gate is of the direct full type, and must be machined to a stub after moulding. This tool is of two-plate design, but if it is made a three-plate tool, a pin gate could be substituted for the full gate, and this method would leave only an unobtrusive mark on the moulding, and require no machining. With either type, the normal press ejection system can be employed. The sprue is the shortest possible, and the flow of material from the central position is even and good.

If it is essential that the sprue appears on the inside of the moulding, a layout similar to that shown at (*B*) will have to be adopted. In this case, the male form is on the fixed side of the tool and the moulding will normally remain on this side when the tool opens. Means can be adopted, however, to ensure that the moulding goes away in the moving female half, as detailed previously, and then ejected normally. On occasions it may be decided to leave the moulding on the male and then eject it from this side, as shown in the illustration, the ejectors being operated by chains from the moving tool. Again, this design is shown as a two-plate tool, but could be a three-plate tool with a pin-gate feed instead of a full gate. The

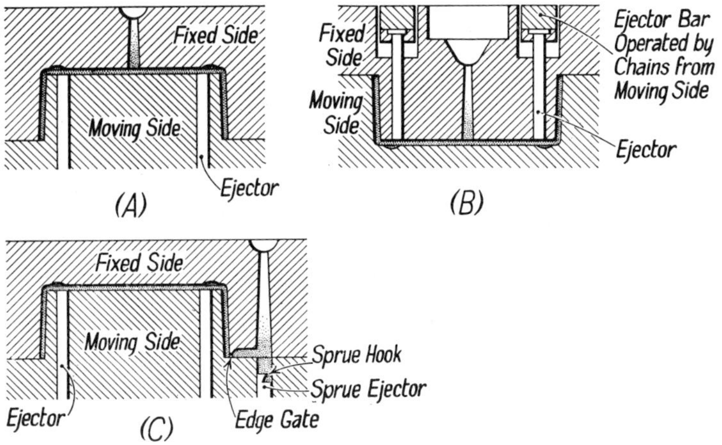

Fig. 3.22. Alternative Feeds and Layouts for Box-type Mouldings.

central gate in either instance gives good flow conditions, but the design approach is not generally as good as (*A*). The disadvantages are:

(i) A much longer sprue is necessary or else the use of an extended nozzle on the press heating cylinder. The former may be slender and difficult to machine, while a moulded broken sprue can be difficult to extract.

(ii) The male core will tend to overheat, due to extra heat being added by the sprue. Thus additional cooling will be necessary, and this can be difficult, especially on small deep cores.

(iii) Ejection difficulties are increased as discussed previously.

At (*C*), the feed is from a side gate, the male form of the tool again being on the moving side and the moulding remaining on the moving side when the tool opens. A gate in this position may be desirable from the visual aspect, but the flow conditions are worse than in (*A*) or (*B*), since the extremity of the mould is farther from the gate than in the first two. The mould will, therefore, require a higher pressure to fill the cavity, and this in turn can lead to increased distortion. Flow lines on the moulding can be worse (with corresponding decrease in moulding strength) and, with some designs, may lead to gas traps in the base of the moulding.

If the mould has a single impression, other difficulties are introduced by virtue of the fact that the centre line of the cavity is offset to the centre line of the platens. This means that the clamping pressure is eccentric, an undesirable condition leading to excessive wear on the press, even although the effects are minimised by the addition of packing blocks to the opposite side*. On occasions, the eccentricity of the tool to the centre line could lead to difficulties in setting up, due to interference with the tie-bars. If the

*Certain presses are available with offset injection units. These allow the use of offset gates without eccentric clamping.

tool is of the multi-impression kind, this eccentricity need not arise. Not infrequently this type of edge-feed layout has to be adopted, particularly on multiple-impression tools where the only alternative is a more expensive hot-runner feed or three-plate centre-shot tool, but type (*A*) should be preferred if possible.

Fig. 3.23 shows various alternative methods of feeding a spindle-knob moulding, the cavity and punch layout being the same in each case.

The design at (*A*) shows a three-plate tool (the back fixed plate not shown) with a pin-gate feed to the centre of the moulding. The gate must be as small as possible since it occurs at a visible portion of the moulding. This position gives optimum flow conditions and maximum strength, since the material flows evenly down the core pin and skirt, venting occurring at the ejector sleeve in the first case and at the flash face from the skirt.

An edge-feed on a two-plate tool is shown at (*B*), this having the advantage that the gate does not occur in a prominent position. This approach is very commonly adopted, particularly for multi-impression tools. In spite of its wide use, however, it is not especially good; the one-sided feed is bad, giving a poor flow pattern which is likely to lead to gas traps on the face of the knob, and a poor mechanical strength in the spindle with possible slight distortion of shape.

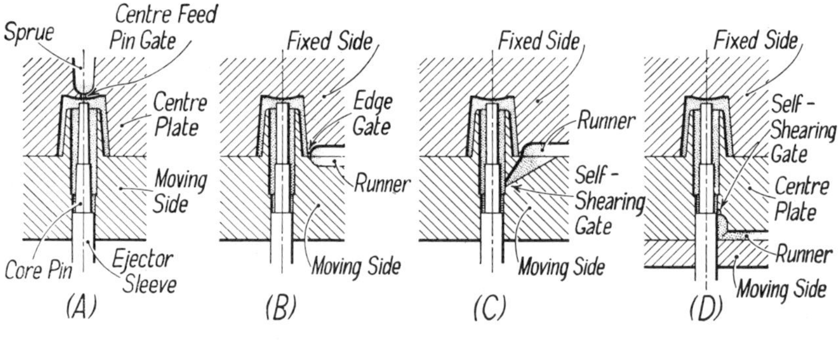

Fig. 3.23. Alternative Feeds for Knob Moulding.

Shown in (*C*) is a self-shearing gate, fed on the edge of the spindle. This gate comprises a conical feed at an angle to the flash face terminating in a pin feed on an unseen face of the moulding. On ejecting the moulding, this feed is sheared off against the edge of the gate, thus presenting the moulding automatically degated. The angle of the feed is such that it can be sprung out when the runner is subsequently ejected. This method is very convenient in production and widely used, but is open to all the objections noted in connection with (*B*).

Another form of automatic gate is illustrated at (*D*). In this instance, a normal gate is fed vertically to the end of the spindle, again being sheared off when the moulding is ejected. This design, however, requires a three-plate tool to extract the runner system. While the feed position is different and the flow pattern is possibly somewhat better, the added complexity of a three-plate tool is not generally warranted.

Alternative disposition of the cavity is shown in Fig. 3.24. The example shows a flat moulding having recessed lettering, such as might be required on an instrument scale. At (*A*) the cavity for the moulding is shown on one side of the tool, and the raised lettering on the other half of the tool. Unless conditions are very satisfactory, there may be a possibility that the moulding may tend to stick in both halves of the tool on opening. On the cavity side this can occur due to the sticking of the gate and the sprue, or

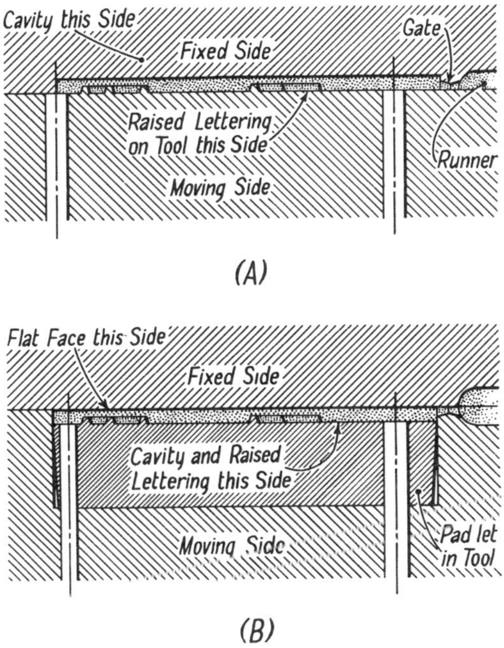

Fig. 3.24. Alternative Positions of cavity for a Moulding with Lettering. The Position (*B*) is to be preferred to avoid Sticking on Tool Opening.

because of negative or insufficient taper, or because of undercuts on the edge left by machining; on the lettering side, such sticking may be due to shrinkage on to the characters. On a flat moulding of this type, any tendency to stick can cause bending and distortion, and in a brittle material such as polystyrene may lead to crazing.

It is preferable to lay out such a tool with both the lettering and the cavity on one side, using a pad to facilitate the machining or hobbing. This is shown at (*B*). In this case, the moulding is sure to remain in the moving tool, whence it can be ejected, a carefully balanced ejector system being used to impart an even ejection and freedom from local bending.

Tool Marks. In any tool layout, the eventual visible effect of the tool construction on the final moulding should be borne in mind. The effect of the tool split line in producing a flash line on the moulding has already been mentioned, but similar witness lines may also be caused by other steel joint lines. With regard to moving cores and similar parts, the moulding material will tend to enter the sliding clearance, and with wear, this tendency will increase. Where static steel blocks make up the tool, the tendency to enter will remain, although if the blocks are well fitted, the line on the moulding may be quite faint. Sometimes, these joint lines are peened over, but the possibility of a witness line appearing on the moulding should be catered for. In this respect also, it is most important to provide that any such joint lines should appear in the line of draw. Thus, if the material does enter the clearance, the moulding can still be ejected, but if there is a possibility of flash occurring at an angle to the line of draw, this will constitute an undercut and seriously interfere with ejection.

CHAPTER IV

MOULD SHRINKAGE, TAPER AND TOLERANCES

Shrinkage Calculations. As the hot plastics material in the cavity cools, it contracts, and the final moulding is smaller than the original mould. Thus when a tool is being designed from a dimensioned component drawing, a shrinkage allowance must be added to give the corresponding tool dimension. The various moulding materials have different shrinkage rates, and these are given in the material manufacturers' handbooks. Some commonly adopted values are given in the table below.

Often a range of shrinkage values is quoted, the exact value depending upon the particular moulding and press conditions.

The shrinkage of the moulding is volumetric, and thus all the component dimensions must have this shrinkage allowance added. It will also be clear that, if a calculated shrinkage allowance is added to a dimension, the actual moulding shrinkage will be based on the original dimension plus the shrinkage allowance instead of on the original dimension only; i.e. a shrinkage on a shrinkage. Thus,

Let S = shrinkage allowance (in. per inch);

d = component dimension (in.);

D = final tool dimension (in.).

Then, $D=(d(1+S+S^2+S^3+S^4+ \ldots$ etc.)

MOULD SHRINKAGE OF VARIOUS THERMOPLASTICS

Material	*Shrinkage alowance (in. per inch)*
Cellulose acetate	0.003–0.007
Cellulose acetate butyrate	0.002–0.005
Nylon	0.010–0.025
P.V.C.—Rigid	0.001–0.002
„ —Flexible*	0.002–0.020
Methyl methacrylate	0.002–0.008
Polystyrene	0.002–0.006
Acrylonitrile butadiene styrene (A.B.S.)	0.003–0.008
Styrene acrylonitrile	0.002–0.005
Polyethylene—Low-density	0.015–0.030
„ —High-density	0.015–0.030
Polypropylene	0.015–0.025
Polycarbonate	0.005–0.007
Acetal	0.025

*Depends on the plasticiser content; the softer the grade, the higher the shrinkage.

Since S is normally very small, any power of S higher than the first may be safely ignored; hence $D=d(1+S)$.

Even if S is comparatively large, for example 0.05 in. per inch nominal, then

$$1+S+S^2+S^3=1+0.05+0.0025+0.000125=1.052625,$$

whence the correct nominal shrinkage to be used should be 0.053 in. per inch. However, the possible shrinkage of a material of this type could be anywhere between 0.045 and 0.055 in. per inch, depending upon the particular moulding and press conditions, and it is pointless to adopt a higher value than $(1+S)$.

Two principal methods are widely used for the accommodation of material shrinkage between tool and component:

1. All the tool dimensions are arrived at by calculation, using the component dimensions as given and then adding the shrinkage to the final result.

2. The shrinkage allowance is first added to the basic dimensions of the component and then from these compensated values the remaining dimensions are derived, and thus automatically provided with their own shrinkage allowances.

Basically there is no substantial difference in either, as is illustrated by the following example.

In Fig. 4.1 are shown two radii of 18.250 in. and 0.742 in. which are to blend. The basic component dimensions are shown in the view (*a*), and to ensure the blending of the two radii, it is necessary to determine h, assuming a shrinkage allowance of 0.015 in. per inch or 1.5 per cent.

The first method is illustrated by view (*a*). Here ABC is a right-angled triangle, so that

$$(18.250-0.742)^2=(h+10.730)^2+(7.150)^2, \text{ whence}$$

$$h=\sqrt{[(18.250-0.742)^2-7.150^2]}-10.730=5.2514 \text{ in.}$$

To this derived value must be added the shrinkage allowance to obtain the corresponding tool dimension; that is the tool dimension

$$=5.2514(1+S)=5.2514(1+0.015)=5.2514(1.015)=5.3302 \text{ in.}$$

Thus, by the first method, the basic component dimensions were used to derive the dimension h, and the shrinkage allowance was then added to this value. By the second method as shown by the view (*b*)

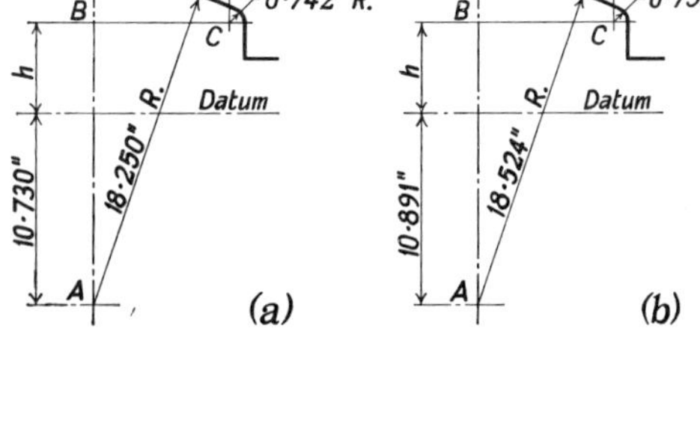

Fig. 4.1. (*a*) Calculating h from the basic Component Dimensions before Shrinkage Allowance is applied; (*b*) Calculating h with Shrinkage Allowance from the Component Dimensions already corrected for Shrinkage to give the Mould Dimensions.

in Fig. 4.1, the shrinkage allowance is first added to each basic component dimension, and then the value for h is derived from these compensated dimensions, the figure so obtained already including its own shrinkage allowance. The figures given in (*b*) are the same basic drawing dimensions as in (*a*) but to which have been added the shrinkage allowances. From these, the value of h is determined. That is

$$h = \sqrt{[(18.524 - 0.753)^2 - 7.257^2]} - 10.891$$
$$= 16.22171 - 10.891 = 5.3307 \text{ in.}$$

The slight difference between the two answers is due to the selection of the third decimal place of the tool dimension after the addition of the shrinkage allowance. Because of this, the second method should always be used, thus ensuring a true blend to the dimensions selected. Also, it must be borne in mind that, to obtain a true indication of the third decimal place in all tool calculations, the minimum number of figures to be used in any logarithmic tables is five, and it is strongly recommended that seven-figure tables should be used.

Component Tolerances. The tolerances shown on the component drawings must cover not only the functional performance of the moulding but also the following three factors in connection with moulding:

(i) *Variations of Shrinkage:* The shrinkage figures quoted by the material manufacturers are based upon a test moulding produced under specific conditions. In practice, however, the shrinkage that occurs is dependent not only upon the particular size and cross-section of the moulding but also upon the particular press conditions. Thus there will be a certain difference between the arbitrary shrinkage allowed and that actually obtained.

(ii) *Mould Draft:* This is the taper provided on the tool to facilitate extraction of the moulding.

(iii) *Tolerances required to cover Tool Making:* It will be appreciated that a tool cannot be made exact to size, and must be allowed a tolerance.

If a component is closely toleranced, it is necessary to accommodate these factors by leaving as much metal on the tool as possible, so that, after initial sampling, suitable corrections can be readily applied to the tool, since it is much easier to machine the metal off to the correct dimensions than it is to add metal.

This is achieved by making all male forms as large as possible and the female forms as small as possible, within the permissible tolerance. One method sometimes adopted is to apply a higher value in the shrinkage range to the male and a lower value to the female. In the case of high-shrinkage material, e.g. high-density polythene, that has a considerable difference of shrinkage in the line of flow from that at right-angles to the flow, due allowance must be made for this effect. Where a precision moulding is required, particularly from high-shrinkage material such as nylon, full provision must be made to allow the tool to be corrected as necessary after sampling, or a simple single-impression tool of the basic form is first made and sampled to establish the actual shrinkage values.

Taper. This is almost always provided on male and female tools in the line of draw. This mould draft is desirable for two reasons:

1. To ensure a clean break from the tool faces as the moulding is rejected.
2. To prevent the effects of machining and polishing marks from forming undercuts.

The necessity of the first of these is in theory disputable, but some taper is desirable for easy release, at least on male forms because of the shrinkage of the moulding on to the form, and is essential on injection moulds. The minimum draft for moulding in general is 1 deg. per side. Unfortunately, it is not always possible to provide this amount of taper, since the component tolerances may not be sufficient. On larger dimensions, the amount of taper may also interfere with the function of the moulding. Reduced draft angles of $\frac{1}{2}$ or $\frac{1}{4}$ deg. per side are employed if such reduction is unavoidable, but should be confined to the essential faces.

The toolmaking tolerance on the mould form varies from company to company but, in general, ± 0.002 in. is allowed for tool making on all dimensions on which the component tolerances are ± 0.006 in. or greater. On closer-toleranced dimensions, the usual toolmaker's allowance is given as ± 0.0005 in. These tolerances are found in practice to be satisfactory for nearly all work up to 12 in. in size. On larger work, the many factors which influence component size in the injection moulding process make closer tolerance work very difficult. On larger work, also, the effect on the mould of thermal expansion from room temperature to moulding temperature should not be overlooked. This expansion can be quite a decisive factor if close tolerances are involved. For example, a tool 20 in. long will expand 0.012 in. in length when heated from 20 to 70 deg. C., since the coefficient of thermal expansion of nickel-chromium steel is approximately 0.000012.

To ensure that sufficient steel is left on the mould for possible modifications, in the moulding of components of close tolerance or using materials of indeterminate shrinkage, a condition of maximum-metal conditions is applied.

Female forms, i.e. cavities, are dimensioned at the bottom (e.g. the bottom of a cavity) and to minimum size.

Male forms, i.e. punches, plugs, pins, etc., are also dimensioned at the bottom, that is at the root of the form or fixed end; and are dimensioned to maximum size.

Taper on the vertical faces, therefore, increases the size of cavities and decreases the sizes of the male forms. These dimensions are generally also the sizes to be aimed at by the toolmaker when machining the form, and thus any slight error in machining can be accommodated by reducing the taper allowed.

The exception to these rules arises when the form is to be produced by hobbing or casting. In these instances, the reverse applies, since the machining is carried out on a male hob or pattern, which in turn is used to produce the cavity. Therefore, the dimension shown on the drawing of the cavity will, in fact, be used to produce the hob or pattern. Since the latter is a male form, the drawing dimensions should be to maximum size instead of to minimum size as in the case of a machined cavity.

To illustrate these points, the following example shows the method of arriving at the tool dimensions required to produce the component shown in Fig. 4.2. This example is chosen to illustrate one of the more difficult cases, the moulding being in nylon, a material of indeterminate shrinkage, and is to be produced to specified limits. Fig. 4.3 shows in plan view only the tool pads to produce this moulding, these being simplified to show clearly the method of dimensioning.

It will be seen that the moulding has been split along a line dividing the 0.547-in. dimension, giving two symmetrical halves. Since the actual shrinkage is indeterminate, a nominal shrinkage factor of 0.025 in. per inch has been adopted. The tool dimensions are so chosen as to provide facilities for eventual adjustment of the tool when the actual shrinkage is determined on sampling (i.e. maximum-metal conditions) as well as allowances for toolmaking tolerance and mould taper. A nominal single taper of 1 deg. can be used as most of the dimensions have a tolerance of ± 0.005 in. A 1-deg. taper over 0.274 in. (i.e. the maximum depth of cavity being half the 0.547-in. dimension) gives 0.0047 in. and, by using an appropriate proportion of the total 0.01 in. tolerance, this taper can be easily accommodated. All dimensions having tolerances of 0.006 in. or less have been taken to four places of decimals, with a correspondingly closer toolmaking tolerance of ± 0.0005 in. instead of the normal toolmaking tolerance of ± 0.002 in. This is illustrated by reference to the moulded bosses of 0.150/0.156 in. diameter, which are marked as important. Since this form is female in the tool, it is necessary to work on minimum size to achieve maximum-metal conditions. Thus

$$\begin{bmatrix}\text{Tool dimension at}\\ \text{bottom of cavity}\end{bmatrix} = \begin{bmatrix}\text{Minimum component size plus toolmaker's}\\ \text{allowance plus shrinkage}\end{bmatrix}$$

$$= (0.1500 + 0.0005) + (0.025 \times 0.1505)$$

$$= 0.1543 \text{ in.}$$

The actual tool drawing would show the toolmaker's tolerance allowed and read 0.1543 in. ± 0.0005 in.

The other principal dimensions are derived as follows:

(1) *Component dimension* 0.133 *in.* ± 0.005 *in.*

From the tolerances given, the component dimension can therefore be 0.128 in./0.138 in. Since the mould form for this dimension is female, maximum-metal conditions are met by making the cavity to minimum size. Additionally, a toolmaking allowance of ± 0.0005 in. must be provided. The modified lower-limit component dimension to include this tolerance is 0.128 in. + 0.0005 in. = 0.1285 in. Similarly, the corresponding upper limit is 0.138 in. − 0.0005 in. = 0.1375 in.

To arrive at the tool dimension, the minimum size is taken, as previously stated, with shrinkage added; thus

$$0.1285 \times (1 + 0.025)^{\star} = 0.1317 \text{ in.}$$

This is the dimension given on the tool drawing with a tolerance of ± 0.0005 in.

The taper on the 0.274-in. depth of the 0.133-in. dimension at one

---

*Although a high-shrinking material is used, the actual component dimensions are relatively small, hence shrinkage calculations are carried to only the first stage.

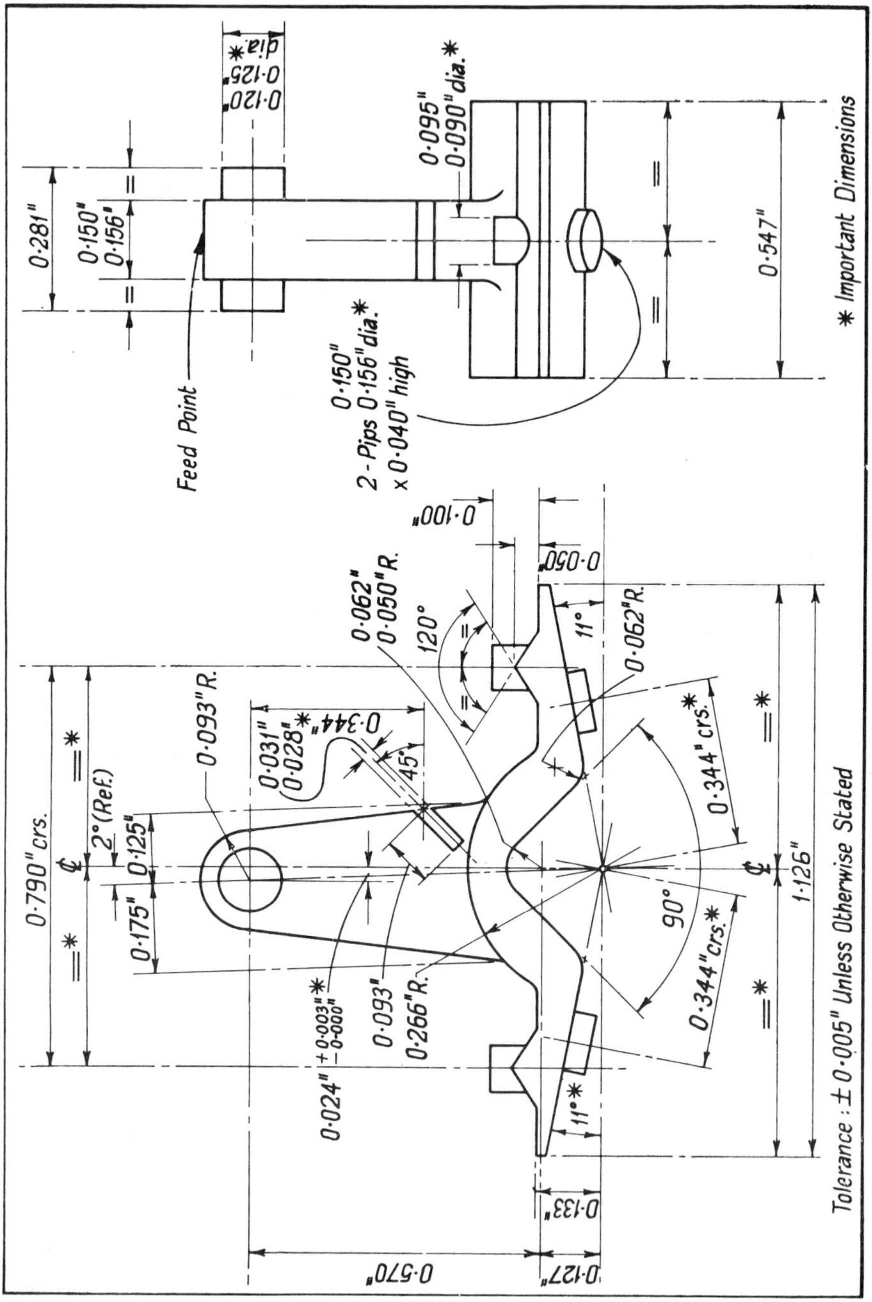
0·120"
0·125" dia.*
0·281"
0·150"
0·156"
0·095"
0·090" dia.*
Feed Point
0·150"
2 - Pips 0·156" dia.*
x 0·040" high
0·547"
* Important Dimensions
0·100"
0·050"
0·062"
0·050" R.
120°
11°
0·062" R.
0·790" crs.
2° (Ref.)
0·125"
0·175"
0·093" R.
0·031"
0·028"
0·344"*
45°
0·344" crs.*
90°
1·126"
0·024" +0·003" −0·000" *
0·093"
0·266" R.
11°*
0·133"
0·570"
0·127"
Tolerance : ± 0·005" Unless Otherwise Stated

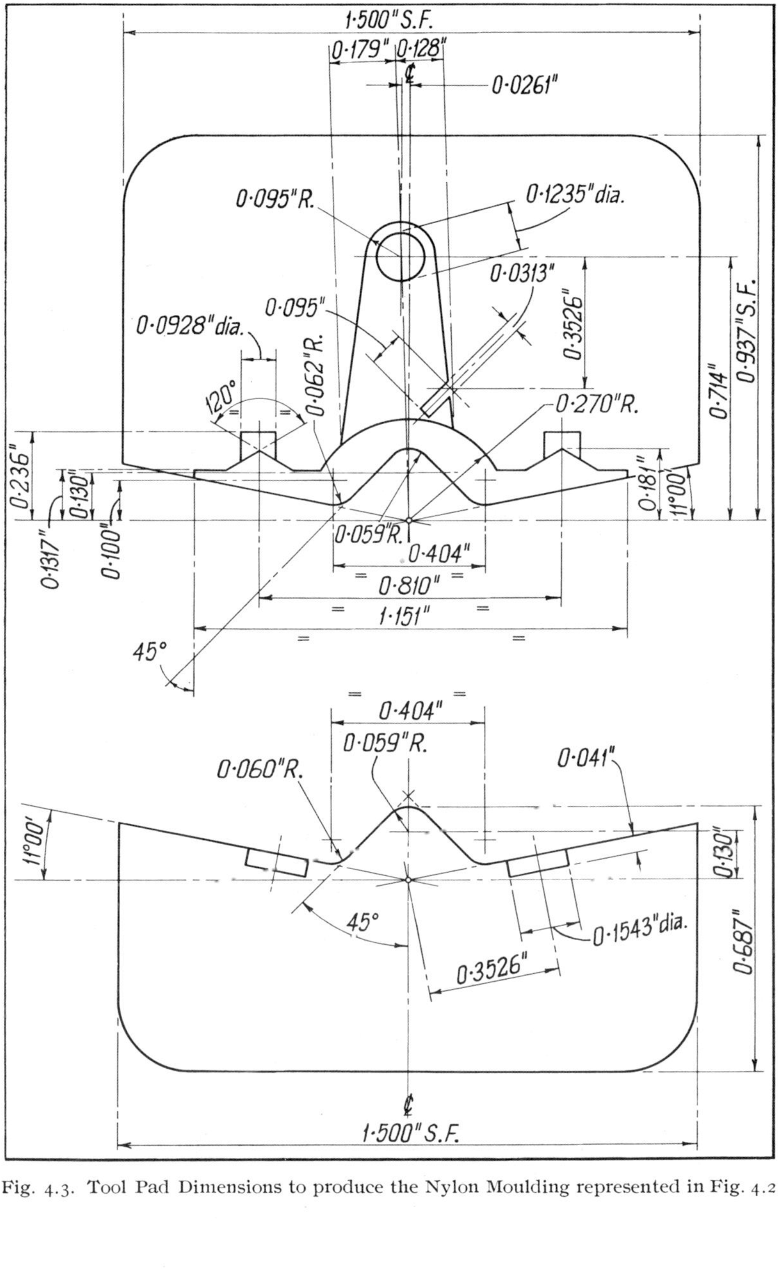

Fig. 4.3. Tool Pad Dimensions to produce the Nylon Moulding represented in Fig. 4.2.

degree is 0.0047 in. This is accommodated between the full maximum and minimum dimensions of the tool, i.e. between

$$0.1285 \times (1+0.025) \quad \text{and} \quad 0.1375 \times (1+0.025)$$

i.e. between 0.1317 in. and 0.1409 in., representing a tolerance of 0.009 in.

However, in addition to the allowance for taper, this 0.009-in. tolerance has also to allow for any variation in shrinkage, unless any tool modification is subsequently carried out.

(2) *Component dimension* 0.127 *in.* ±0.005 *in.*

This relates to centres and hence shrinkage can be added to only the mean dimension. Thus

$$\text{tool dimension} = 0.127 \times (1+0.025) = 0.130 \text{ in.}$$

(3) *Component dimension* 0.100 *in.* ±0.005 *in.*

The tool dimension here again relates to female form. It is not convenient, however, to dimension the tool in the same manner as the component drawing in this instance, since this would involve setting-off the 0.100-in. depth of the boss from the face of the webs which, on the tool, are tapered. On the pad drawing, therefore, this dimension is set-off from the parallel datum line. The 0.100-in. dimension is therefore added to the modified component dimension 0.130 in. as derived in section (1) from the component-drawing dimension 0.133 in. to give the modified component dimension 0.230 in. To this is added the shrinkage, in the previous manner, to give a tool dimension of 0.236 in. The taper on the depth of the split 0.100-in. dimension is too small to consider, and is therefore ignored.

(4) *Component dimension* 0.028 *in.*/0.031 *in.*

The tool for this form is male. The tolerance given is 0.003 in. which is less than the normal toolmaking tolerance of ±0.002 in., therefore the reduced toolmaking tolerance of ±0.0005 in. is employed, and the tool dimension given to four places of decimals. The application of this reduced toolmaking tolerance therefore gives a modified component dimension of 0.0285 in./0.0305 in. Since the tool form is male, the largest dimension is used and the shrinkage added, hence:

$$\text{tool dimension} = 0.0305 \times (1+0.025) = 0.0313 \text{ in.}$$

The toolmaking tolerance of ±0.0005 in. taken from the total component tolerance of 0.003 in. leaves 0.002 in. to accommodate taper over the split 0.150 in./0.156 in. depth of the 0.028 in./0.031 in. dimension; i.e. 0.002 in. for taper over a depth of $\frac{1}{2}$(0.150/0.156 in. + shrinkage) = 0 deg. 44 min. maximum taper.

This information is included, with other taper allowances, on the final tool drawing, but is omitted in the illustration. The reduced toolmaking tolerance applying to this dimension is also stated, either specifically or in the drawing tolerance table.

(5) *Component dimension* 0.093 *in.* ±0.005 *in.*

The tool form is again male, and therefore maximum dimensions are used. Here, however, the depth of mould form that is subject to taper is slight, and this together with toolmaking tolerances and maximum-metal allowances can easily be accommodated within the component ±0.005-in. tolerance. The mean value of 0.093 in. is therefore taken and the shrinkage added, giving tool dimension = 0.093 × (1 + 0.025) = 0.095 in.

(6) *Component dimension* 0.024 *in. plus* 0.003 *in. minus* 0.000 *in.*

This dimension relates to centres and is treated as in section (2). The mean value of the dimension is 0.0255 in. and when shrinkage is added,

$$\text{tool dimension} = 0.0255 \times (1 + 0.025) = 0.0261 \text{ in.}$$

On the component drawing, this dimension is marked as important and hence, to ensure due attention in toolmaking, the tool dimension is given to four places of decimals.

(7) *Component dimension* 0.344 *in.* ±0.005 *in.*

This dimension also relates to centres. The mean value is 0.344 in. and when shrinkage is added as in section (6) a tool dimension of 0.3526 in. is given. Again the dimension is marked as important and the figure is carried to four decimal places.

(8) *Component dimension* 0.570 *in.* ±0.005 *in.*

This centre dimension is more conveniently related to the bottom datum line, for toolmaking, i.e. 0.570 in. + 0.127 in. This gives the component a modified mean value of 0.697 in. The tool dimension is derived as in section (2) by adding shrinkage, to give a tool dimension of 0.714 in.

(9) *Component dimension* 0.062 *in.*/0.050 *in.* (*radius*)

The tool form is male and is therefore made large. However, the large tolerance of 0.012 in. is more than sufficient to provide for shrinkage variations on this small dimension. Therefore, 0.002 in. is taken for toolmaking tolerance plus a further 0.002 in. off the allowance for shrinkage variation, to give a modified component dimension of 0.062 in. − 0.004 in. = 0.058 in. The tool dimension is then obtained by adding shrinkage: $0.058 \times (1 + 0.025) = 0.059$ in. The tool dimension at top limit as derived above is 0.059 in. and the full lower limit similarly derived is 0.053 in. The taper on the $\frac{1}{2}$ (0.547 in.) depth of this dimension at 1 deg. amounts to 0.0047 in., and this is accommodated in the 0.006 in. difference between top and bottom limits.

(10) *Component dimension* 0.266 *in.* ±0.005 *in.* (*radius*)

This form is female and is treated as shown in section (1). This gives a modified component dimension of 0.263 in./0.269 in. Shrinkage added to the lower value of 0.263 in. gives a tool dimension of 0.270 in. Taper is also accommodated in the same manner as in section (1).

(11) *Component dimension* 0.175 *in.* ±0.005 *in.*

The tool form is female but is otherwise similar to that discussed in section (5), where the depth of taper is shallow. The mean value is therefore taken to give a tool dimension of 0.179 in.

(12) *Component dimension* 0.125 *in.* ±0.005 *in.*

This is identical with that of section (11) and gives a tool dimension of 0.128 in.

(13) *Component dimension* 0.095 *in.*/0.090 *in. diameter*

The tool form is female and minimum sizes are used. The component tolerance is low, hence a reduced toolmaking tolerance of ±0.0005 in. is employed to yield modified component dimensions of 0.0905 in./0.0945 in. The tool dimension which is obtained by adding shrinkage to lower limit has the value 0.0928 in. Since the form is circular, the taper is integral.

(14) *Component dimension* 0.120 *in.*/0.125 *in. diameter*

Here the procedure is identical to that in section (13). The modified component dimension is 0.1205 in./0.1245 in. The tool dimension is 0.1235 in. and the taper is again integral.

(15) *Component dimension* 0.790 *in.* ±0.005 *in.*

This dimension relates to centres and is treated as in section (2). The tool dimension obtained by adding shrinkage to the mean dimension is 0.810 in.

(16) *Component dimension* 1.126 *in.* ±0.005 *in.*

This tool form is female and its treatment is similar to that of section (1). With a toolmaking tolerance of ±0.002 in., the modified component dimension is 1.123 in./1.129 in. The tool dimension is obtained by adding shrinkage to the lower value to give 1.151 in. The taper on the $\frac{1}{2}$ (0.547 in.) depth of this dimension at one degree is again 0.0047 in. which is obtained within the 0.006-in. allowance between top and bottom limits.

(17) *Component dimension* 0.062 *in.* ±0.005 *in. radius*

This dimension is determined in the same manner as that considered in section (10). The modified component dimension is 0.059 in./0.065 in. Shrinkage added to the 0.059-in. value gives a tool dimension of 0.060 in. The taper on the $\frac{1}{2}$ (0.547 in.) depth is again obtained from the 0.006-in. allowance between top and bottom limits.

(18) *Component dimension* 0.150 *in.*/0.156 *in.*

This female form producing the two bosses on the underside of the webs is treated in the usual manner [see section (13)], with the reduced toolmaking tolerance of ±0.0005 in. The modified component dimension is 0.1505 in./0.1555 in., and shrinkage is added to the lower limit to yield a tool dimension of 0.1543 in. The dimension also relates to a circular form in the line of draw and hence the taper is integral.

(19) *Component dimension* 0.050 *in.* ±0.005 *in.*

Form is again female and the method used in section (1) is applicable. This gives a modified component dimension of 0.047 in./0.053 in., and a tool dimension of 0.048 in. It is not convenient, however, to dimension the tool in the same manner as the component, so that to assist toolmaking, this dimension is referred to the datum, by adding the corresponding tool dimensions of the 0.133 in. and 0.050 in. component dimensions; hence

tool dimension = 0.133 in. + 0.048 in. = 0.181 in.

As has been stated, the foregoing example illustrates the procedure for mouldings to be produced to high accuracy, or in materials of high, intermediate, or varying, shrinkage. In less accurate mouldings, or for materials of small shrinkage, the usual practice is to take the nominal dimension and obtain a tool dimension by adding the shrinkage allowance directly to it.

Distortion of Form to produce a Component within Specified Limits whilst accommodating Taper, Shrinkage and Toolmaking Tolerance. In the example shown in Fig. 4.4, it is assumed that a hole 0.5 in. ±0.003 in. diameter is to be moulded in a component 0.125-in. thick. The hole, therefore, has upper and lower limits of 0.503 in. and 0.497 in., and this is shown on the figure by outside and inside dotted lines. Toolmaking tolerance and taper have to be provided between these two limits.

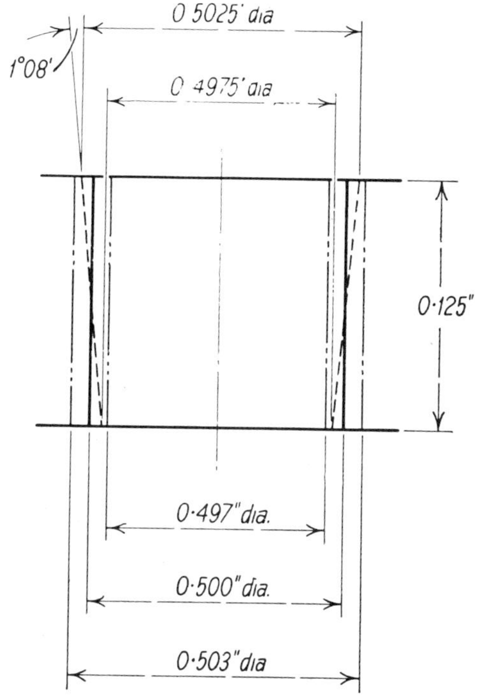

Fig. 4.4. Distortion of Form to meet Specified Limits whilst accommodating Taper, Shrinkage and Toolmaking Tolerance.

The toolmaking tolerance has to be subtracted from the top limit and added to the lower limit and, if the toolmaking tolerance is taken at ±0.0005 in., this gives a value for the core pin of 0.5025 in./0.4975 in. To this must be added shrinkage. If maximum-metal conditions are adopted, the largest diameter of the pin will be taken. Thus, core-pin diameter = (0.5025 in. + shrinkage) ± 0.0005 in. This dimension will be that of the bottom of the pin, at its mounting point, and will decrease in diameter along its length by virtue of the taper that must be provided. This taper must be accommodated between the top and bottom limits of the pin, i.e. between (0.5025 in. + shrinkage) and (0.4975 in. + shrinkage).

The maximum taper angle per side of core pin will therefore be given by

$$\tan\ \theta = \frac{(0.5025+S)-(0.4975+S)}{2(t+S)} = \frac{0.5025-0.4975}{2(0.125+S)}$$

where: θ = angle of taper per side;
S = total shrinkage;
t = length of pin = thickness of component.

If shrinkage allowance is 0.004 in. per inch, then

0.125 in. + shrinkage = 0.125 (1 + 0.004) = 0.1255 in., and

$\tan\theta = (0.5025-0.4975)/(2\times 0.1255) = 0.0199203$; whence

$\theta = 1$ deg. 8 min.

If this taper is greater than that allowed as the nominal single taper applying elsewhere in the mould, this lesser taper should also be used for the pin, since it will result in a larger pin and hence improve the maximum-metal conditions.

A final consideration on taper applies to the draft to be provided on shut-off faces operating in the vertical plane. Here, the taper is required to separate the tool faces rapidly, to minimize wear and consequent scoring. Taper in these conditions should be the maximum possible and values of from 3 to 10 deg. are common.

Hardening and Plating. If the moulded component is a precision moulding, it is imperative that the mould is tried and samples approved before any final hardening and plating of the mould takes place, in case any modification or adjustment is required. Although the mould may be subsequently stripped, annealed, modified and then rehardened, it is difficult to prevent mould distortion and consequent further component errors. It is thus better to obtain the correct sizes before final treatment. Allowance for chromium plating is commonly 0.0005 in., but on many moulds this may be ignored.

Gate Position. The size and form of feed and gate have considerable influence on the degree of shrinkage obtained, a restrictive feed system causing excessive shrinkage. Moreover, the position of the gate can influence the dimensional accuracy and distortion of the moulding. For example, feeding into the central portion of one side of a long narrow moulding, or a slender tubular moulding, will almost always cause a distortion that is concave to the feed.

CHAPTER V

PRINCIPAL TYPES OF INJECTION MOULD

Single-impression Two-plate Tools. This form of tool is generally the simplest type of mould, comprising essentially two die plates which carry the female cavity and the male punch respectively, the ejector system if provided, and the packing and tool backplates.

In Fig. 2.3, page 48, a tool of this type is shown in simplified form, the sprue being of the stalk type and leading directly into the body of the moulding. The nozzle of the heating cylinder engages in the spherical seating of the sprue bush, while the location spigot engages in the hole provided in the fixed platen of the press. The location spigot and sprue bush are normally made separate items, and the sprue bush is invariably hardened. At (*A*) and (*B*) in Fig. 3.22, page 97, are shown diagrammatically the die plates in a single-impression mould, with stalk-type sprue. One die plate carries the female tool, this either consisting of a single block of steel or, in more elaborate tools, being built-up from a number of blocks or die inserts set in a bolster.

The second die plate is machined to form the male core, and both plates may be drilled for water cooling. The ejector system often consists of a number of ejector pins, which, when moved forward, eject the moulding from the tool.

The ejector pins themselves are most economically made as round pins, but their form is often dictated by the requirements of the moulding; thus many ejectors are made in the form of sleeves, rectangular blades, half-round pins or stepped blades. These, and other forms of ejector are discussed in a later chapter.

An ejector system incorporated in a two-impression tool is shown in Fig. 5.2, and this form of construction would be equally applicable to a single-impression mould. The other detail features shown covering the location spigot, guide pins and bushes, ejector packing blocks and tool backplates are also similar to those used in a single-impression tool.

The simple single-impression two-plate tool is very extensively used, but more complex tools are built up from this basic form by the addition of other members such as insert carriers, threaded pins and moving cores, together with others that may be dictated by the actual moulding.

The feed in this type of mould may either be direct into the moulding by a stalk sprue, as shown in Fig. 2.3, and at (*A*) and (*B*) in Fig. 3.22, or by an edge feed from the outside edge of the cavity as shown at (*C*) in Fig. 3.22. The gate in this type may be of various shapes, such as rect-

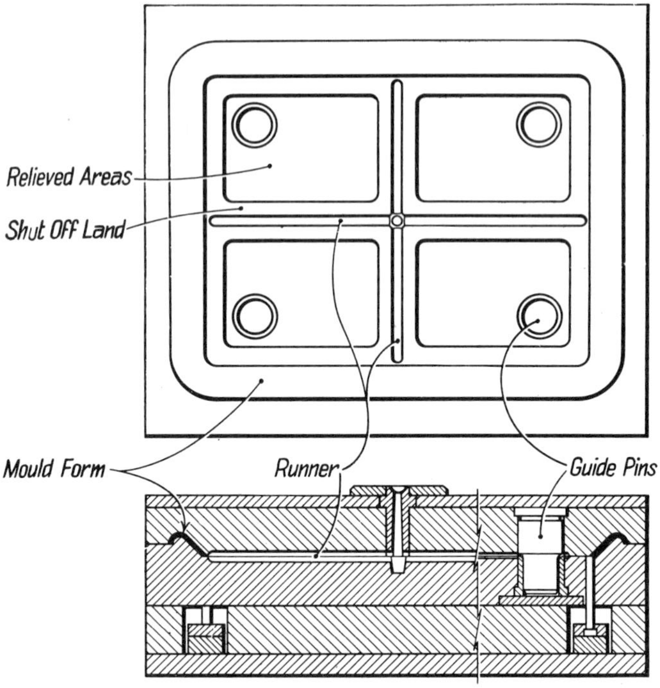

Fig. 5.1. Single-impression Two-plate Tool for producing a Frame-type Moulding and with Runner System and Guide Pins inside the Mould Form.

angular, half-round, fan, or tab. For such a single-impression edge-feed tool, some degree of offset on the mould is required, since the sprue must be on the centre line of the press. However, in the case of mouldings with a central opening, the sprue and runner system may be provided within the central shut-off area, thus maintaining the mould in the central position. Such mouldings include box- or bowl-type mouldings with central base openings; rings, television or radio frame-type escutcheons; and scale blanks with face openings. Fig. 5.1 illustrates a mould for a television surround of the kind which comprises a frame moulding with a large central opening. In this, the mould is central, the sprue leading through the opening of the die and then being connected by means of runners within the central area to feed into the moulding on the inside edge by means of one or several gates. This method, besides eliminating mould offset, also has the advantage over a moulding fed from the outside rim in that the provision of multiple gates can more readily be effected. This may be important where the section thickness is small and the path of flow long. In such cases, the feed from a single gate is long and may require excessively high injection pressures to ensure a full moulding. If multiple gates are provided, however, the length of flow from each gate is greatly reduced, while positions of weld lines can be more readily controlled. A moulding illustrating this feature is shown in Fig. 8.36 on page 255.

Another advantage, on the larger types of mould, is the ability to provide guide pins also within the shut-off area. This means that the tool can be made considerably smaller than if the dowels had to be provided in the mould face outside the moulding area. The relieving of the central area of the mould should be noted. This concentrates the clamping force on the land areas and hence the flash line.

This kind of tool is quite common for the medium and large sizes of moulding. For small mouldings, a diaphragm or disc gate can accomplish a similar result, with improved flow conditions.

With these the sprue is led to the central area of the mould, where sufficient gap is left between the face of the punch and the cavity to provide a diaphragm runner. The diaphragm extends to the edge of the inside of the mould form, where its thickness is reduced to form a rim gate communicating between the diaphragm and the mould cavity. On injection, the diaphragm is filled with the plastics material, which enters all parts of the mould simultaneously through the rim gate, thus ensuring an even flow and obviating all weld lines. After moulding, the diaphragm is separated from the completed moulding by stamping or cutting out the rim gate.

Another variation of the single-impression tool which is used in special circumstances is the portable tool. In some mouldings, the depth of the component is too great to allow removal of the moulding, because of insufficient die-opening space. To overcome this difficulty, the male tool is mounted so that it will engage a slide fitted to the press platen, the slide generally being arranged vertically so that the tool is dropped in from the top of the press. The portable male tool is retained in the bottom half of the slide so that it accurately locates within the female tool when the press closes. This is best achieved by providing a location on the diameter of the male tool itself. It is arranged that the moulding remains on the male half of the tool, which is then withdrawn completely from the press. Once outside the press, the moulding is removed from the tool by means of special ejector jigs, if necessary. (An illustration of a two-impression portable split tool is shown in Fig. 5.15, page 134.)

This method, however, can be used only if the weight of the male tool is reasonable. Nevertheless, in the case of heavier tools, it is sometimes arranged that the male tool can be withdrawn horizontally into a set of special carrier rails arranged at the side of the press. When the press opens, the male tool is withdrawn sideways to engage in the external rails, so that the moulding can be removed in the manner previously described. This method, however, tends to be rather slow in operation, and the external equipment may obstruct the operator's access to the press.

Multiple-impression Tools. The basic reason for the employment of multiple-impression tools is either to reduce component cost by producing more than one moulding in the same press, thereby reducing labour time, or to obtain a greater output in a given time. As to the number of impressions, 2-, 4-, 6-, 8-, and 12-impression tools are common, and some tools exist with a hundred impressions, or more. Multiple-impression tools can produce a number of the same component, or a number of different components, in the same material and colour. Most of these tools are two-plate moulds, but three-plate construction is also used where necessary.

A typical two-plate two-impression mould is shown in Fig. 5.2, with main details of construction as given in the following list, which also indicates the accepted nomenclature.

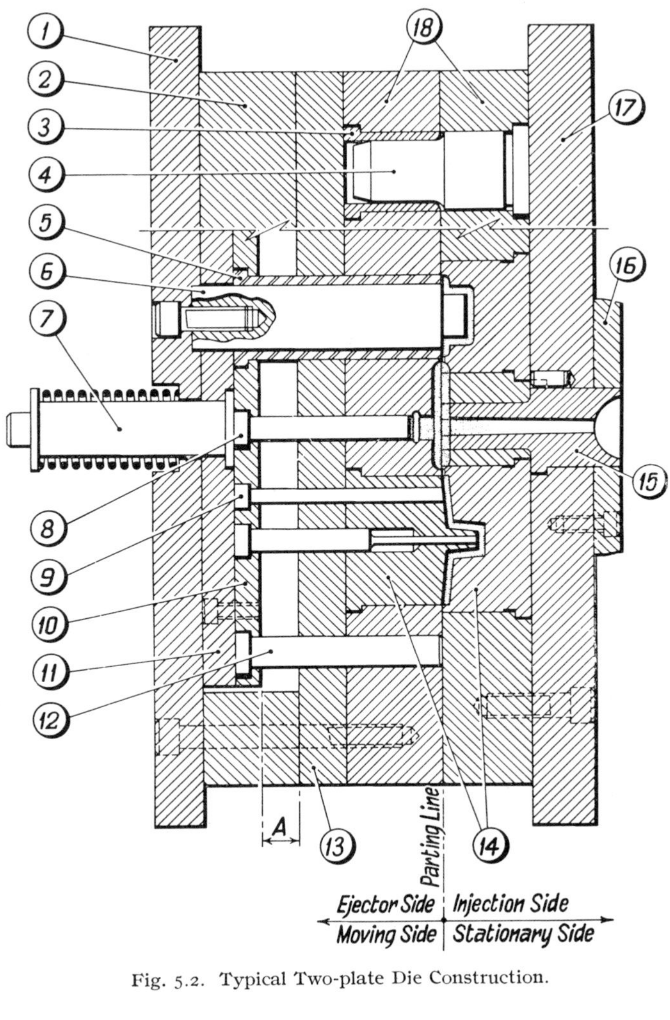

Fig. 5.2. Typical Two-plate Die Construction.

TYPICAL TWO-PLATE DIE CONSTRUCTION WITH NOMENCLATURE

(1) Backplate (mild steel).
(2) Packing to provide ejector space *A*; this must be sufficient to eject the moulding clear of the male form (mild steel).
(3) Guide bush (nickel-chromium steel, hardened and tempered).
(4) Guide pin (nickel-chromium steel, hardened and tempered).
(5) Sleeve-type ejector (nickel-chromium steel, hardened and tempered).
(6) Male core pin (nickel-chromium steel).
(7) Ejector-operating and return-spring mechanism (mild steel).
(8) Sprue hook stripper (nickel-chromium steel, hardened and tempered).
(9) Ejector pin, plain (nickel-chromium steel, hardened and tempered).
(10) Ejector locking plate (mild steel).
(11) Ejector backplate (mild steel).
(12) Ejector return pins (nickel-chromium steel, hardened and tempered).
(13) Die-insert backing plate (nickel-chromium or mild steel).
(14) Die insets (nickel-chromium steel with heat treatment depending upon the number and type of moving parts).
(15) Sprue bush (nickel-chromium steel, hardened and tempered).
(16) Sprue locating plate (mild steel).
(17) Die-insert backing plate (nickel-chromium or mild steel).
(18) Die plates (nickel-chromium steel).

This mould is for producing two dissimilar mouldings, edge-fed from a round runner cut half-and-half in each die plate. The sprue puller hook is provided to extract the sprue from the sprue bush and to retain the runner and sprue on the moving side of the tool, for eventual ejection with the mouldings. The two cavities are machined as die inserts set in a die plate, and are positively retained by the heads of the die inserts being clamped between the die plate and the backplate. The sprue bush is similarly retained, rotation being prevented by means of the dowel. The punch for the wide-skirted knob is in the form of a male die insert set in the moving plate. In the case of the other impression, forming the "top-hat" knob, it is intended to effect ejection by means of a sleeve ejector. This prevents a die insert from being used for the male form and hence a core pin mounted on the backplate is used. When the press opens, the sleeve ejector moves forward along the core pin and strips the moulding off the male form.

For the wide-skirted knob, ejection is effected by means of a central ejector pin which removes the moulding from the boss, and by ejector pins set round the skirt (for clarity, only one is shown in the illustration). These outer ejectors are provided to prevent distortion of the moulding which might arise if ejection were from the centre only.

It will be noted that the central ejection pin has been reduced in diameter to enter the restricted diameter of the centre boss. This has been done in preference to using a full-length pin of the final diameter, which would be too slender and might buckle. All the ejectors carry button heads, and these are clamped between the ejector locking (retaining) plate and the ejector backplate.

The ejector may be positively locked or may be allowed a slight amount of float sideways to assist location. The ejector-operating pin is spring-loaded and projects rearwards, through the mould backplate. When fitted to the press, this passes through the platen hole to make contact with the stationary ejector pin of the press as the press opens. On the closing of the mould, the return of the ejector gear is assisted by the ejector return pins (only one is shown in the illustration). As the die faces come together, these make contact with the non-form faces of the stationary die plate, and hence push back the ejector bar.

The two backplates of the tool are extended beyond the main body of the mould, in order to provide space for clipping or direct bolting to the press.

The layout of cavities in multi-impression tools requires considerable care to achieve the best results. Considerations to be borne in mind include:

(i) Optimum disposition of the cavities to achieve minimum overall tool size;

(ii) Attainment of minimum length of runners to cavities;

(iii) Correct layout of cavities to obtain balanced clamping.

Application of the first of these is illustrated by Fig. 5.3, from which it may be seen that the overall tool size of layout (*B*) is less than that of (*A*),

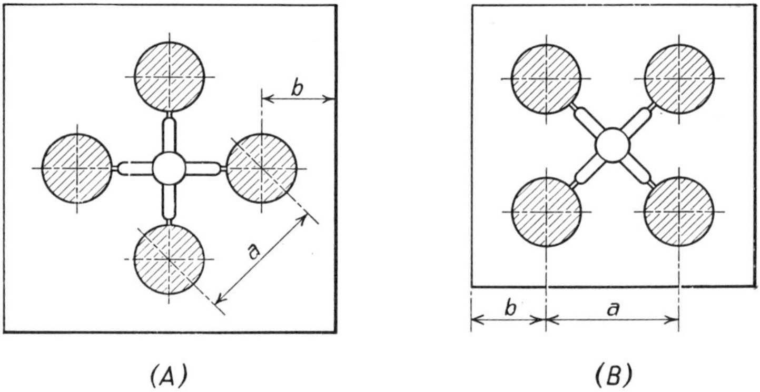

Fig. 5.3. Alternative Disposition of Cavities to reduce Overall Tool Size.

although the dimensions between cavities and from the edge of the tool are the same in both instances.

The truth of this remark may be shown as follows:
Area of plate $A=(a\sqrt{2}+2b)^2$; area of plate $B=(a+2b)^2$; so that the area of plate A is greater than that of plate B by the amount $(a\sqrt{2}+2b)^2-(a+2b)^2=a^2+1.656ab$.

Layouts to reduce the length of runner are shown in Fig. 5.4. The design at (*A*) is poor for the following reasons:

1. Due to the excessive length of runner, a high pressure drop will result from the rapidly cooling moulding material as it travels along the runner. This will give rise to a high pressure in the sprue vicinity, and a much lower pressure at the end of the runner. The cavities at the centre of the tool will thus fill first, whereas those at the extremity of the runner will fill last, at lower cavity pressure; or, if conditions are adverse, will not

fill completely. This condition is normally dealt with in any multi-impression tool with cavities feeding off a common runner by adjusting the gate size and length of land to compensate; but excessively long runners may make this much more critical and difficult to achieve.

2. If the number of impressions is large, the layout of the tool shown at (*A*) can result in a tool of excessive length, particularly when additional allowances have been made for members such as guide pins.

3. With a long tool of this nature, the clamping pressure at the ends of the tool is well distant from the centre line of press action, and this, under difficult conditions, could lead to local flashing of the tool.

The foregoing difficulties are overcome by repositioning the cavities as shown at (*B*), in which runner lengths are shorter, the tool is more compact, and the clamping pressure better distributed.

Attainment of balanced clamping is important for correct operation of the tool with minimum wear and freedom from flashing. The mould-opening force arising from cavity pressures should be so disposed as to

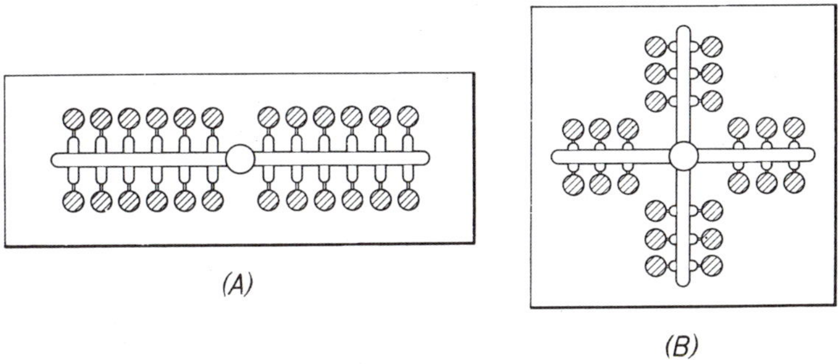

Fig. 5.4. Alternative Disposition of Cavities to reduce Length of Runners.

be evenly opposed by the clamping force acting on the press action centre line. Obviously, if the opening force is one-sided to the clamp (i.e. out of balance), a mechanical moment will be set up which could cause the tool to open on that side and lead to flashing. This problem is illustrated in Fig. 5.5 which depicts a multi-impression tool producing five dissimilar components. The original layout is shown at (*A*). The cavity pressure is assumed equal in each impression, while the areas in each cavity are as given in the table on page 120. Since the cavity pressure is constant, the mould-opening force of each impression will be in direct proportion to the individual projected areas.

The opening moment of each impression is given by
(Area of cavity) × (Distance of centre of pressure from centre line of tool).

Equating moments about the centre line *aa*:

$$(2.768\times 2)+[1.750\times(-0.625)]+[0.452\times(-2.250)]+[1.768\times(-1.500)]+(3.510\times 1.750)$$
$$=5.536-1.094-1.017-2.652+6.143=6.916 \text{ in.}^3.$$

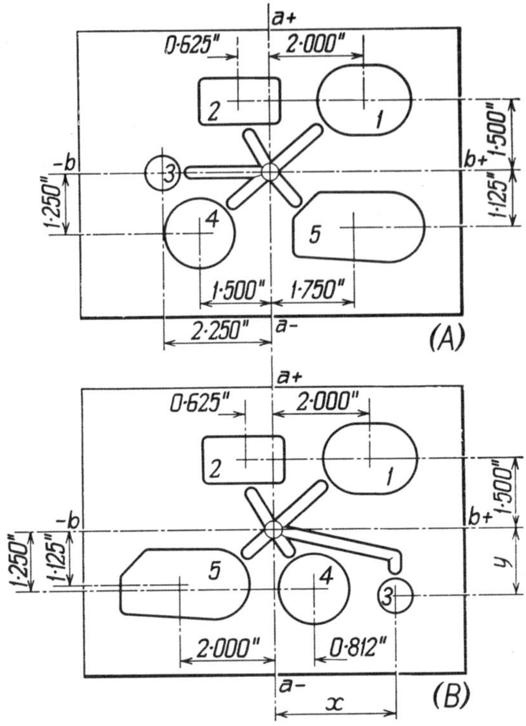

Fig. 5.5. Balancing of Live Cavity Pressure in Multi-impression Tools.

If it is assumed that the cavity pressure is 4 ton/sq. in., then this figure gives an out-of-balance force of $6.916 \times 4 = 27.664$ ton-in.

Similarly, equating moments about *bb*:

$$(2.768 \times 1.500) + (1.750 \times 1.500) + (0.452 \times 0) + [1.768 \times (-1.25)] + [3.510 \times (-1.125)]$$

$$= 4.152 + 2.625 - 2.210 - 3.949 = 0.613 \text{ in.}^3.$$

At 4 ton/sq. in. cavity pressure, the moment of the out-of-balance force about *bb* is $4 \times 0.613 = 2.452$ ton-in.

Such out-of-balance forces can lead to difficulties, but can be reduced, and perhaps eliminated, by repositioning some of the cavities. For the

CAVITY LAYOUT (*A*), FIG. 5.5.

Impression Number	Projected area of cavity (sq. in.)	Distance of the centre of pressure of cavity from centre line of tool From *aa*. (in.)	From *bb*. (in.)
1	2.768	+2.000	+1.500
2	1.750	−0.625	+1.500
3	0.452	−2.250	0.000
4	1.768	−1.500	−1.250
5	3.510	+1.750	−1.125

present example, this has been done for the cavities Nos. 3, 4, and 5 at (*B*) in Fig. 5.5. Here, the positions for cavities Nos. 4 and 5 have been assumed by inspection, the tool being finally balanced by calculating the position of impression No. 3, as follows:

CAVITY LAYOUT (B), FIG. 5.5

Impression Number	Projected area of cavity (sq. in.)	Distance of the centre of pressure of cavity from centre line of tool — From *aa*. (in.)	From *bb*. (in.)
1	2.768	+2.000	+1.500
2	1.750	−0.625	+1.500
3	0.452	x	y
4	1.768	+0.812	−1.250
5	3.510	−2.000	−1.125

In the above table, the dimensions x and y are determined by taking moments about *aa* and *bb* respectively.

Equating moments about *aa* to zero,

$$[2.768\times 2]+[1.750\times(-0.625)]+[0.452\times x]+[1.768\times 0.812]+[3.510\times(-2.0)]$$

$$=5.536-1.094+0.452x+1.435-7.020=0.452x-1.142=0.$$

Hence $x=(1.142)/(0.452)=2.527$ in.

Similarly, equating moments about *bb* to zero,

$$0.618+0.452y=0, \text{ i.e.}$$

$$y=(-0.618)/(0.452)=-1.367 \text{ in.}$$

The foregoing illustrates one scheme of repositioning cavities to reduce out-of-balance forces; various other arrangements of cavity disposition could, of course, be made to achieve the same object.

Although some small out-of-balance force may be tolerated in practice, it is essential that the greatest care is exercised that these do not become excessive. Otherwise, flashing can result which can be rectified only by employing less than optimum injection pressure. This can particularly arise if the projected area of the mouldings is close to the maximum possible for the clamping force available on the press employed.

In Fig. 5.6 is shown a further example of the balancing of cavity pressure, together with a corresponding reduction of overall mould size. The two impressions are identical.

In the layout shown at (*A*), it is clear that the tool is considerably out of balance due to the juxtaposition of the maximum projected-area portions of each cavity. If the cavities are repositioned as shown at (*B*), the pressure conditions will be symmetrical and in balance, while overall mould size will be reduced. The modified layout requires a longer runner, but this is justified in that it maintains the gate position in the same place in both cavities and in the preferred position, i.e. the widest portion of the cavity.

Reduction of out-of-balance forces is equally important for single-impression tools. This is illustrated by Fig. 5.7, where the repositioning

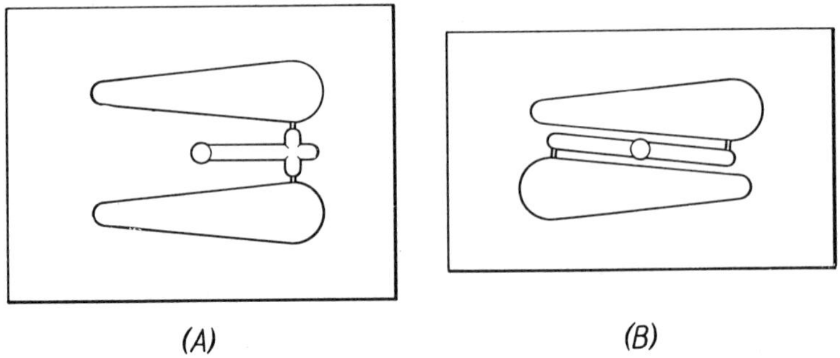

Fig. 5.6. Rearrangement of Two-impression Tool to balance Cavity Pressure and to reduce Mould Size.

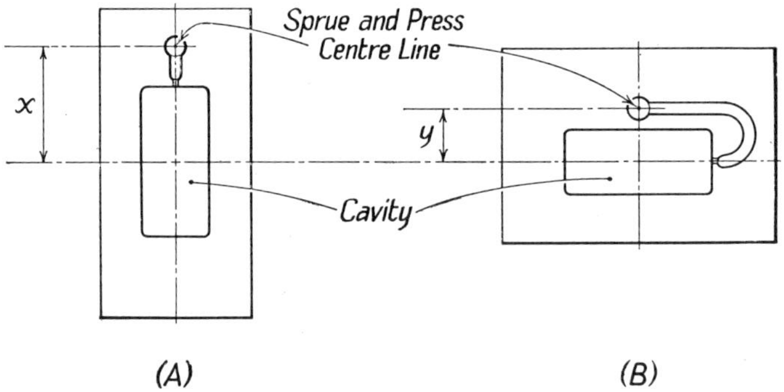

Fig. 5.7. Rearrangement of Cavity to reduce Out-of-balance on Offset Single-impression Mould.

of the cavity and the use of a longer runner in (*B*) produces a much lower opening moment than the original design (*A*), the dimension y being less than the dimension x.

In certain cases it is desired that certain impressions in a multi-impression tool should be "shut-off". This can arise if certain impressions are found to be defective or not capable of moulding satisfactorily. More frequently, however, this requirement occurs when the tool comprises a set of dissimilar components and it is desired to make certain pieces and not others; e.g. components consisting of either left- or right-hand pieces of an assembly. If the cut-off is to be semi-permanent, it is the practice to weld up the gate, but if the cut-off is temporary, merely to afford a selection in production, it is usual to include in the runner system a plug that can be turned so that the runner may be allowed to continue or be shut-off. More frequently, two interchangeable plugs are employed, one being blank for shutting off the runner, the other being shaped to continue the runner. However, when this method is considered, due attention must again be paid to the effect on the clamping-pressure balance.

Three-plate Tools. In addition to the two plates, one on each of the fixed and moving platens, as found in the two-plate tool, the three-plate tool has a third plate which is known as the floating or centre plate. This latter carries the gate, some of the runner system, and a requisite portion of the mould form. In the tool-open position, this third plate is separated from the other two, allowing the mouldings to be ejected on one side and the sprue and runners on the other. The chief uses of this type of tool are:

(i) Centre-feeding of components in multi-impression tools; e.g. a multi-impression knob tool with a pin-gate feed in the centre of each knob. Mouldings produced in a mould of this kind are shown in Fig. 3.16.

(ii) Centre-feeding of components with restricted gates in single-impression tools; e.g. a central pin-gate feed into the base of a bowl.

(iii) Area feeding of components with multiple restricted gates; e.g. multiple gates into the surface of a tray.

(iv) Edge-gating of single-impression or unbalanced multiple-impression tools without producing an offset mould; in large moulds this is important.

It is evident that none of the tools just listed could be made as a two-plate mould, since it would be impossible to remove the sprue-and-runner system—hence the third plate. Three-plate moulds are always more expensive than two-plate moulds, sometimes considerably. They also tend to be somewhat slower in production due to the necessity for the operator to clear the moulded sprue-and-runner system when the mould is open. This generally has to be done manually unless special provision has been made to eject the sprue.

Three-plate tools are frequently used for the production of small or medium sized components. Due, however, to the increasing massiveness and weight of the floating plate, they are less often used on the larger types of moulding, and rarely on very large mouldings.

In Fig. 5.8 is illustrated diagrammatically a three-plate, pin-point-gate, two-impression mould which is representative of the three-plate moulds employed for small components. The various elements of the tool are:

(1) Backplate (stationary side).

(2) Floating centre plate, or third plate. This plate must separate sufficiently from the backplate to allow the sprue-and-runner system to be extracted easily. The distance between the plates is governed by the length of the stripper bolts (7), and it should be noted that dimension a must exceed or not be less than dimension a_1.

(3) Stripper plate. In the tool shown, ejection of the mouldings off the male cores is effected by means of a stripper plate. The ejector system adopted in any given tool, however, would depend on the component.

The tool in the closed position is shown in the upper view, and fully open in the lower. After injection, the coolings will shrink on to the male cores and away from the cavity.

The initial movement of the moving platen parts the mould between the plates (3) and (4), this being assisted by the mouldings staying on the male cores. When the male cores clear the female cavities, the heads of the stripper bolts (8) engage with the injection side of the stripper plate (3). Thus dimension b must be greater than dimension b_1. The plate (3) then strips the mouldings from the male cores.

During this further movement, the adjusting nuts on the outrigger rods (9) engage on the back face of the plate (5) and then commence to move the plate (2) forward, giving a further parting line between the plates (2) and (1).

The initial movement of (2) thereby pulls the sprue from the sprue bush by virtue of the undercuts in the feeds down the male cores. The feed-stripper plate (6) is thus trapped between the plate (2) and the top runner system because of the same undercuts.

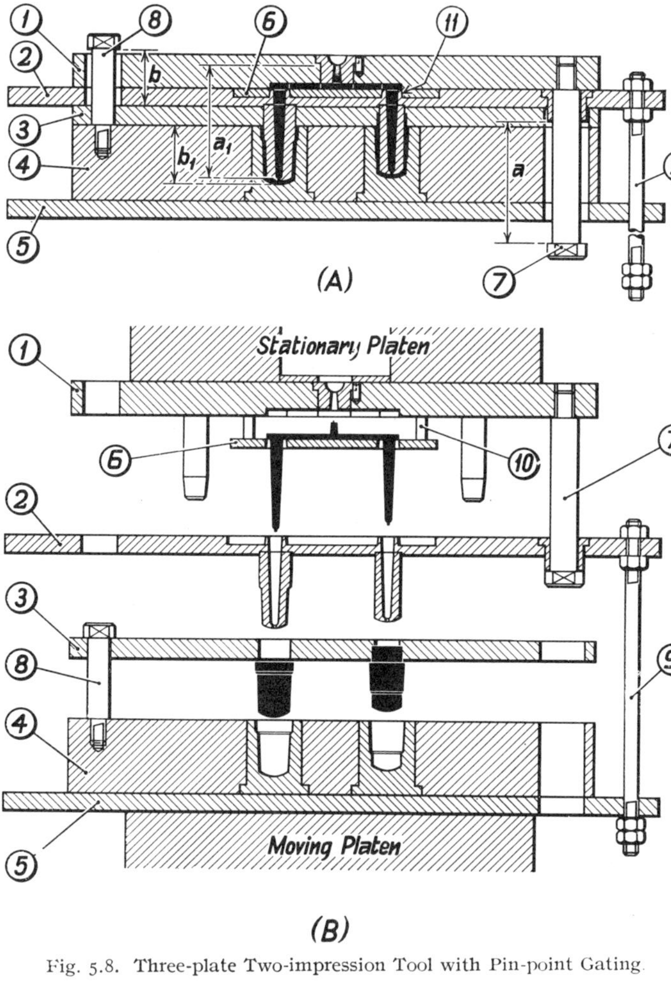

Fig. 5.8. Three-plate Two-impression Tool with Pin-point Gating.

However, the plate (6) has only a limited movement that is determined by the length of the stripper bolts (10). When the heads of (10) engage with the ejection side of the plate (6), the movement of this plate is halted and the plate pulls the feed from the male cores, overcoming the resistance of the undercuts. The total movement of (2) is restricted to the length *a* between the underside of the heads of the stripper bolts (7) to the bush face. To clear the runner system, the dimension *a* must exceed a_1. The runner system is allowed to fall clear from (6), because the face (11) is in the form of an open-ended slot, the open end being designed always to face downwards.

It should be noted that the actual sequence of the tool is of little importance, providing that the final position of the plates is controlled so that the respective distances are sufficient to allow the ejection of the moulding and the feed system. The undercuts on the feed down the male core are not normally machined in until after sampling, since normal machining of the feed often leaves marks that are sufficiently pronounced to act as undercuts.

It will be clear that three-plate tools require a greater opening stroke on the press than corresponding tools of two-plate design. It is thus important that the availability of the necessary stroke is checked against the distances allocated to the various moving parts.

(4) Die plate.

(5) Backplate (moving side).

(6) Sprue-ejection plate.

It is quite common practice to remove the sprue-and-runner system from the mould manually, or with the aid of a hooked wire. However, means can be provided to eject the feed automatically, one method being shown in Fig. 5.8. The feed stripper plate is recessed into the plate (2). engaging in the tapered faces (11) to ensure location. The plate is guided upon the pins (10) which allow the plate to move forward a predetermined distance as shown in the lower view. The runners to each cavity pass through the plate by the opening formed by the tapered faces (11), this opening being in the form of an open-ended slot with the open side downwards towards the floor. In operation, as the tool opens, the main sprue is pulled from the sprue bush by the pull of the two cavity sprues, undercuts being provided if necessary. This, in turn, causes the plate (6) to slide forward along the pins (10). When the plate has made its full movement, further opening of the tool causes the cavity sprues to be pulled from the cores until they are clear. The sprue and runners are then free, and drop from the tool through the slots in the plate (6). (An alternative mechanism is considered in Chapter VIII.)

The pins supporting the third plate must be of adequate strength for this purpose and, to this end, it is necessary to ascertain that the pins are of sufficient size to maintain the stress in the pin within safe values and that the deflection of the pins under the weight of the plate in the fully extended position is not excessive.

STRESS IN PINS. It is advisable to assume that all load is taken by two pins only, since, if alignment is not perfect, this condition can arise.

Let d = diameter of pin (in.); l = the effective length of pin (in.); f = the safe tensile stress of steel in the pin (lb./sq. in.). Also, let W be the weight supported by the two pins so that the maximum bending moment on each pin is equal to $\frac{1}{2}Wl$, which is equal to Zf, where Z is the sectional modulus of the pin, i.e. $Z = \pi d^3/32$.
Hence, $\frac{1}{2}Wl = f\pi d^3/32$, so that $d = [16Wl/\pi f]^{\frac{1}{3}}$.

Example: Let $W = 300$ lb.; $l = 6$ in.; $f = 5{,}000$ lb./sq. in. Then

$$d = [(16 \times 300 \times 6)/5{,}000\pi]^{\frac{1}{3}} = 1.22 \text{ in., say } 1.25 \text{ in.}$$

In many cases, the pins supporting the third plate also act as guide pins, hence the duty of aligning the tool will exceed that of supporting the third plate, and so pins of greater diameter than that shown by the foregoing calculations may be specified.

DEFLECTION IN PINS. The maximum permissible deflection can depend on several factors; but it should not exceed 0.02 in. and positive stops should be provided at the end of travel.

Referring to Fig. 5.9, let

xx = the horizontal centre line between the upper and lower pins;
p = the distance between the centre of the upper pins and the line xx;
I = the moment of inertia of each pin about a horizontal diameter, such as AA;
d = diameter of pins (in.).

Then the moment of inertia (or second moment of area) of the four pins about xx is given by

$$I_{XX} = 4(I + \tfrac{1}{4}\pi d^2 p^2) = 4[(\pi d^4/64) + (\pi d^2 p^2/4)]$$
$$= (\pi d^4/16) + \pi d^2 p^2 \quad \ldots \quad (1)$$

If it be assumed that the centre of gravity of the block coincides with the end of the pin, this will represent the worst condition, and the maximum deflection S will be:

$$S = Wl^3/3EI_{XX}, \text{ whence, substituting for } I_{XX} \text{ from (1),}$$
$$S = (Wl^3)/[3E(\pi d^4/16 + \pi d^2 p^2)] \quad \ldots \quad (2)$$

where W is the weight of the plate.

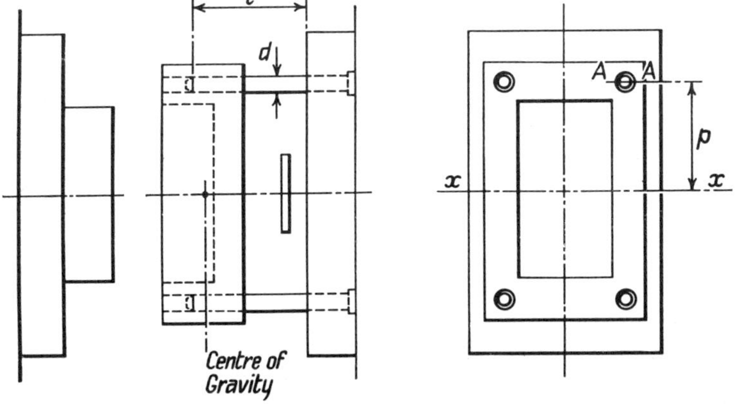

Fig. 5.9. Deflection in Pins.

In many instances, this deflection is very small and may be ignored.

In Figs. 5.10A and 5.10B is shown the application of a three-plate tool to the multiple feeding of pin-point gates on a tray-type moulding. Views (*a*) and (*b*) illustrate respectively the side section and plan of this moulding from which it will be noted that the four runners (2) diverge from the sprue to enter four pin-point gates (3) to the moulding (15). The runners are cut into the stationary plate *A*, and terminate in the third or floating plate *B*. The plate *C* is attached directly to the moving platen. The third plate *B* is carried upon the guide pins (1), these also acting as dowels for the whole tool. When the tool opens, the sprue is broken free by means of the sprue hook, while the multiple gates are parted by means of the stalk hooks (4). In this design, no provision is made for the complete automatic stripping of the sprue-and-runner system and, after initial freeing in the manner described, the feed system is removed manually after the tool has opened. The male form on the plate *B* is located with respect to the female by means of the taper pads (6). These are provided

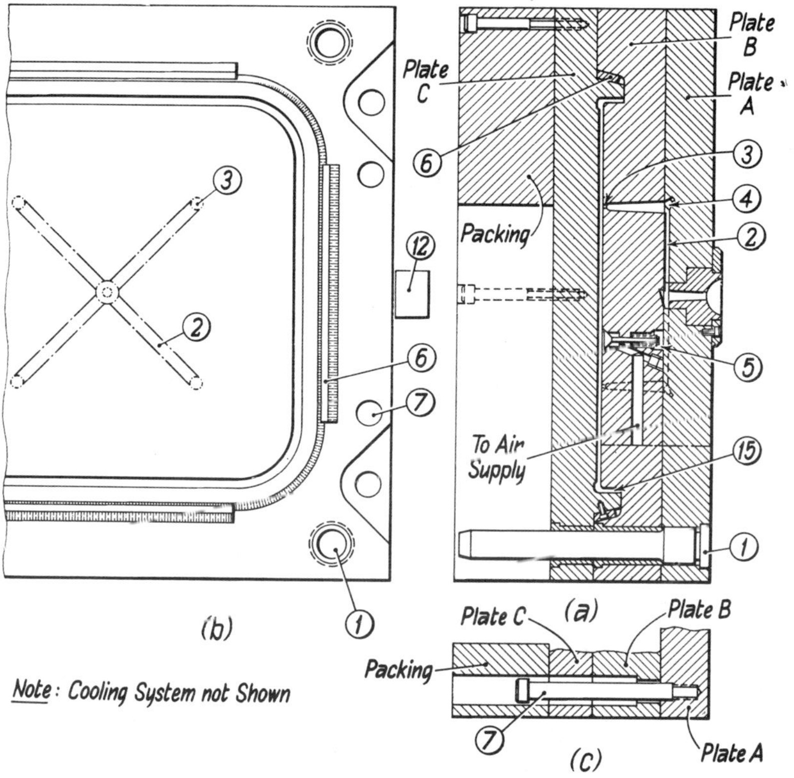

Fig. 5.10A. Application of a Three-plate Tool to the Multiple Feeding of Pin-point Gates on a Tray-type Moulding.

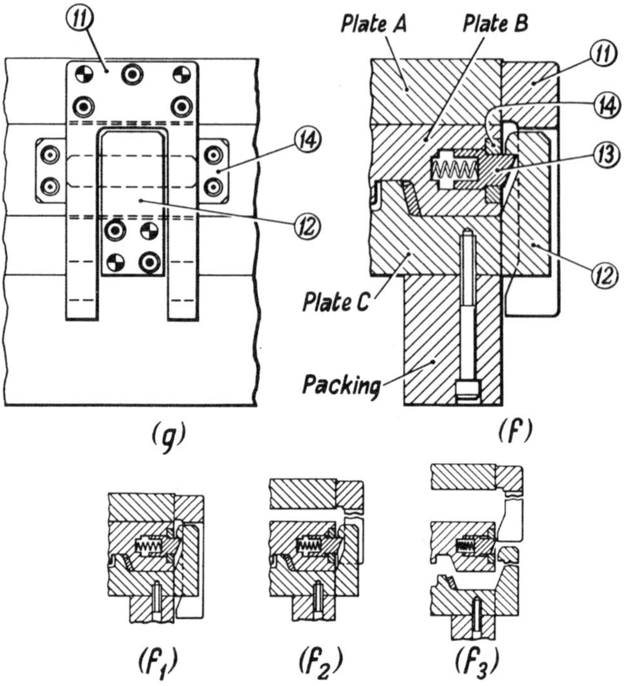

Fig. 5.10B. The Catches used for the Mould shown in Fig. 5.10A, and their Operation.

for wear, and in order that adjustment may be made if necessary to the relative positioning of the two halves of the form. An air valve (5) is provided, this comprising a taper seat valve which is held down by means of a compression spring. When air is admitted, behind the valve, the valve opens slightly and allows compressed air to permeate between the moulding and the surface of the male form. Since the depth of the moulding is shallow, it may then be readily removed by hand from the tool, assisted by the breaking of the vacuum and the slight air pressure supplied by the air ejector. View (*c*) shows one of the four shoulder screws (7) which prevent further movement of the third plate after it has opened the designed amount. One of the catches which operate the relative opening of the plates is shown in the sectional side view at (*f*) in Fig. 5.10B, and in plan view at (*g*). This device comprises a rectangular pawl (13) which is retained in a recess in the third plate by the cover plate (14). The catch is forced outwards by means of several compression springs.

Mounted rigidly to the plate *C* is the catch (12) whose free outer end closely engages the pawl (13) and keeps the two plates in close contact. Fastened to the plate *A* is the second member (11), which has two projections lying either side of the locking piece (12). The free-end portions

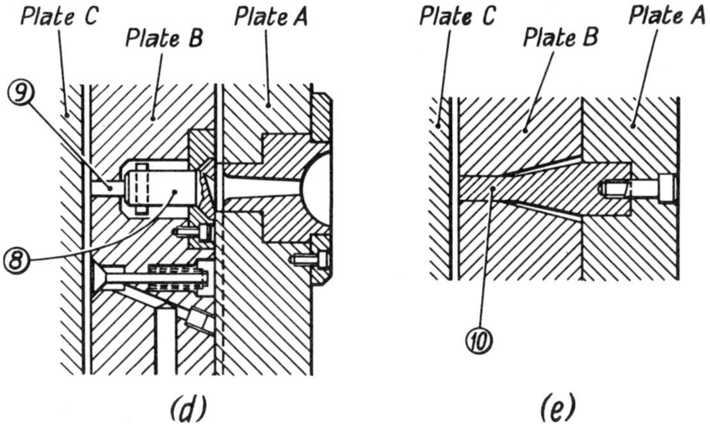

Fig. 5.10c. Venting Arrangements of the Mould in Fig. 5.10A. (See Chapter 8.)

of the projections are profiled to form an incline plane which can eventually engage the inclined angle of the catch (13). The stages of operation of the catch are shown in views (f_1), (f_2) and (f_3). In the first view, the assembly is shown in the closed position, ready for moulding. After moulding has been completed, the tool commences to open with the plate *A* remaining stationary since this is on the fixed platen of the press. The plates *B* and *C* move away, since *B* is retained in contact with *C* by action of the locking member (12) which engages the protruding catch (13).

At the same time, the inclined face of the member (11) approaches the incline of the catch (13), since (11) is stationary, being attached to the fixed plate *A*, and the catch (13) is moving away, being mounted on the plate *B*. At the designed point the two inclined faces of (11) and (13) make contact and the catch (13) is forced inwards against spring pressure. At the point where this catch is sufficiently depressed to discontinue contact with the locking member (12), further movement of the plate *B* is prevented by the shoulder screws (7). Since the plates *B* and *C* are no longer connected through the catch plate, *B* becomes stationary and the moving plate *C* continues opening to allow removal of the moulding.

The venting arrangements on this tool relating to views (*d*) and (*e*) are described on page 257.

Split Moulds. There are many varieties of split mould, and correspondingly numerous methods of operation. This type of tool is employed where some feature of the moulding forms an undercut relative to the line of draw, and the mould must be parted in a second direction before the component can be extracted. The action of the splits is frequently at right-angles to the line of draw, but this is dependent upon the component, and angles of split can vary. In many cases, the whole form is contained in the splits, but, in others, only certain portions, the remainder of the form being extracted in the line of draw. This latter generally occurs where only a particular feature on the moulding forms an undercut.

The chief forms of split tool fall into the following categories:

(i) Splits on the stationary side of the tool and operated by cams or pins on the moving side.

(ii) Splits on the moving side of the tool and operated by cams or pins on the stationary side.

(iii) Splits operated by the ejector mechanism.

(iv) Portable split tools that are operated manually or by means of a jig.

The first of these is illustrated by Fig. 5.11, in which the component has rims forming undercuts at the shoulder and top. The splits are mounted in slides on the fixed (injection) side of the tool, and are usually urged apart by suitable compression springs, or by outside tension springs;

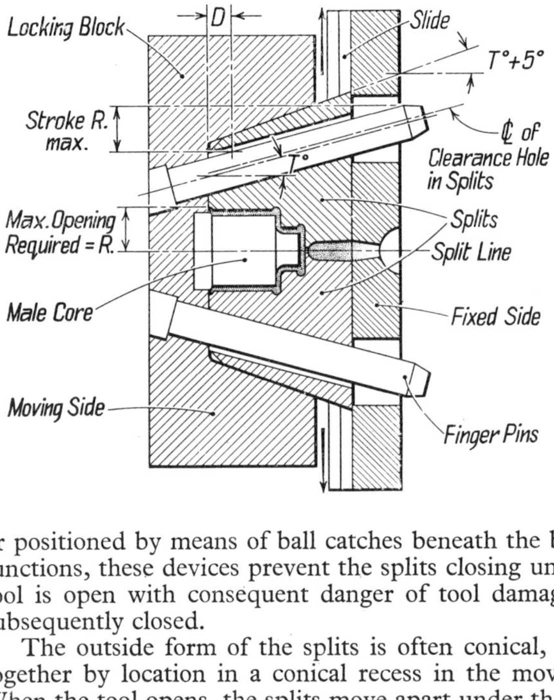

Fig. 5.11. Split Tool—Stationary Side.

or positioned by means of ball catches beneath the blocks. Among other functions, these devices prevent the splits closing under gravity when the tool is open with consequent danger of tool damage when the press is subsequently closed.

The outside form of the splits is often conical, the splits being held together by location in a conical recess in the moving part of the tool. When the tool opens, the splits move apart under the control of the angle finger pins mounted in the moving portion of the mould. During mould closing, the reverse action takes place, the splits being moved together by the finger pins. In the design shown, the holes in the splits for the finger pins are provided with clearance. This means that the tool will part for a distance D (Fig. 5.11) before the splits begin to part. During this period of delay, the male core commences to retract from the moulding while the remainder of the component is firmly held. This means that with many types of moulding no other ejection mechanism is necessary, resulting in a simpler tool. It will be noted that, in the following type of split mould producing the same component, an additional sleeve ejector is required. The amount of opening *R* provided on each split must not be less than is necessary to clear the undercut on the component.

The second type is illustrated in Fig. 5.12, in which the splits are mounted upon the moving side of the tool. The tool shown at (*A*) produces the same component as the tool in Fig. 5.11. In the present case, however, the delay *D* provided by the clearance of finger pin and hole serves no purpose beyond normal operating clearance, and no freeing of the male core occurs. Hence ejection must be provided by means of a sleeve ejector, and on the tool closing, some means must be provided to return

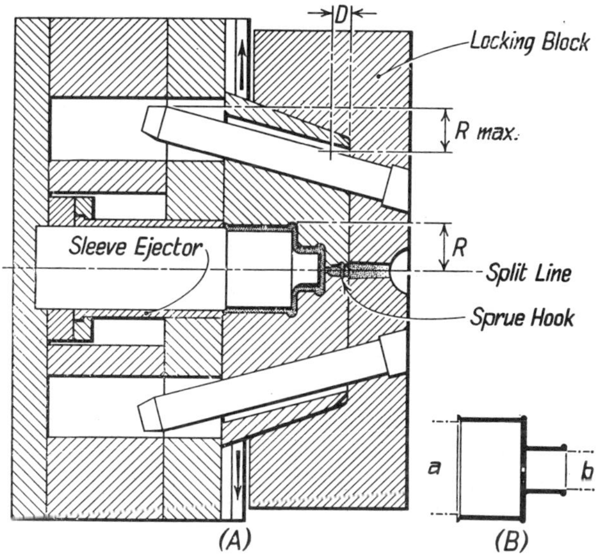

Fig. 5.12. Split Tool—Moving Side.

this ejector before the splits come together (see page 182). However this type of tool is usefully employed on components similar to that shown at (*B*). Here, two male cores are required at *a* and *b*, mounted respectively to the moving and fixed sides of the tool. If a tool with the splits mounted on the stationary side were employed, the *a* core would be stripped, but the moulding would still need to be ejected from the *b* core, and this ejection would require chain or link operation. However, if the tool is of the second type with splits on the moving side, the delay provided by the clearance between finger pins and holes allows the *b* core to be stripped before the splits part, while the moulding is subsequently ejected from the *a* core, the sleeve ejector being operated directly from the press ejector gear. The provision of a sprue puller hook on this type of split tool will be noted. This is necessary to extract the sprue with the moulding as the press opens. On the system shown in Fig. 5.11, no sprue hook should be provided, as the sprue remains with the moulding on the fixed side and must be freely removable.

An alternative kind of finger cam for use with these tools is shown in Fig. 5.13. This is the dog-leg type giving an increased delay. As previously, *D* is the delay portion of the opening stroke of tool during which no opening of the splits occurs; and *R* is the amount of opening provided on each split, which must not be less than that necessary to clear the undercut on the component.

It is important that the angle of all finger cams should be smaller than the angle provided on the locking block; especially if the delay time is small or non-existent. If the angle is the same or greater than that of the block, the splits will attempt to open more rapidly than the angular clamping faces withdraw, with the likelihood of serious damage to the tool. The angle of the finger pins should not exceed 30 deg. from the vertical.

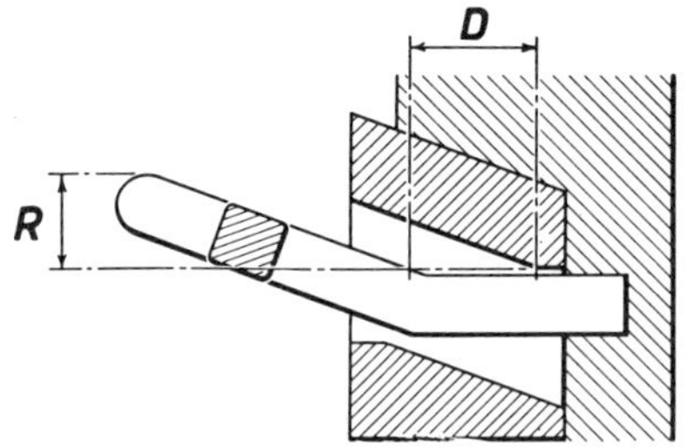

Fig. 5.13. Dog-leg Finger Cam.

In many tools, only a portion of the component is undercut, and it may be inconvenient to place the entire tool in splits—especially with larger tools. In such cases, the undercut portion can be freed by introducing a separate core block, which can be removed to allow ejection of the component in the usual way. Such a core can be operated by pneumatic or hydraulic means, but, in many cases, mechanical movement of the block is effected from the ejector gear. An illustration of one such method is indicated in Fig. 5.14. In this, the moulding (1) has a local undercut section (2) formed by the core (3). The sideways movement of this core is controlled by the rod (4) which slides in and out of the block (5). One end of the rod is secured to the core block and the other end terminates at the roller (6), which engages a hardened plate (7) secured to the tool ejector gear.

On the opening of the tool, the fixed half of the mould parts from the moulding, leaving the component in the moving tool and with the undercut core block still engaged. Towards the end of the press opening stroke, the ejector gear begins to move forward, ejecting the moulding in the press line of draw. At the same time, the forward action of the ejector gear forces the rod (4) outwards, its angular disposition resulting in the roller (6)

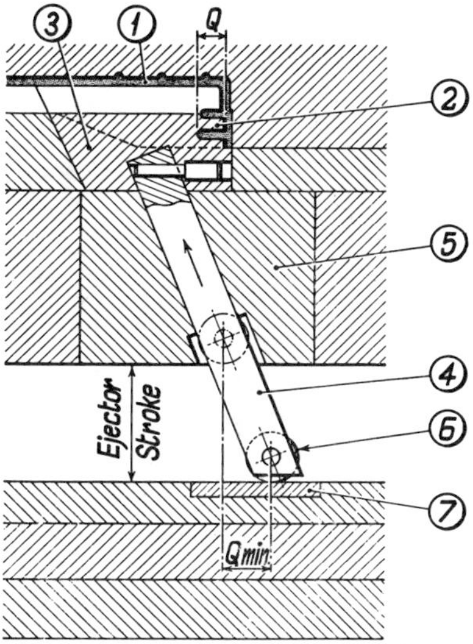

Fig. 5.14. Local Undercut formed by Core Block and operated from Ejector Gear.

traversing the plate (7). Since the core block is anchored to the rod, this travels forward at the same time. During the same period, the block also moves outwards due to the angle of action of the rod. This outward motion withdraws the undercut portion of the core block from the moulding and hence frees the component.

Where mouldings are required in limited numbers or involve difficult inserts, it may not be economic or convenient to manufacture an automatic split mould, and hence the employment of portable split tools is common. These are slower in operation, since the splits are removed bodily from the press and subsequently opened manually or by means of a jig. They are generally somewhat cheaper, however, since the cam opening mechanism is omitted.

These tools, too, are particularly convenient if the component includes an insert which is difficult to load into a conventional tool. This especially applies if the insert has to be loaded between the splits. Such inserts can readily be loaded into one half of the split mould, on the bench, the second half retaining the inserts before the tool is remounted in the press. Access to the tool for component removal is similarly facilitated, thereby removing a difficulty which can arise with extensive flashing round complex or uneven inserts.

One of the many types of portable mould is illustrated in Fig. 5.15. This is a two-impression tool for producing square-coil formers. The cage (3) is mounted on the moving platen, and its taper-sided slot is closed at one end by the plate (5). The splits (2) slide in the slot and are retained in the cage by the plates (1). They are positioned relative to each other by the dowels (7). In operation, the two splits are inserted into the cage from the top, and as the press closes the slight projection of the splits

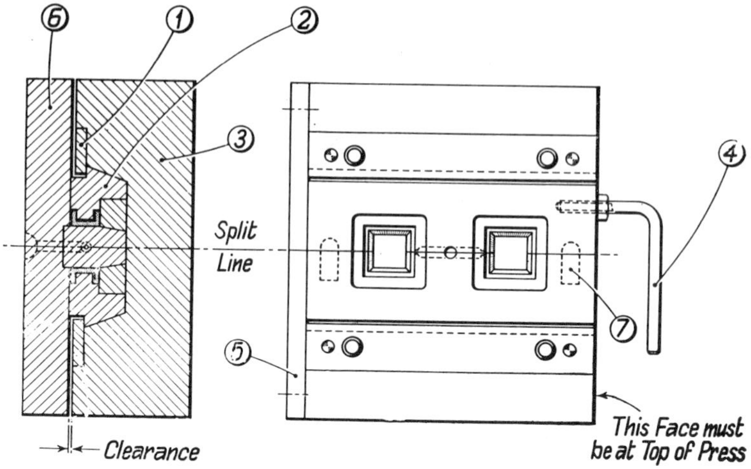

Fig. 5.15. Portable-type Tool.

beyond the cage receives the clamping pressure, and this beds home the tool and forces the splits towards each other by the pressure exerted by the angle faces of the slot. When the tool is opened, the splits are removed from the press by the handle (4). Wherever possible, the tool should be arranged so that the splits are entered from the top. This ensures that the weight of the splits brings them fully against the stop, and thus guarantees that the splits are aligned to the injection half of the tool.

Tools with Sprung Undercuts. The necessity to employ a split tool or tool with collapsing core can be avoided in certain instances if the undercut can be sprung from the tool; i.e. if the undercut is not too great or difficult and the moulding material is not too rigid. Materials which, in favourable circumstances, can readily be stripped from the tool include:

Toughened polystyrene; high- and low-density polythene; nylon; P.V.C., flexible and semi-rigid grades; cellulose acetate; cellulose butyrate; cellulose propionate, polypropylene and A.B.S.

Experience is the chief factor in determining whether a given undercut can be extracted in this manner, since its success is very dependent upon the particular design of the component, the severity of the undercut and the material employed.

A typical male-tool undercut intended for stripping is illustrated in Fig. 5.16. This could otherwise only be handled in a tool with a collapsing core. The view at (*a*) shows the cylindrical component with an enlarged view of the internally undercut section. It should be noted that, to ease stripping, not only is the undercut itself radiused but large radii are provided at 1 and 2 to ease extraction of the male tool.

At (*b*) is shown one method of stripping the component. When the tool opens, the moulding remains on the male, due to the undercut, but once it is free of the female it is free to expand outwards as it is sprung over the raised portion of the male. Relief is generally provided between the stripper sleeve and core, as shown in more detail in Fig. 6.17, page 171.

An alternative method is illustrated at (*d*) in Fig. 5.16 in which the stripper sleeve is replaced by a stripper plate. In practice, however, the sleeve would be chosen, since the moulding and male tool are cylindrical and the moulding is not too large.

Sleeve ejectors also allow greater scope for correction of positional errors and wear, particularly on multi-impression tools. Normally, however, a stripper-plate construction is simpler and almost a necessity for profiles other than circular. To overcome scoring on all stripper-plate tools, the mating faces of plate and core are tapered with an angle of not less than 5 deg. per side. If possible, both ejector sleeves and plates should be hardened.

It is obvious that, during the ejection stroke, the ejector sleeve or plate must be able to pass over the raised portion of the male tool forming the undercut. Moreover, clearance must be provided between this portion of the tool and the sleeve, to ensure that, if the moulding fails to strip, the material distorted by the ejector has room to pass without damaging the tool. The recommended clearance between ejector and maximum point of undercut is $\frac{1}{2}$ (undercut), and this is indicated in view (*c*) where the undercut dimension is given as u. The particular value that is effective depends upon the kind of material and its elasticity, and the above rule cannot be exactly applied in all cases.

This clearance requires that the rim of the moulding be fairly wide, and such a thick rim requires, in turn, that the moulding be centre-gated at the top. Any attempt at edge-feeding may cause the moulding to fill at the thick rim first, before the sides are filled, thus producing gas traps and bad flow lines.

An alternative form of ejection to the sleeve or plate is the large centre ejector, which may be either a large fixed plain pin as shown at (*e*), or a

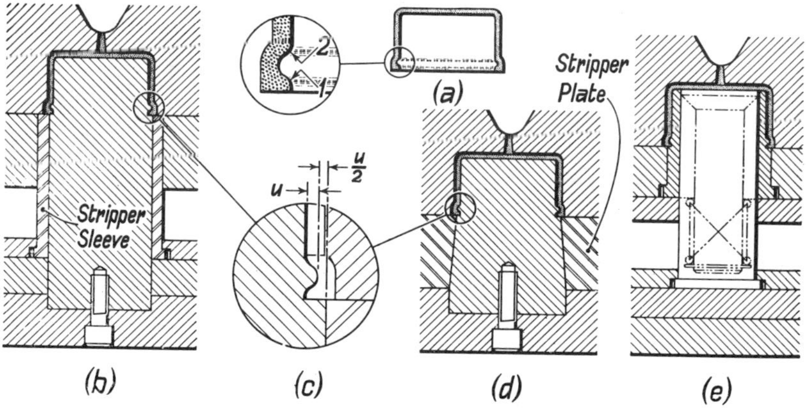

Fig. 5.16. Stripping Moulding from Male Tool.

spring-loaded valve-seated air ejector as shown chain-dotted in the same view. These centre ejectors may be of as large a diameter as possible; if they are too small they cause the top of the moulding to dish upwards, thereby distorting the sides of the moulding outwards and tightening the undercut on the tool.

Attempts to eject a non-rigid moulding by replacing a sleeve or plate ejector by a series of peripheral ejector pins can lead to difficulties, as these may cause distortion of the moulding between the ejector pins.

The necessity to strip mouldings from a female tool probably does not arise so frequently. The method is not as easy as stripping from a male, and hence it is normally overcome by the use of a split tool. However, occasions do arise when this technique can be employed, resulting in a cheaper tool. At (*a*) in Fig. 5.17 is illustrated a component with an exterior undercut which could not be produced in a split tool, since the undercut extends above the skirt. This would prevent the splits opening, hence

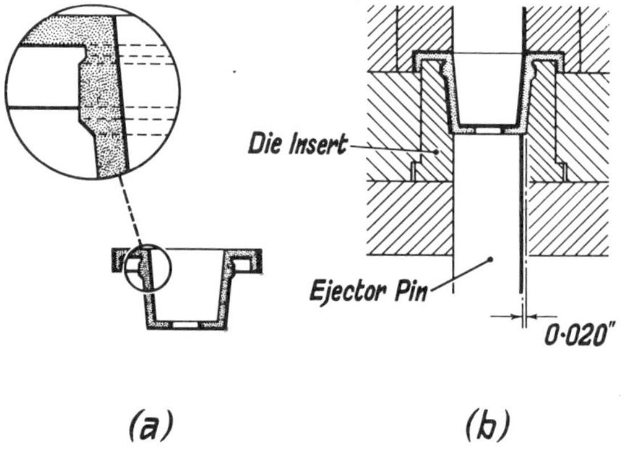

Fig. 5.17. Stripping Moulding from Female Tool.

stripping must be employed. The generous chamfer provided on the undercut is indicated in the enlarged view. This assists the sliding out of the moulding from the corresponding recess in the female tool. At (*b*) is shown the tool scheme suitable for moulding and stripping this component, with the male tool on the injection side. When the tool opens, the male core is withdrawn from the moulding as the female tool moves away, the component being retained in the female by the undercut.

On ejection, the moulding is moved upwards and out, the moulding being free to collapse inwards slightly to allow the undercut to ride free of the die recess. The ejector pin should be as large as possible to prevent distortion of the base of the moulding, but should have a 0.02 in. clearance between the pin and the minimum diameter of the cavity to prevent scoring the cavity wall.

In general, it is not possible to strip such deep undercuts from female tools as it is from male tools. This is due to the fact that mouldings can be more easily freed when stretched and expanded over the undercut than when the moulding has to be compressed to achieve ejection and the moulding collapsed inwards to free a female undercut. In both cases, however, it is important to arrange that the moulding is free to expand or contract to free the undercut at the time of ejection.

Hot-runner Moulds. Also known as a sprueless or runnerless mould, the hot-runner mould has as its chief objective the obviating of the whole or part of the moulded feed system which is inherent in the conventional type of mould. With single-impression moulds, this means the shortening or elimination of the sprue, and with multi-impression moulds, the elimination of both sprue and runner system. In the conventional tool, this feed system is cooled at the same time as the moulding, and is removed from the tool at the same time, to be subsequently reground. In hot-runner moulding, the whole or part of this feed system is maintained at an elevated temperature, so that the moulding material in the feed system is kept hot ready for the next cycle. This has several advantages: the provision of faster cycles, the elimination of finishing operations in gate removal, and the reduction of scrap or reground material arising from the sprue or runner. Tools of this type are more suitable for large production quantities, since tool cost is generally higher. Hot-runner moulding is achieved by various methods which are usually as follows:

(i) By extending the heating nozzle of the injection cylinder through the tool until it is in direct contact with the mould cavity. This method is indicated at (*A*) in Fig. 5.18. In the case of multiple-impression tools, several heated extension nozzles can be taken from a heated manifold mounted directly on the injection cylinder, each nozzle leading direct to one of the cavities, as shown at (*B*) in Fig. 5.18. Unless the nozzles are closely pitched, however, difficulties may arise in passing through the normal-sized hole in the fixed platen. The single extension nozzle tool is the most useful of this type. By using this, advantages are obtained in ease of heating, reduction of insulation problems, and facility of access by movement of the injection unit carriage from the fixed platen.

(ii) The nozzle of the injection cylinder is heated and extended to pass into the tool and terminates in a conventional sprue, but of considerably shorter length, as shown at (*C*) in Fig. 5.18. This produces a moulding with a stub sprue, which can sometimes be left on the moulding.

(iii) Hot-runner moulds in which the material from the heating cylinder is led into a heated manifold from which diversionary heated nozzles extend into the cavities. This method is normally employed on multi-impression tools, but may also be used on multi-gate single-impression tools or where the feed is to be carried to a single gate on the moulding remote from the centre line (thus avoiding an offset tool). The nozzles may extend direct to the cavity, to give a sprueless mould as shown at (*D*) or may be foreshortened to give a stub sprue.

The operation of tools of this type depends upon maintaining the moulding material within the nozzle at a sufficiently high temperature to prevent freezing, but not sufficiently high to allow the nozzle to dribble into the cavity when the tool is opened. Similarly, cooling must not cause

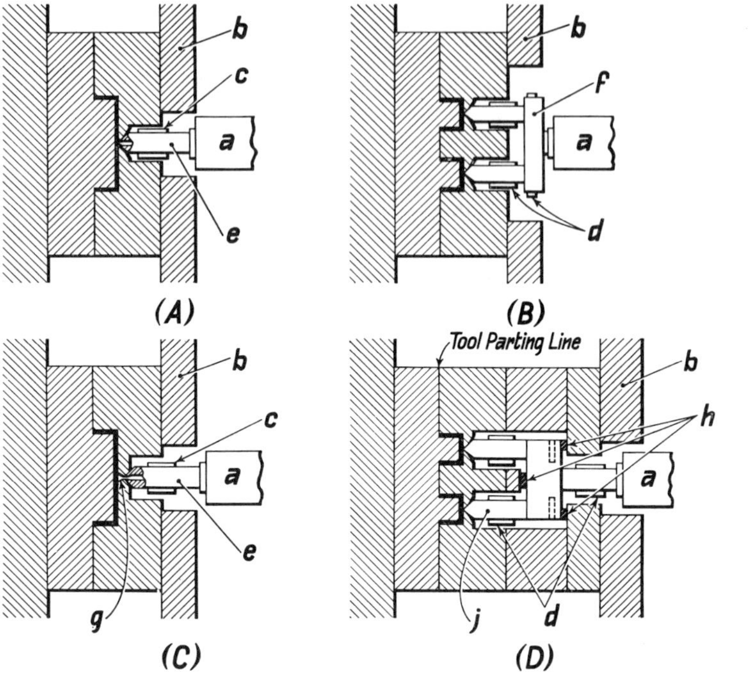

Fig. 5.18. (*A*) Sprueless Moulding with Direct-contact Extension Nozzle; (*B*) Sprueless Moulding on Multi-impression Tool with Multiple Extension Nozzle; (*C*) Extension Nozzle with Stub Sprue; (*D*) Sprueless Moulding with Hot-runner Mould. The various Members and Parts are denoted as follows: *a*, Heating Cylinder; *b*, Stationary Platen; *c*, Extension Nozzle Heater; *d*, Heaters; *e*, Extension Nozzle; *f*, Manifold and Multiple Extension Nozzles; *g*, Stub Sprue; *h*, Insulation Spacers, *j*, Hot-runner Manifold and Nozzles.

the nozzle to freeze, but must be sufficient to allow the moulding in the gate zone to set.

These requirements mean a strict control of nozzle heating, minimum contact area between nozzle and tool, and close control of the mould temperature. Continuous maintenance of the moulding cycle is essential, since erratic cycle times upsets the critical heat balance, especially with sprueless moulds.

The first requirement is met by the provision of adequate controllers on the separate manifold and nozzle-heating system, and it is usual to place a thermocouple as close as possible to the nozzle end. Reduction of heating losses is best attained by provision of relief areas where the nozzle contacts the tool*, and adequate insulation between the extension

*For the sake of simplicity in Fig. 5.18 full metal-to-metal contact is shown between nozzle and tool. In order to prevent excessive heat loss however, the connection will be one of the following:

(a) Provided with only the minimum seating between the nozzle and tool—illustrated in the direct nozzle of Fig. 5.20. *or* (b) Effected through the use of a pre-chamber—as shown in Fig. 8.6, page 235. This applies to the methods as shown

nozzle or hot manifold runner and the remainder of the tool. Mould-temperature control requires that the cooling channels are provided as close as possible to the nozzle area, and that the temperature of the cooling medium can be properly controlled. In some cases, differential cooling between the areas adjacent to the nozzle and the remainder of the tool may be necessary, since cooling at one temperature may be satisfactory for preventing the nozzle freezing but may be too high for an overall mould temperature to allow sufficiently rapid setting of the moulding. Thus separate circuits for these two areas are warranted.

A typical direct-injection nozzle is shown in Fig. 5.19. Here the nozzle is insulated effectively from the cavity by the provision of a clearance of 0.03 in./0.06 in. which fills with plastic melt on injection. The nozzle is carried just short of the cavity, a further gate being cut through to the

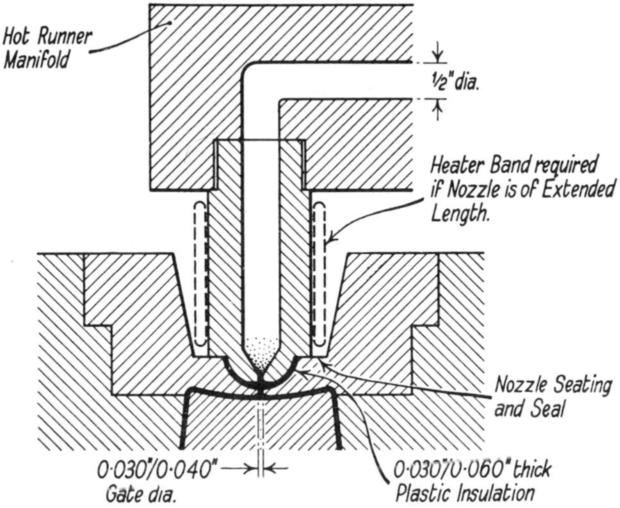

Fig. 5.19. Direct-injection Nozzle.

cavity. On injection, any chilling of the insulated film across the nozzle orifice is limited by the insulation provided and is swept away by the following shot.

It will be noted that the nozzle orifice is kept small to prevent dribbling and allow closer temperature control over the plastic material in the mouth of the nozzle.

Earlier attempts in the U.S.A. and Great Britain to employ a flat-ended nozzle which formed part of the mould cavity surface were not

in Fig. 5.18 (*A*), (*B*) and (*C*); *or* (c) Insulated by a plastic seal—as shown in Fig. 5.19; *or* (d) Broken each cycle by employing the sprue break. This applies to the methods as shown in Fig. 5.18 (*A*), (*B*) and (*C*).

The last two methods are probably the most common, while the use of a pre-chamber is particularly popular for single-impression tools.

particularly successful. However, the stub sprue nozzle method is successfully used.

The critical maintenance of temperature balance at the nozzle of the sprueless mould is eased if the moulding is produced with a stub sprue, and this compromise is often preferred. In the case of multiple-impression tools, particularly, such moulds still save the whole of the normal runner system with the advantage of easier working conditions.

The design of hot-runner or sprueless moulds requires attention to several detail points of construction, in addition to those of heating and cooling.

Of first importance is the control of any leakage of moulding material within the tool. Such leakage can primarily occur at the junction of nozzle and tool, but may also arise at joints between the extension nozzles and the hot-runner manifold, and between the manifold and the injection cylinder. In the design of the tool, it is useful to consider such possibilities and, firstly, to ensure that such leakage can be readily detected, and, secondly, to provide means of leading away such leakage outside the tool without fouling the heater system or interfering with the operation of the mould. If the tool becomes blocked by undetected and trapped leaking material, considerable work and delay will be involved. To this end, close and tight fitting of the parts of the hot-runner system is essential, including welding or solid construction where possible. The minimum number of fitting parts should be employed.

To prevent backward leakage of the nozzle, not only must the fit of the mating faces be good, but sufficient pressure must be maintained between nozzle and tool. This should not be carried, however, to such an excessive degree that the nozzle is hobbed into the tool with probable resulting distortion of the mould-cavity surface around the gate. For the same reasons, care must be taken in a multi-impression tool that all the extension nozzles bed home on the tool at the same time, since otherwise leakage will occur on some, or else excessive pressure will have to be applied to the longer nozzles to ensure the seating of the shorter. Such discrepancies of nozzle length can arise if the extension nozzles are not properly made or fitted, or are made up of several fitted parts.

With large manifolds, the effect of thermal expansion must be considered, since this can cause distortion of the nozzle, where one end is held in the seating and the other end is moving outwards as the manifold heats up. One method of avoiding this is to provide a flat face to the nozzle which is held firmly against a flat seating. The orifice in the seating is made larger than the nozzle orifice. Thus when the manifold expands, the flat faces slide over one another without shutting off the passage. This problem is equally minimised by use of the plastic seal illustrated in Fig. 5.19.

In the case of floating manifolds, the hot runner system is carried by the injection cylinder, the nozzles making contact with the tool. The pressure of contact is transmitted to the heating cylinder, no substantial contact or support being otherwise provided between mould and hot-runner system. This minimizes heat insulation problems, but means that tool and hot-runner system are two separate parts. It is often preferred to build the hot-runner system into the tool as a unit. Thermal insulation is arranged by providing an air space between the hot-runner manifolds and the extension nozzle, except for local points where contact and mounting is provided by means of pads. These pads are set behind the hot-runner manifold (between hot runner and backplate) opposite each extension nozzle, and a single pad on the front of the hot-runner manifold opposite the heating cylinder entry. The first are to sustain and resist the pressure of the extension nozzles and the cavities, and the latter to prevent forward movement of the hot runner and to withstand any pressure transmitted from the injection unit. These pads may either be of hard asbestos-cement or else of steel.

In addition to the air insulation gap, provision should be made for ample water cooling in the die plate adjacent to the hot-runner manifold. Assembly of the hot-runner system in the tool must be exact, to ensure proper contact between the nozzles and the tool.

In the case of tools employing an extension nozzle direct from the heating cylinder as shown at (*A*), (*B*) and (*C*) in Fig. 5.18, access to the heaters for maintenance can readily be achieved by withdrawing the injection unit or moving the stationary platen.

In the case of moulds employing an integral hot-runner system, similar ready access to the heaters must also be provided, since to remove and dismantle the tool in order to replace a burnt-out element can be vexatious. For manifolds not subject to excessive heat losses a heater loading of 15 watts per cu. in. of metal in the hot-runner manifold is usually sufficient. However, values up to 30 watts per cu. in are often employed.

In Fig. 5.20 is shown a four-impression hot-runner mould for producing two pairs of matched spheres in polythene, one half of each sphere carrying a moulded thread. For illustration purposes the figure shows two different methods of feeding these mouldings from a hot-runner manifold, the upper cavity being fed with a stub sprue, and the lower showing a sprueless direct-injection nozzle.* It will be noted that the hot-runner manifold (Fig. 5.20A) is in the form of a cross (minimum metal in the manifold is obviously desirable), heating being provided by means of cartridge heaters.

The manifold is mounted against the backplate, being separated by insulation spacers, its forward position being maintained by a centrally raised portion which abuts against the access plate. Apart from these location pads, the hot-runner manifold is insulated all round by an air gap. The extension nozzles themselves are heated by means of electrical band heaters[1, 2]. Ready access to the nozzles and heaters is achieved by the

---

*In the illustration the nozzle/mould seating is shown as the minimum metal-to-metal contact type. If a direct gate was required however, the most common practice would be to utilise the plastic-seal type shown in Fig. 5.19.

[1]In tools where nozzle heater-bands are not employed, the nozzles are sometimes made from beryllium copper. Where numerous nozzles are concerned (e.g. large multi-impression tools) maintenance of optimum temperature conditions at every gate is difficult. In these cases it is good practice to provide each nozzle heater with individual temperature control.

[2]Long nozzles from the manifold are necessary in Fig. 5.20 in order to extend the depth of the male core. Whenever possible, however, these nozzles are kept as short as possible.

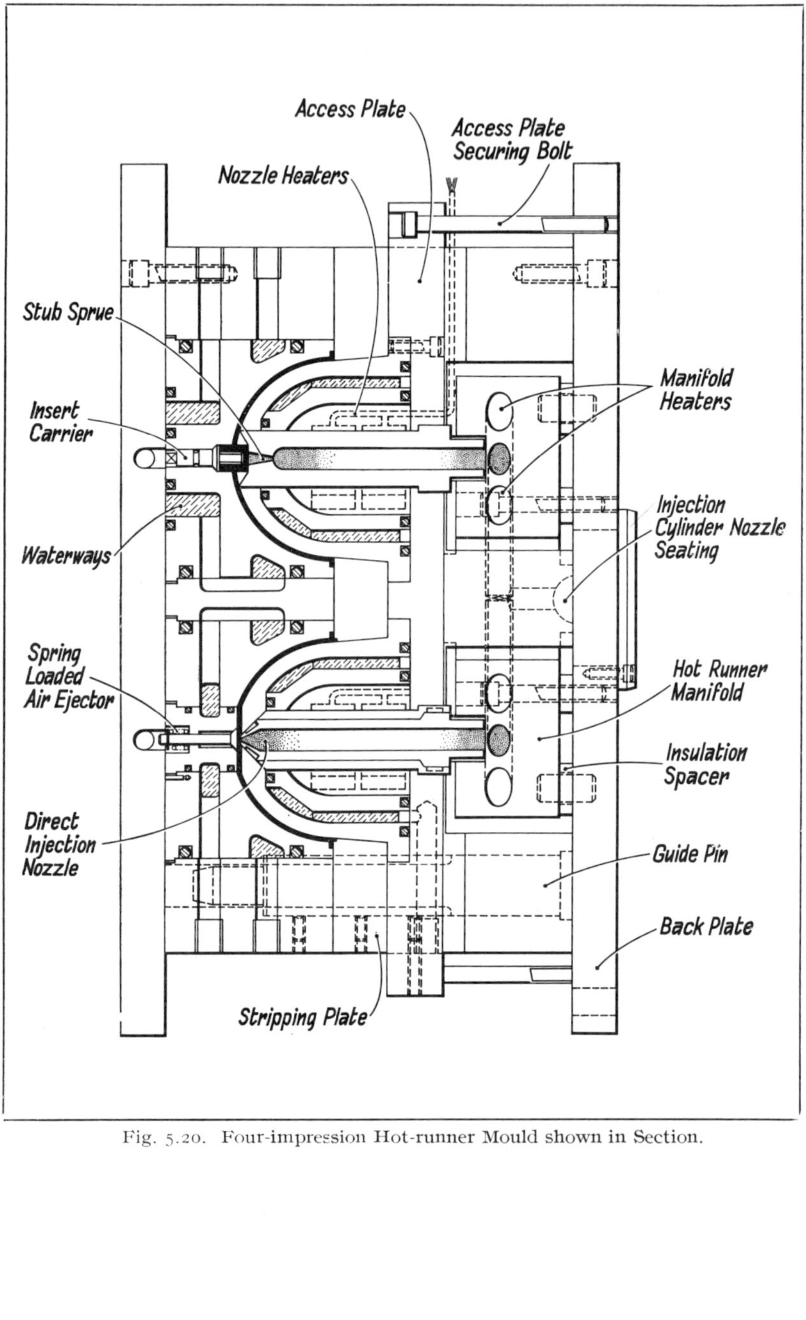

Fig. 5.20. Four-impression Hot-runner Mould shown in Section.

provision of an access plate. The latter is bolted back against the stationary platen back plate in normal use, and retains the male form. Should it be necessary to inspect or maintain the heaters, or should a leakage occur, the access plate bolts are removed and reversed to bolt the plate to the moving platen backplate. Thus, when the press is opened, this plate with the male form is separated from the remainder of the tool on the stationary side. This then leaves the hot-runner manifold and packing on the fixed side with the extension nozzles protruding, ready for immediate attention. Maintenance can thus be rapidly executed without the need to remove and dismantle the tool. The provision of ample cooling water should be noted,

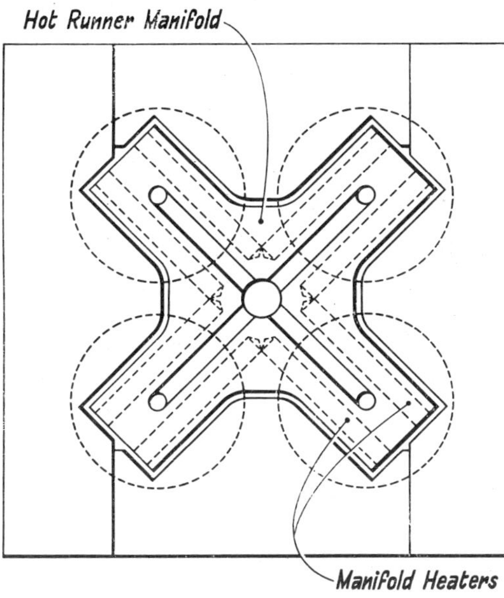

Fig. 5.20A. Plan View of Hot-runner Manifold shown in Fig. 5.20.

the channels being machined within the blocks and sealed from one another by means of rubber O-rings.

A variant of the hot-runner tool, which was originally developed to take advantage of the good heat-insulating properties of polythene, is the insulated-runner tool. In this mould, the runners in the three-plate form of tool are considerably enlarged in diameter. The third plate is locked to the stationary plate. When the plastics material is injected, the outer skin in the runner freezes, but the inner core remains molten for the next shot. In subsequent cycles, the initial outer skin on the runner remains in place as an insulator, the fresh material continually flowing through the hot core as long as the cycle is maintained. By this means, the necessity to remove the runner system inherent in the three-plate tool method is avoided. At the same time, this approach avoids the close temperature balance that has to be maintained with a hot-runner tool. The provision of a miniature heat torpedo in the gate itself, where the insulating effect of the outer skin is reduced, is advantageous. The principle of the insulated-runner tool as applied to a multiple-impression tool is illustrated in Fig. 5.20B. A more widely used technique is employed with a single-impression centre-gated two-plate tool. Here, instead of the usual stalk sprue leading direct into the cavity, the sprue is made much larger (0.5 in. to 0.75 in. diameter) and with a reverse taper. Instead of leading direct into the moulding, this sprue communicates through a short gate of 0.07 in. to 0.09 in. diameter, to the centre of the moulding. The gate has a reverse taper, so that the sprue is extracted with the moulding. The enlarged sprue and the restricted gate are machined in the body of the sprue bush. By this means, restricted gating with almost direct feed conditions can be simply achieved on a simple two-plate tool. The reverse

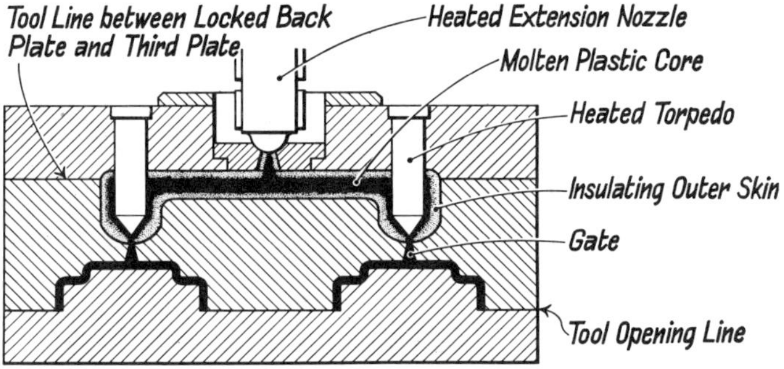

Fig. 5.20B. Principle of Insulated-runner Tool.

taper on the sprue enables this to be cleared rapidly by separating the nozzle from the sprue bush, should a freeze-up occur; see Fig. 8.6, page 235.

Valve-gated Moulds. Various methods have been employed to utilize the high compression energy of the hot plastics material in the injection cylinder when under pressure, these usually being known as variants of the precompression technique. In conventional moulding, the material commences to flow into the mould as soon as the injection pressure begins to rise, whereas, in the precompression system, this filling of the mould is delayed until full injection pressure is developed. Then, the sudden opening of the mould cavity allows the rapid decompression of the melt to fill the cavity at a much higher rate than normal. This technique was developed along various lines by a number of concerns and some applications form the subjects of patent claims. In some, the valves to restrain the flow until precompression is achieved are built into the injection unit and utilize standard moulds;* in others, standard injection

*One system of this type is described in "Sequential Impact Moulding" by G. D. Gilmore and Judson Decker, *Modern Plastics*, Vol. 36, No. 3.

machines are used with special moulds. A valve-gated mould of the latter type has been developed in the U.S. by Columbus Plastic Products Inc. and W. R. Grace and Co.*

Reference to Fig. 5.21 will show diagrammatically the method of operation. The tool is a hot-runner mould (only the female tool is shown in the figure) in which the runners are led directly into the mould cavity, terminating in a conical seating which is open to the cavity and forms a gate. A valve rod can be moved forward to seal off the gate, or withdrawn to allow the passage of the material from the runner to the cavity. The valve rods are operated by means of external pneumatic or hydraulic cylinders. This type of gate produces a sprueless moulding.

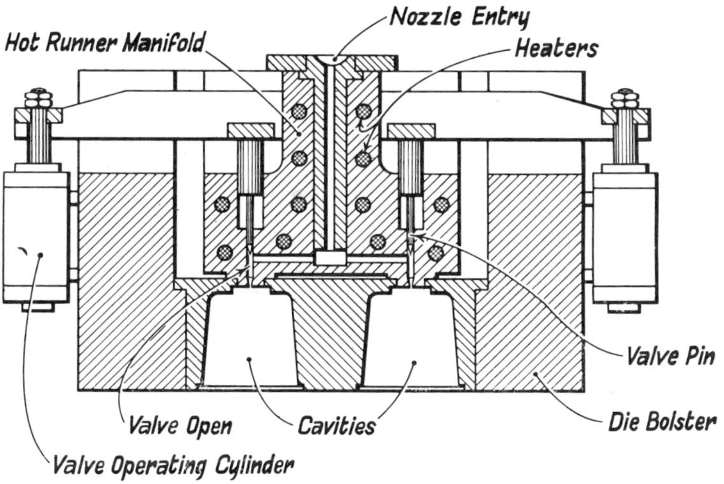

Fig. 5.21 Valve-gated Mould.

During the mould-open period, the gate valves are shut. The injection ram is then brought forward to raise the material to full injection pressure, against the closed valves. When the mould is closed, the valves are rapidly opened, thus releasing the full injection pressure to the cavity. This allows instantaneous decompression of the material which commences to fill the cavity at an extremely rapid rate.

Complete filling of the mould is achieved by the follow-up of the injection ram. The amount of material fed into the cavity by decompression alone is dependent upon various factors, but can be substantial. After filling, the valves are closed and the ram returned to take up the next charge of material while the moulding is cooling. Due to the high velocity of the decompressed material, care must be taken in the mould design to

*"Valve Gating of Injection Moulds", A. Spaak and G. Kelly, *Modern Plastics*, Vol. 36, No. 1.

provide as smooth a flow path as possible, without sharp corners. The hot-runner manifold and cooling arrangements must be properly designed, as in a normal direct-injection hot-runner mould, except that it is claimed that the heat-balance problem at the gate is less in a valve-gated mould, due to the positive sealing of the gate.

The advantages of this type of moulding are stated to be: reduced cycle times, due to reduction of idle time by the initial compression of the material in the injection cylinder during the mould-open time instead of after the mould has been closed as in the conventional process; and a lower overall injection pressure or longer flow path, because of the rapid decompression of the material. Provision can be made to close the valves as soon as the mould fills, providing packing pressure is not required, and this in combination with the lower injection pressure can effect a saving in weight on the completed moulding.

This type of mould is considerably more expensive than conventional tools and in spite of the advantages claimed has only enjoyed limited adoption. To an extent its achievement of lower pressure/longer flow-path has, in any event, been equally served by the advent of screw injection machines.

Multi-colour Moulding. This type of moulding comprises the production of a first moulding in a specific colour, the first moulding then being used as an insert or preform for insertion in a second tool. In the second moulding operation, material in contrasting colour is injected around the first moulding, except for certain areas on the original moulding which are left clear to form a character or design. The process is used for the production of such parts as key tips and radio knobs, usually in two colours but sometimes in three. A separate press is normally used for each colour, the mouldings being passed, mounted on a tool plate, from the first press to the second, as they are produced*.

Fig. 5.22 illustrates diagrammatically a multi-impression two-colour set-up employing two presses to produce number and letter key tips.

At (*A*) is shown the first tool, producing the inner moulding and character; an enlarged view of the cavity and the core pin being shown at (B_1), (B_2) and (B_3). The cavities are in the form of die inserts with hobbed characters (1), these being set in a bolster. The core pins (2) pass through the transfer plate (3).

The face of the core pin is machined to form a number of feeder ribs which shut off against the bottom face of the cavity. The plastics material flows along these ribs to fill the character. Where the core pins pass through the transfer plate, two or more slots are cut into the transfer-plate bore. The splines (7) are raised on the male core to engage these slots, and in the tool-closed position these splines stop short of the face of the transfer plate to form a recess. These recesses produce moulded lugs projecting from the base of the inner (first) moulding. The height of the lugs is designed to be the difference between the depth of the first (inner) moulding plus the depth of character, and the depth of the second (outer)

*A few presses are however, manufactured having two separate injection cylinders on the same frame.

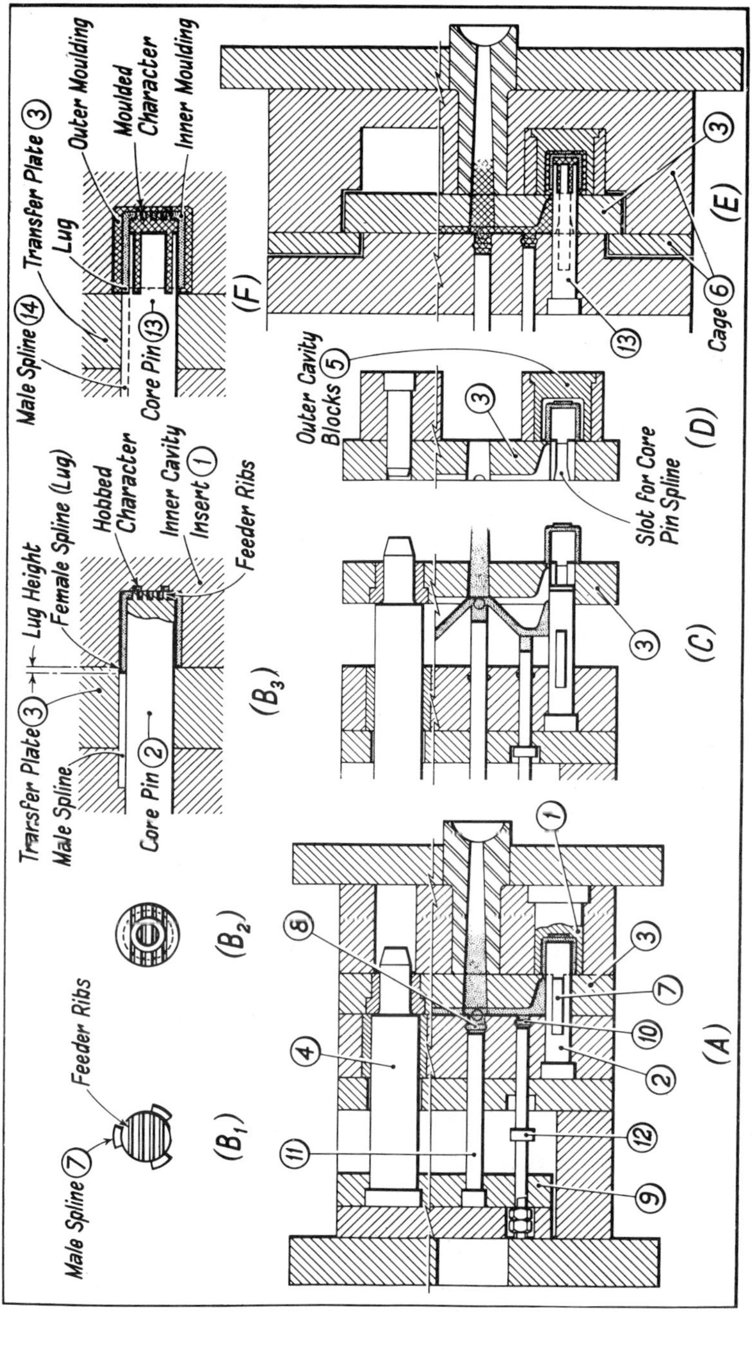

Fig. 5.22. Two-colour Injection Mould.

cavity. The feed is led from the back of the transfer plate along the core pin and communicates with the cavities through a restricted gate.

As the tool opens on completion of the moulding, the latter remains on the core, the sprue being pulled from the sprue bush by the sprue puller hook (8). On the commencement of ejection, the ejector bar (9) moves forward, together with the location pins (4). These in turn cause the transfer plate to move forward, stripping the mouldings from the core pins and breaking off the gate, the runner still being retained by the hook (10). When the mouldings are freed from the core pin, they are still retained in the transfer plate by the moulded lugs.*

As soon as it is free, the transfer plate, together with the mouldings, is removed bodily from the tool. As the transfer-plate ejector pin (4) moves forward, the sprue ejector pin (11) also moves forward. Since, however, the runners are still retained by the hooks (10), the movement of the pin (11) causes the runners to bend, since these are still warm and flexible. By this bending action, the runners are more easily freed from their spread-eagled position across the several core pins. Since the runners are partly formed round the diameters of the core pins, removal will be found more difficult without this action. Ejection of the runner system is finally completed by the ejector bar striking the fixed collar (12) and ejecting the hook (10). The fully ejected position is shown in the view (*C*).

After the transfer plate is removed from the first press, the plates carrying the second, or outer, cavities (5) are located in position on it. This view is shown in (*D*).

This complete assembly is then slid into the cage (6) on the injection side of the second tool. As the tool closes, the male cores (13) pass through the core-pin location holes in the transfer plate and enter the cavities and the bore of the first moulding. The core pins have splines (14) matching those on the first moulding which produced the moulded lugs. These splines pass through the slots provided in the transfer plate. As the tool closes, the face of the splines engage with the bottom of the moulded lugs, and continued tool movement causes the lugs to be ejected out of the transfer-plate slots and the whole of the first moulding to be carried forward until the face of the character is shut off hard against the bottom of the cavity. The closed position of the second tool is shown in view (*E*), with the enlarged view of the cavity and core pin in view (*F*).

The core pins are machined at their ends to mould the slots necessary for the moulded key tips to be mounted on the key lever of the business machine or typewriter.

The forward movement of the inner moulding removes the wall of the first moulding from covering the gate and thus allows the second injection to take place. This material, of a different colour, covers the outside of the first moulding and the top—except where the character is shut off against the cavity face. Entrance of the material to fill the centre of such characters as "O" is effected by the material flowing up through

*Various other methods are also used to retain the moulding on the transfer plate, including fixed blades projecting from the transfer plate into the rim of the moulding.

the feeder ribs moulded on the face of the inner moulding. The interior of the first moulding and the bottom are also filled with the second colour.

When the tool opens, the male cores are withdrawn from the mouldings, which are retained in the cavities by the transfer plate. The gate is broken off at the same time, the sprue and runners being retained by the puller hooks and moving away with the tool. They are ejected in the same way as the first moulding. The transfer plate, complete, is removed from the cage, and the cavity blocks (5) are knocked off and the mouldings extracted.

In the manufacture of tools of this type, it is important that all the hobbed cavities are of even length. Too long a first moulding or too shallow a second cavity will cause a distortion of the first moulding as the second tool closes, while the reverse condition will prevent the inner moulding character from shutting off against the bottom of the second cavity and thus allow the second colour to cover or partially cover the character.

CHAPTER VI

TOOL CONSTRUCTION

Complete Tool. The overall dimensions of the finished tool must be such that the mould will fit the intended press. These essential dimensions are primarily those of the closed height of the tool, the tool opening necessary to extract the moulding, and its width and depth. The maximum and minimum closed tool height, opening stroke, and the tie-bar clearance on the particular press are noted from the press data sheet as explained in a previous chapter.

The final tool design must fall within these limitations, and thus the final permissible sizes must be borne in mind as detail tool design proceeds.

The opening stroke of the tool must be clearly sufficient to allow the ready extraction of the moulding, and this generally means that the opening stroke of the tool and press must be equal to not less than twice the height of the component (i.e. length of the male punch plus length of component), plus working clearance. This amount can be reduced only in special cases, where the component is extracted sideways, or where the tool is portable and removed or partly removed from the press before ejection.

The closed die space on injection presses can be adjusted within its range either by varying the position of the moving platen or by adjusting the clamping stroke, depending upon the type of press. Most presses have fairly large minimum closed die heights and many moulds have insufficient height to fit. In these cases it is not good practice to increase the designed height of the tool by purposely increasing the thickness of the backplates, etc., but rather to make up the height by means of mild-steel packing blocks behind the tool. As an alternative, standard packing "stools" can be made up of various heights, and these can be used with different moulds.

The overall sizes of the tool must, of course, allow the mould to be inserted between the tie bars of the press so that it may be mounted on the platens. This problem can be eased somewhat by providing that the halves of the tool may be fitted separately, thus reducing the overall size, or else by canting the tool to pass through the tie bars at an angle.

It is obvious that all respects of the fitting and operation of the tool relative to the press should be fully satisfied before tool-making is commenced.

The Cavity. The female tool can be designed and made in various ways:

(1) By machining entirely from the solid steel;

(2) By machining from the solid steel and subsequently adding other features by means of separately fitted pads and inserts;

(3) By making up the cavity from an assembly of separately machined blocks.

Wherever possible, solid cavities are employed because of the greater strength and freedom from flash lines or tool print lines. The criterion is essentially whether the form can be readily machined from the solid; if it can, then it is the most satisfactory method of procedure, and may be used for both small and large moulds. Where cavities can be wholly or

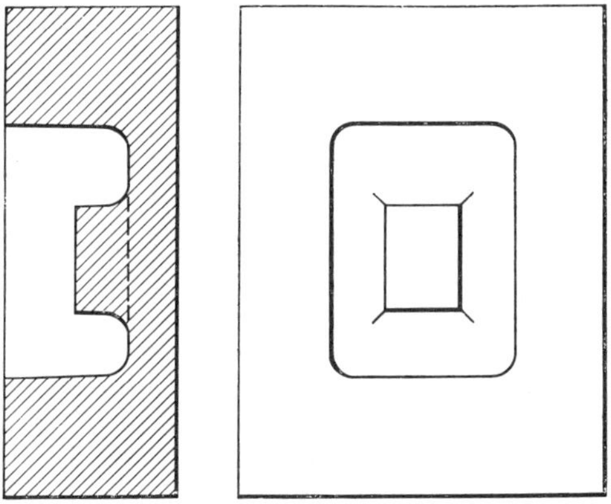

Fig. 6.1. Cavity cut from the Solid.

largely produced by turning, machining from the solid is clearly to be preferred. Where moulds are hobbed, the cavity is produced directly into the solid.

A number of cavity forms that are suitable for machining from the solid are indicated in the following illustrations.

In Fig. 6.1 is shown a single-impression cavity of simple form with all the corners radiused. The proportions of length and breadth to depth are such that the cavity may be easily machined in the solid by utilizing a standard-type tapered ball-ended cutter.

A similar type of cavity is illustrated in Fig. 6.2, except that the corners at the bottom of the cavity are square instead of radiused. This cavity also can be machined from the solid, by the use of a tapered end-mill, the square end of the cutter producing the square bottom corners.

In Fig. 6.3 is depicted an asymmetrical cavity with a raised portion

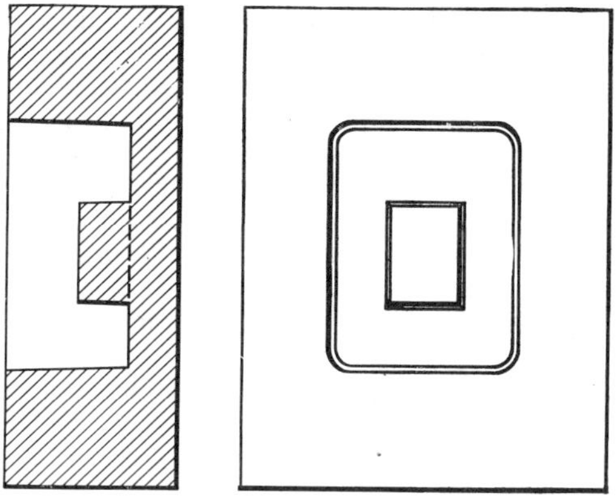

Fig. 6.2. Solid Cavity with Sharp Bottom Corners.

at the bottom. At first sight, it might be considered that this should be produced by letting in a pad to form this raised area, but since all the intersections to the main wall are radiused, it is evident that the cavity may readily be machined from the solid, and no advantage would be gained by a composite construction. It is probable that the cavity would be too large to hob and, in any event, it would prove difficult to do so, because of the uneven load. Since it is usually easier to machine a male form to dimensions than a female, it is frequent practice in such cases first to produce the reverse male form in an easily machined material such as laminated sheet. From this, a hard non-shrink plaster cast is taken to produce a plaster replica of the cavity. This in turn is used as the master to produce the final steel cavity in a duplicator or copying mill.

In many cases, however, complete machining of the cavity from the solid is impossible or difficult, and hence it has to be made up from one or

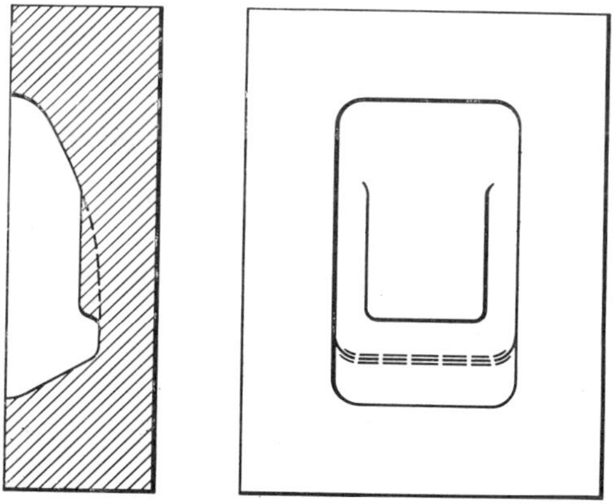

Fig. 6.3. Solid Cavity with Raised Area.

more parts. Into quite how many components the tool should be broken down is often a matter of judgment, but the number should be kept to the minimum, so that fitting work is reduced and print lines on the moulding, arising from the penetration of flash into the joint lines* of the steel blocks, avoided as far as possible. At one end of the scale, the cavity may be completely machined from the solid but with an added pad or pin let-in, or it may be wholly made up from numerous blocks mounted in a retaining chase. Apart from easing machining, cavities principally made up from blocks are used on some large moulds to avoid the necessity of machining out large masses of metal.

A cavity having sharp corners both at the bottom and on the sides is depicted in Fig. 6.4. This would be produced by machining the greater part of the cavity by normal means and then producing the profile and sharp wall corners by slotting. Taper may be produced by a compound

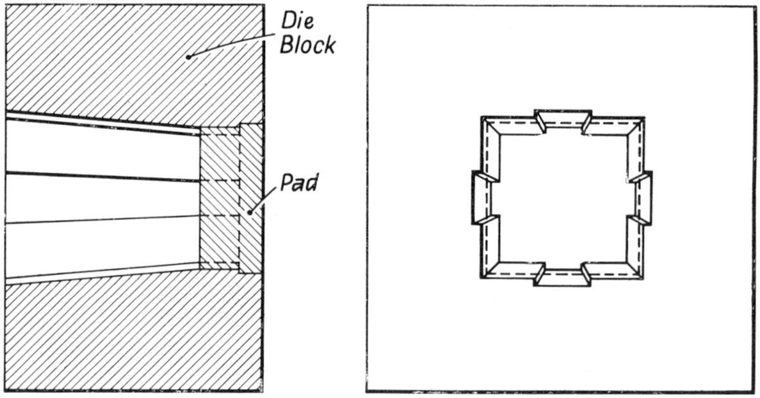

Fig. 6.4. Construction of Cavity with Sharp Corner Side Walls.

setting of the slotting machine. To obtain a sharp corner at the bottom of the cavity and to facilitate chip removal, the cavity is taken right through the die block and the bottom pad fitted afterwards. The shoulder retaining the pad should be noted. The complete cavity is, of course, eventually screwed down to a backplate, which is not shown in the illustration.

This type of construction would also be used if the bottom of the cavity

*In spite of the most careful machining and fitting, the possibility that these joint lines will show on the component should always be borne in mind. They should, therefore, be placed in a position which will not irretrievably ruin the appearance of the moulding if print lines do show. To avoid this difficulty, the joint is often deliberately made a feature of the design by allowing the pad to form a raised or recessed area in the moulding, with the joint line as the step. A common case occurs with pads carrying embossed or engraved letters, where the pad itself can form a moulded panel on which the lettering appears. The effect of print lines on unseen undersides of mouldings can usually be ignored.

were to carry detailed machining or engraving; by making the pad separately, such work is facilitated. In other instances, different mouldings are required where only a portion of the component is changed, the remainder of the moulding remaining the same; e.g. a radio cabinet with a different front panel design. In such cases, an interchangeable bottom pad can be employed in a manner similar to that illustrated.

The cavity shown in Fig. 6.2 can be produced from the solid because only the bottom corners are sharp, and not the corners of the side walls. The round side-wall corners could be produced by the rotary action of an end-mill, but could not be less than the radius of the smallest practical cutter, and if they were to be sharp, the cavity would have to be produced by a method similar to that just given. In general, the use of sharp corners formed by machining from the solid should be avoided or kept to a minimum, since these could set up points of stress concentration and, hence, weaken the mould.

An alternative method of making up the cavity given in Fig. 6.4 is shown in Fig. 6.5. In this design, the four profiled sides of the cavity form are made up as four separate blocks which are assembled together and fitted into an outer chase. It should be noted that the two small blocks must be keyed into the other pair to prevent inward movement. The bottom of the cavity is made up by a tapered centre block. This must be assembled from the front and hence must be screwed down to the backplate to prevent any tendency to lift. Use may be made of the taper

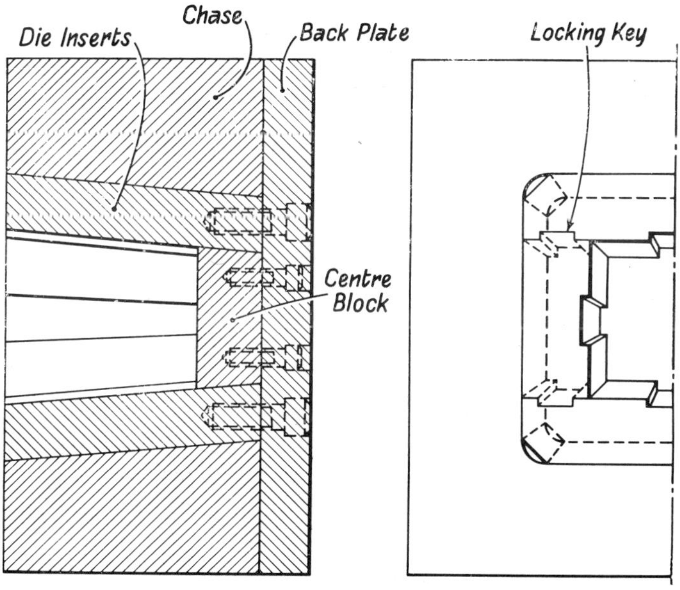

Fig. 6.5. Cavity with Sharp-corner Side Walls made up from Separate Die Inserts.

on the centre block to lock down the side blocks, but, if the taper is small, it is preferable to bolt these down also. The chase must be of sufficient thickness around the blocks to prevent deflection under moulding-cavity pressure. The advantages of this form of construction over that shown in Fig. 6.4 are:

(i) Machining of the form is easier; the individual blocks can be readily end-milled and ground;

(ii) Polishing of the mould faces is easier;

(iii) Subsequent design modification can often be more readily executed.

The disadvantages are:

(iv) A chase has to be manufactured unless a standard unit is suitable and available;

(v) The mould is usually larger due to the combined thickness of the inserts and chase.

The chase shown in the illustration is tapered on all four inside faces. This has the advantage that when the inserts are pulled back, the taper ensures that all the joint lines between the inserts are closed and tightened. Some chases have a taper face at one end only, the remainder being straight-sided, a type that is employed when end-wise tightening of the die blocks or inserts is desired. The taper angle is commonly from 10 to 15 deg. per side. The provision of taper faces, however, increases the cost of the chase, and straight-sided chases are usually employed, particularly where only a few blocks are involved.

Another cavity made up from blocks to ease machining is shown in Fig. 6.6. Here the regular shape of the cavity is broken by a sharp-sided raised step. This mould could be produced from the solid by employing a small ball cutter and machining as close to the step as possible. The final radius left would than have to be removed by hand. This method would be long and tedious, and the preferred method would be to break the form into its three natural segments. These blocks can then be readily machined to the form and their abutting faces ground. The outsides of the blocks are then machined to fit tightly into a chase or bolster plate. The two end blocks are provided with a location shoulder engaging in a recess in the chase, to prevent upward movement. Additionally, each end block is recessed to receive a location shoulder provided on the centre block. Thus the two end blocks are held back by the bolster, while the end blocks retain the centre block. The corners of the end blocks need not match the corners of the chase, but may be provided with clearance by either of the two methods shown. The assembly is completed by the addition of a backplate (not shown).

One disadvantage of this construction is that any taper required on the step, in the line of draw, would have to be added afterwards, probably by hand. If taper is omitted, the moulding may be marked or scored on ejection.

The blocks in this particular cavity are set in a straight-sided chase, but it is also common practice where a number of blocks are provided in a bank, to make one end block with a taper face to engage in one end of the chase recess. By this means, all the blocks are forced tightly together endwise, when the taper block is screwed down to the backplate.

Cavities of complex bottom form and varying depth constitute another application where the cavity may be built up. In such cases, the provision of taper on the end faces of the cavity may be awkward due to machining difficulties, and these may be avoided if separate blocks are added to form

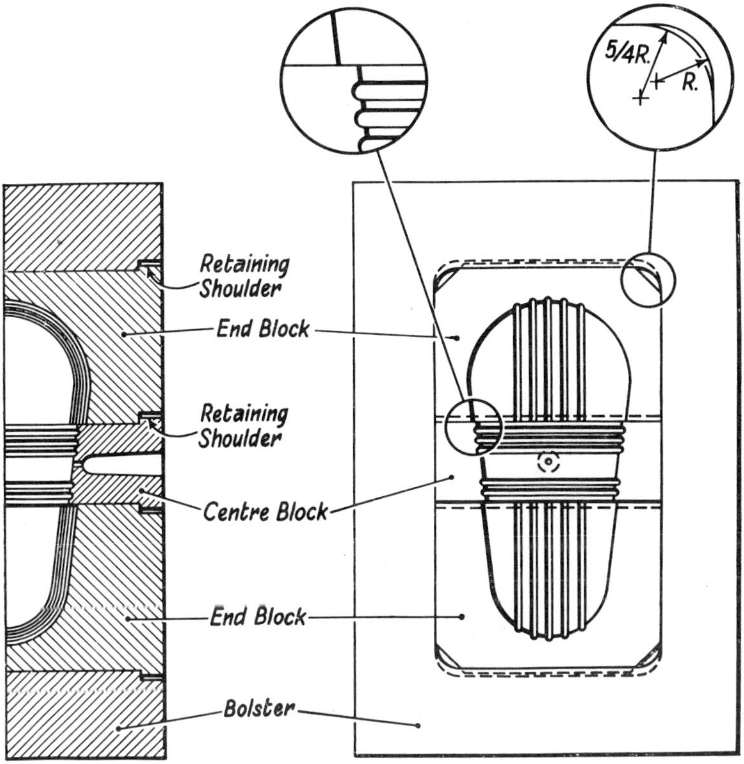

Fig. 6.6. Construction of Cavity from Blocks.

these ends. An illustration of one end of a typical cavity of this kind is given in Fig. 6.7, and it is assumed, as is the normal case, that the component will require a straight line across the end face with a single dimension to this surface. Moulding taper must be allowed on this face and machining this may be considered in a number of ways:

(1) The cavity may be machined with a parallel cutter to produce a vertical end face. Taper is then added by filling, running on to a straight line *XX* scribed across the top face of the mould. This will give an

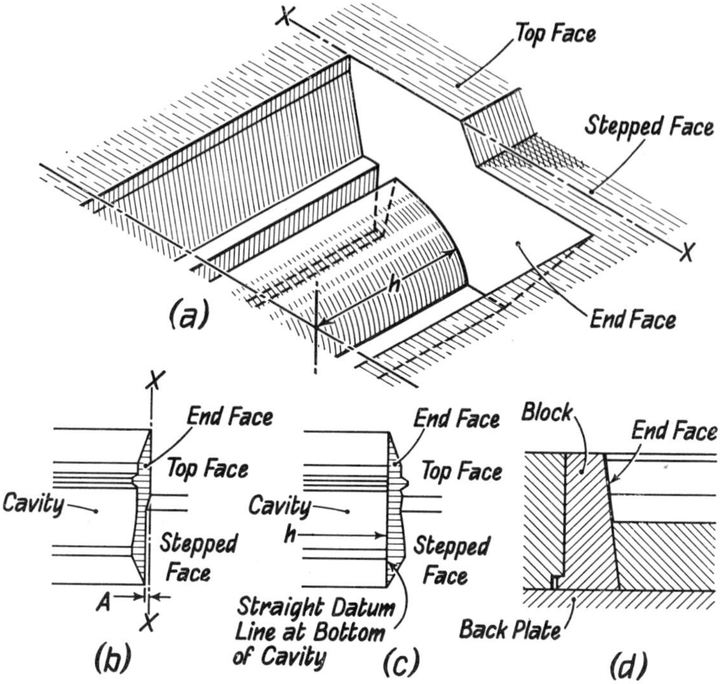

Fig. 6.7. Methods of Making Complex Cavity.

appearance, in plan, as shown at (*b*). In this view, the offset *A* is the effect of the tapered end face intersecting the stepped face at the lower depth than the intersection with the top face.

(2) The cavity may be machined with a taper cutter of the same angle as the taper to be machined, and the machining taken to depth by means of profile templates. This method is the most satisfactory, but difficult to execute, since, as the cutter is altered in depth, the cutter diameter at the mould top face and stepped-face level changes, and hence the whole cutter must be moved in and out to maintain the straight line *XX*. In plan view, the final appearance is again as shown in (*b*), and this approach produces the most satisfactory component.

(3) To avoid the difficulty of constantly adjusting the position of the cutter to maintain the straight line across the top face of the mould, another method can be adopted. This is to maintain a constant dimension *h*, view (*a*), from the centre line to the bottom of the cavity. The taper cutter is then only adjusted for depth, templates being employed. This method is much simpler than those already discussed, but the changing diameter of the taper cutter at the mould top and stepped-face level as the depth of the cutter is adjusted will produce an uneven line at the intersection of mould face and end face. In plan this would appear as shown in

view (*c*), and this effect is often not acceptable on many components.

(4) A further method is to maintain constant dimensions at the top and bottom of the cavity: i.e. working between the bottom straight line datum from dimension *h*, in view (*c*), and the top straight line *XX*, in view (*b*). This is not very easily attained except by hand filing.

All the foregoing difficulties can be avoided if the end face is formed by a separate block as shown in the side view (*d*). The block forms an end face on which the taper can readily be machined, and forms clear-cut lines. All the advantages of methods (1) and (2) are achieved, and the plan view is identical to that shown at (*b*).

A worked example of a built-up cavity is shown in Fig. 6.8. The cavity includes a raised block (1) at its base having an angled top face, and a rectangular protrusion (2) with sharp corners rising beyond this. To

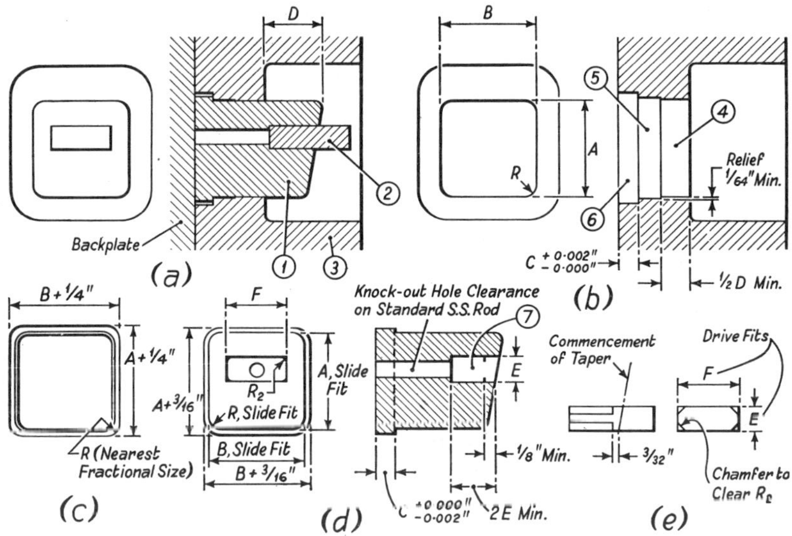

Fig. 6.8. Built-up Cavity.

facilitate machining of the cavity, the block (1) is let in from the back of the die block (3). Although this block is machined separately, the blade (2) is also made an individual item in order to ease the machining of the angled top to the block (1). The assembly of the completed cavity is shown at (*a*).

The details of the recess in the die block (3) to receive the pad (1) are shown at (*b*). The close fit of the pad need not be carried down the whole depth of the recess, but only sufficiently far to ensure a good fit and that the bearing load on the face opposing any side thrust arising from cavity pressure is sufficient to prevent the block being "hobbed" sideways into the wall of the recess.

The depth of the fitting face (4) should not be less than half that of the area exposed to cavity pressure. As normally the length of the recess and of the core plug are the same, the fitting depth is $D/2$ minimum. This reduces the bearing load to a maximum of twice the cavity pressure which, in normal conditions, should be satisfactory for a bolster or chase of nickel-chromium steel. The remainder of the recess which receives the parallel portion of the plug (5) should be relieved $\frac{1}{64}$-in. all round. The depth of recess (6) to receive the shoulder of the plug should be made to the limits +0.002 in. and 0.000 in. to ensure that the plug does not remain proud of the back surface. The height of shoulders or inserts of this nature is usually $\frac{3}{16}$ in. to $\frac{3}{8}$ in. The length and width proportions of this recess are given in the view (*c*).

The details of the main core plug (1) are shown at (*d*), and those of the small core (2) in the view (*e*). Since the sharp corners of the latter extend down to the angled face of the main core plug, this requires that the sharp corners on the core plug (2) be carried part way into the body of (1). The mouth of the recess (7) must therefore be cut with square corners to receive the sharp corners on (2). This is carried down for a depth of $\frac{1}{8}$ in., after which the recess has the radius of the cutter used to machine the cavity. Similarly, the sharp corners on the plug (2) are carried $\frac{3}{32}$ in. below the intersection of the angled face of (1) and thereafter are chamfered to clear the radius in the remainder of the depth of the recess (7). A knockout hole is provided which is a clearance on a standard silver-steel rod, in order that the core plug (2) can be driven out of its recess. Means such as this should be provided for the removal of any fitted part in a tool in order to facilitate repair or modification.

It will be noted that all tool joint lines have been made vertical and in the line of draw. This is to be strongly recommended for all general practice, since any plastics material penetrating the joints will not then prevent ejection.

The usual taper allowances must, of course, be added to the cavity and plug form.

In considering the construction of a cavity made up from blocks, due provision must be made for cooling. Cooling holes cannot be merely bored through the assembled blocks and across joint lines, however tightly these may be clamped together, since leakage at the joints will very likely occur. If a bored cooling hole must pass across a joint, proper sealing with rubber seals must be provided. An alternative method sometimes applicable is to pass a copper tube completely through an oversize hole or into an open-sided channel, sealing the pipe in position by casting in a low-melting-point alloy. In general, however, cooling can often be satisfactorily provided by cooling the chase or the bolster itself, and not the blocks. Since the chase is solid, cooling holes can be bored without difficulty.

Cavity Pads in Multi-impression Tools. Multi-impression tools are frequently constructed by providing each cavity as a separate die insert, the inserts being let into the bolster. They are inserted from the back of the bolster, and located by a shoulder on the die insert which

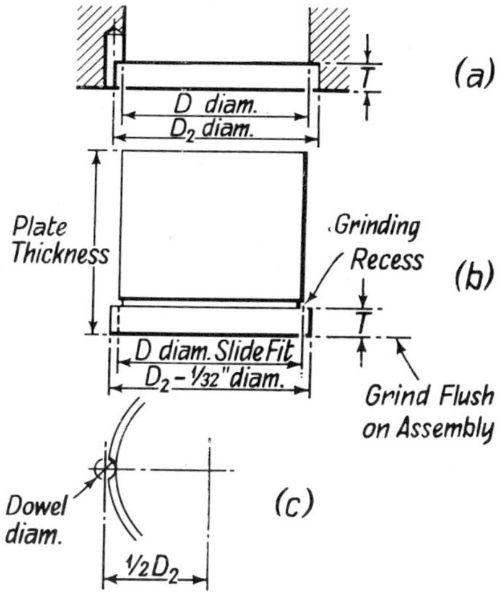

Fig. 6.9. Assembly of Die Insert.

engages a corresponding recess in the back of the bolster. The insert is finally locked by a backplate secured to the rear of the bolster. A typical assembly is shown in Fig. 6.9, where (*b*) shows the die insert itself, and (*a*) the corresponding hole and recess in the bolster. It will be noted that the location shoulder on the die insert is ground flush with the rear face of the bolster after assembly and prior to the fastening of the backplate. Rotation of the insert can be prevented by fitting a blind dowel half through the location shoulder as indicated at (*c*). Alternatively, if a number of inserts are in line, a long key can be recessed into the backplate to engage with flats that are ground on each insert location shoulder.

Strength of Cavities. The calculation of the true strength of mould cavities is almost impossibly complex for practical application, even for those of regular profile; hence it is usual to base calculations on simple basic forms and to allow an ample margin of safety.

The cavity is subjected to high internal pressure from the injected moulding material, and the essential determination is to calculate the wall thickness necessary to keep the deflection of the cavity under load below a specified maximum. While the stress level achieved in the mould must remain within safe limits, it is the actual physical deflection of the walls of the mould that is of prime importance in the operation of the tool and, if these are maintained within the specified limits generally, the stress values will also be satisfactory.

For square or rectangular cavities of normal proportions,* several

*These methods reasonably apply only where the length of the cavity wall exceeds that of cavity depth. As the depth approaches the length or exceeds it, its effect as a cantilever on the total deflection should properly be taken into account.

simple approaches are possible, including the following:

(i) Calculations based on considering each wall of the cavity as a fixed beam with uniformly distributed load;

(ii) Calculations based on considering each wall of the cavity as a freely-supported beam with uniformly distributed load;

(iii) Calculations based on considering the cavity as a uniformly internally-loaded portal frame.

(iv) Calculation based on considering each wall of the cavity as a freely-supported uniformly-loaded rectangular plate.

The effects of clamping pressure and the retaining effect of the back-plate are ignored.

A safe maximum cavity pressure on which these calculations are based is 9,000 lb./sq. in.* The maximum value for deflection should be from 0.005 in. to 0.01 in., the smallest value in this range being taken, consistent with the general proportions of the mould. Deflections of this order are acceptable for cavities cut from the solid, but special care should be exercised where cavities are made up from blocks, particularly solid cavities with a separate base plug. Here, the deflection should not allow the clearance between the separate pieces to increase to such a degree that the entry of flash can occur. The clearance between the blocks should not be allowed to rise above 0.003 in. to 0.004 in. for polystyrene or acrylic, or above 0.001 in. for nylon.

The following calculations are necessarily based on solid cavities, and it should be noted that, with built-up moulds, any loose fitting or local lack of rigidity or strength may give considerably inferior results. Also, it will be appreciated that the larger the deflection, the more plastics material is taken to fill the mould. Again, as the material cools and shrinks, the cavity pressure will fall and the sides of the mould attempt to return to straightness. This, in turn, may cause the moulding to be nipped between the steel faces of the male and female tools and hamper mould opening.

This factor can be particularly serious when the component is formed between the steel faces of one-half of the mould so that the deflection causes the component to become trapped, thereby making ejection difficult. The greater the amount of deflection, the greater the difficulty.

The first method is based on the standard fixed beam formula:

$$y = wL^4/384EI \quad \ldots \quad (6.1)$$

where y = deflection of side wall (in.);
w = load on inside of cavity (lb. per in. of span);
= cavity pressure in lb./sq. in. × depth of cavity (in.);
L = inside length of cavity wall (in.);
E = modulus of elasticity (steel = 30×10^6 lb./sq. in.);
I = moment of inertia of cavity wall = $dt^3/12$;

*In some cases, a limit is set by the available clamping pressure to the maximum cavity pressure that can be achieved. Beyond this limiting value, flashing will occur. For example, if the projected area of the moulding is 300 sq. in., and the clamping pressure is 750 tons, then the maximum cavity pressure is $(750 \times 2240)/300 = 5{,}600$ lb./sq. in. See also footnote on page 64.

d = total depth of cavity wall (in.); and
t = thickness of cavity wall (in.).

The necessary wall thickness to permit a specified maximum deflection can therefore be obtained by transposition of equation (6.1),

$$\text{i.e.} \quad t = [wL^4/32yd]^{\frac{1}{3}} \qquad (6.2)$$

This calculation is first made for the long side of the cavity. For the short sides a similar calculation can be made, or the same wall thickness used as was determined for the long side, since this latter will give an even smaller deflection.

By using (6.2), the minimum wall thickness against a given deflection is obtained. In practice, however, the sides of a mould cavity do not completely fulfil the basic requirements of end fixing upon which the fixed-beam formula is based, and the method can lead to considerably higher actual deflections in service. If formula (6.2) is used, therefore, it should be confined to moulds in which the two halves locate in each other in such a manner that the male tool prevents the opening of the cavity, by acting as a tie (i.e. similar to the mould shown in Fig. 5.10).

Another approach is to consider the sides of the cavity as a freely supported beam, according to the basic formula

$$y = 5wL^4/384EI \qquad (6.3)$$

Thus the wall thickness for a given deflection y is given by

$$t = [5wL^4/32yd]^{\frac{1}{3}} \qquad (6.4)$$

This formula is commonly used, giving the greatest wall thickness for a given deflection of all the three methods described. In this respect, therefore, it provides the greatest margin of safety, and in many moulds a greater wall thickness above that strictly necessary is often of little consequence.

However, a more rational approach is offered by considering the mould as a portal frame. Both the previous methods deal with the problem by considering each side of the cavity independently as a uniformly loaded beam, with fixed ends and freely-supported ends respectively. Neither is true, since the ends are neither fixed nor free; nor does any one wall of the cavity deflect independently of the others, i.e. the effects of the corners must be taken into account.

The outward deflection of one wall of the cavity will set up a moment about each corner at the ends of the wall, and this in turn will tend to cause the walls at right-angles to the first wall to deflect inwardly. Since, however, these are attempting to deflect outwards under cavity pressure, the net effect is that the outward deflection of each wall is reduced by the opposing corner moment that is produced by the adjacent walls.

This is illustrated in Fig. 6.10, where the cavity wall considered as a uniformly-loaded simply-supported beam gives a bending moment of $wL^2/8$, this being resisted by the corner-fixing moment M_1. The bending moment causing the actual deflection of the wall is therefore the difference between these two, i.e. $(wL^2/8) - M_1$. The cavity is hence considered as a uniformly internally-loaded portal frame, with the usual assumptions

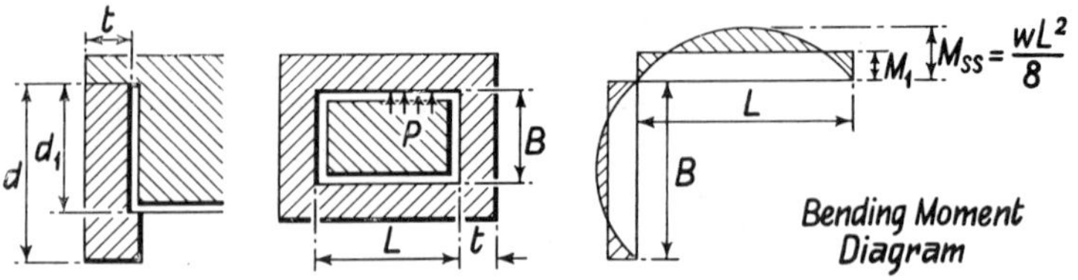

Fig. 6.10. Portal-frame Method of Determining Cavity Strength.

that the corners remain square and fixed in space. The effect of clamping pressure and the retaining effect of the backplate are again ignored.

Let,

p = cavity pressure (lb./sq. in.), say 9,000 lb./sq. in.;
L = length of long side of cavity (in.);
B = length of short side of cavity (in.);
d = total depth of cavity wall (in.);
d_1 = depth of cavity wall subjected to pressure (in.);
t_L; t_B = thickness of cavity walls (in.);
I_L; I_B = moment of inertia of cavity walls (in.^{4});
w = load on inside of cavity (lb. per in. of span) = pd_1;
M = the bending moment at mid-span;
M_1 = bending moment due to the influence of the load on the adjacent side;
E = modulus of elasticity = 30×10^6 lb./sq. in.;
y_L; y_B = deflection of side walls.

The bending moment at any point along the span is equal to $M_{SS} - M_1$; where M_{SS} = bending moment due to the cavity pressure minus the bending moment due to the beam treated as simply supported. Hence, $M = (wL^2/8) - M_1$.

From this, the net slope of the beam at the corner is given by

$$dy/dx = wL^3 - 12M_1L)/24EI_L \qquad (6.5)$$

Similarly, the net slope for B at the corner is given by

$$dy/dx = (wB^3 - 12M_1B)/24EI_B \qquad (6.6)$$

From the assumption that the corner remains a right angle, these slopes must be equal, i.e.

$$(wL^3 - 12M_1L)/24EI_L = (wB^3 - 12M_1B)/24EI_B \qquad (6.7)$$

From (6.7)

$$M_1 = w(I_BL^3 + I_LB^3)/12(I_BL + I_LB) \qquad (6.8)$$

If $t_L = t_B$, then $I_L = I_B$, and

$$M_1 = w(L^3 + B^3)/12(L + B) \qquad (6.9)$$

Now, $y_L = (5wL^4/384EI_L) - (M_1L^2/8EI_L)$, so that

$$t_L = [(1.5L^2/dEy_L)\,(5wL^2/48 - M_1)]^{\frac{1}{3}} \qquad (6.10)$$

Similarly,

$$t_B = [(1.5B^2/dEy_B)\,(5wB^2/48 - M_1)]^{\frac{1}{3}} \qquad (6.11)$$

The value of M_1 is calculated from (6.8) or (6.9), while y_L and y_B will be assumed from the permissible conditions.

All these formulae apply fundamentally to plain-sided rectangular cavities, but calculations taking into account non-regular forms, local recesses, etc., would be too complex for practical use. In these cases the cavity must be considered as a rectangle which is based on the largest dimensions.

Example. To determine the wall thickness to allow a maximum deflection of 0.01 in. with cavity walls 15 in. long and 6 in. long, and a cavity pressure of 9,000 lb./sq. in., the cavity being 10 in. deep.

Here $L=15$ in.; $B=6$ in.; $d=d_1=10$ in.; $p=9{,}000$ lb./sq. in.; $y_L=0.01$ in.; $E=30\times10^6$ lb./sq. in.; $w=9000\times10=9\times10^4$ lb.

(i) Treated as a fixed beam from formula (6.2)

$$t_L=[(9\times10^4\times15^4)/(32\times30\times10^6\times10^{-2}\times10)]^{\frac{1}{3}}=3.61 \text{ in.}$$

(ii) Treated as a freely-supported beam by formula (6.4)

$$t_L=[5\times46.8]^{\frac{1}{3}}=6.16 \text{ in.}$$

(iii) Treated as a portal frame, assuming that wall sections on all sides are equal, i.e. $I_L=I_B$, and $t_L=t_B$, from formulae (6.9) and (6.10),

$$M_1=[(9\times10^4)/12]\,[15^3+6^3)/(15+6)]=128.4\times10^4 \text{ in. lb.};$$

$$t_L=[(1.5\times15^2/10\times30\times10^6\times10^{-2})\{(5\times9\times10^4\times15^2/48)-(128.4\times10^4)\}]^{\frac{1}{3}}$$
$$=4.7 \text{ in.}$$

These three examples illustrate different methods of determining the thickness of the cavity walls. Method (i) gives the thinnest wall, but an open mould cavity does not strictly fulfil the conditions of a fixed beam, and this method is only applied where the cavity is fully tied by an interlocking male tool. Method (ii) gives the thickest wall, hence the largest margin of safety. In some cases, however, an excessive wall thickness can prevent the tool fitting on the specified press. Method (iii) is more rational, since it takes into account the interdependence of the walls on each other, but the method of calculation is more complex and depends on the proportions of the tool providing a completely rigid corner which remains square under load.

A quite different approach that can sometimes be employed is to consider each side of the cavity as a freely-supported uniformly-loaded rectangular plate. To fulfil these conditions of support, the base of the cavity must be either solid or rigidly registered into the side, while the open end of the cavity must be completely and firmly tied across by substantial registers in the male tool. Then, from a formula derived by Timoshenko and Lessells, the approximate maximum deflection is given by the equation

$$y=Cpd^4/Et^3 \qquad \text{.. .. ..} \quad (6.12)$$

where C is a constant having the values as shown at the top of the next page for the ratio L/d, L being the longest main dimension of cavity wall (normally the length of the wall), and d the shortest main dimension of the cavity wall (normally the depth of the cavity).

L/d	C	L/d	C	L/d	C
1.0	0.044	1.5	0.084	2.0	0.111
1.1	0.053	1.6	0.090	3.0	0.134
1.2	0.062	1.7	0.096	4.0	0.140
1.3	0.070	1.8	0.102	5.0	0.142
1.4	0.078	1.9	0.106	—	—

Hence by (6.12)

$$t = [Cpd^4/Ey]^{\frac{1}{3}} \quad \ldots \quad (6.13)$$

Example: Applying formula (6.13) to the previous example where $L=15$ in.; $y=0.01$ in.; $p=9{,}000$ lb./sq. in.; $d=10$ in.; and $E=30\times10^6$ lb./sq. in.,

the ratio $L/d=1.5$, hence from the table, $C=0.084$, so that $t=[0.084\times 9000\times10^4/30\times10^6\times0.01]^{\frac{1}{3}}=2.94$ in.

Deflection of Cylindrical Cavities. The increase in radius of a cylinder under internal pressure is given by Lamé's equation,

$$\mu=(rp/E)\,[\{(r^2+R^2)/(R^2-r^2)\}+m] \quad \ldots \quad (6.14)$$

where μ = increase of inside radius (in.);
r = original inside radius;
R = original outside radius;
m = Poisson's ratio (= 0.25 for steel)

Example: Let $r=3$ in.; $R=5$ in.; $p=9000$ lb./sq. in.; $E=30\times10^6$ lb./sq. in.; $m=0.25$.

then $\mu=(3\times9000/30\times10^6)[\{(5^2+3^2)/(5^2-3^2)\}+0.25]$
$=0.0021$ in.

Hence, increase in internal diameter is $2\times0.0021=0.0042$ in.

The Male Tool. Most male tools are made from the solid since they are much more easily machined than the female, and the solid construction is preferable for strength.

The various approaches to producing and mounting a solid form should be considered.

In Fig. 6.11 is illustrated a solid male form with its backplate, machined

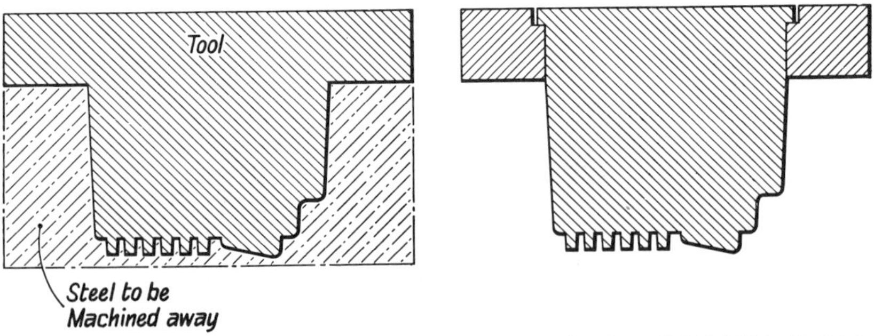

Fig. 6.11. [*Left*] Male Tool machined from Solid Block. Fig. 6.12. [*Right*] Male Tool let through Mounting Plate.

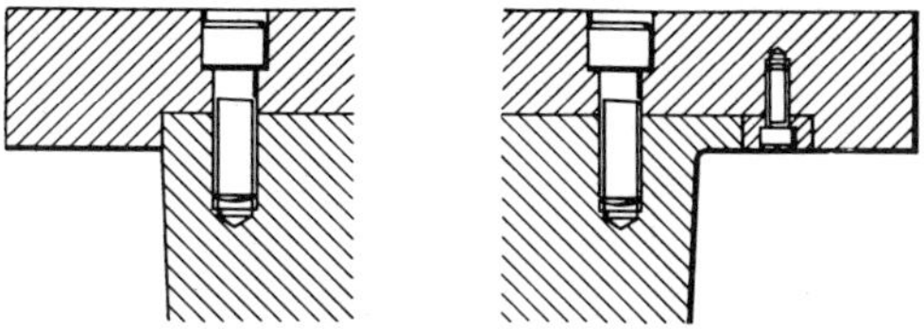

Fig. 6.13. [*Left*] Male Tool recessed into Backplate. Fig. 6.14 [*Right*] Male Tool recessed into Backplate with Positioning Strip.

from one block of steel. The chain-dotted area shows the volume of steel that has to be machined away. This may not be objectionable in a small tool, but in a large tool this excess stock removal would be lengthy and wasteful.

A solid male tool with a separately fitted mounting plate is shown in Fig. 6.12. Here the male is let through an aperture in the mounting plate and located by means of a machined shoulder that fits into a recess in the plate. Finally, a solid backplate (not shown) is secured to the back of the mounting plate, to lock the male in position. This method is acceptable for small or medium-sized tools, but for large tools the expense of machining the large aperture in the mounting plate and the close fitting of the punch is generally uneconomical.

A very satisfactory method of mounting the male in all but tools with long slender punches of small base area is shown in Fig. 6.13. This construction provides maximum rigidity with the minimum of machining. The fit of the form into the backplate recess should be of a quality to prevent penetration by the moulding material.

An alternative arrangement is shown in Fig. 6.14, where the punch is located in the recess by separate positioning strips, these allowing adjustment of the male relative to the female if necessary. In practice, this method is not largely employed on injection tools. It is unwise to place the positioning strip against a vertical wall on the punch, to give a sharp bottom corner as in Fig. 6.11, since the fit of the strip is generally insufficient to prevent leakage.

The construction shown in Fig. 6.15 is very widely used since it

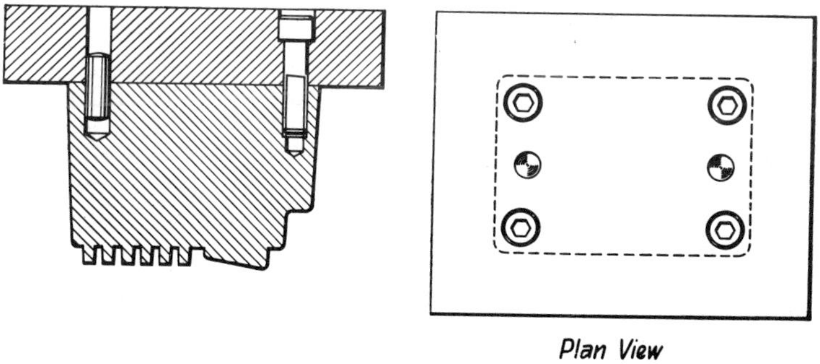

Fig. 6.15. Male Tool with Screw and Dowel Location.

represents minimum cost. This form of mounting, however, is not good practice since excessive side pressure produced by flow conditions in the cavity can frequently shear or distort the dowels. It is preferable to absorb any side thrust over a wide face of solid steel as indicated in Fig. 6.13.

If the punch carries small areas of detailed machining or interchangeable portions, these may be made as separate items and secured to the main form. If positioning or wear blocks are to be provided in the tool, these are invariably fitted to the male, since they are then more easily accessible for adjustment or repair.

As previously shown, the male tool is made from the solid where possible, but in some cases construction from individual blocks cannot be avoided. This latter is true when the male carries deep narrow recesses. These are difficult to machine and polish, and if the polishing is defective, the fragile ribs on the moulding will hold up on ejection and break. This may be avoided by building up the punch from a number of blocks, which are selected to expose the recesssed surfaces, thereby affording easy machining and polishing. Such blocks are provided with location shoulders, so that they interlock and retain each other on the backplate. (Similar retaining shoulders are shown on the blocks for the cavity in Fig. 6.6.)

The assembly of blocks in the male tool is partly recessed into the backplate, as in Fig. 6.12, and key blocks are screwed to the back of the plate. The assembled blocks must fit tightly around the recess profile, otherwise leakage of moulding material will occur at the block joints, or the joints will produce an objectionable print line on the moulding. Taper end blocks are not generally employed on male tools.

Construction of male tools from blocks is usually confined to punches of small depth. If the length of the punch is considerable, difficulties can arise in securing the blocks together, especially at the top remote from the backplate.

Normal cooling is provided in the male. If the punch is not deep, it is adequate to provide cooling channels in the backplate, but if it is long or large, the punch itself must be cooled. For punches made up from blocks, the cooling is generally confined to the backplate, because of the difficulty of sealing off the water at the block-joint lines.

Ejection. Whenever possible, the stroke on the ejector should clear the moulding completely from the form, and this can often be achieved. In any event, such local items as bosses should be fully ejected, since manual removal can otherwise very easily cause distortion of the moulding about such points. Another instance concerns male forms with little or no taper, such as spindle holes for radio knobs. Here the moulding will shrink on to the core in the normal manner, and manual removal can again cause distortion. For deep-drawn components, however, the required ejector stroke would be excessive, and the ejectors must merely free the component from close contact with the mould and allow easy withdrawal off the form, by hand.

The amount of ejection with respect to the mould taper should be greater than the amount of component shrinkage. Thus, where a moulding has a $\frac{1}{2}$-deg. taper per side (approximately 0.017 in. per inch inclusive) in a material such as polystyrene, assuming an "on mould" shrinkage of half the full shrinkage (say, 0.002 in. per inch), the minimum ejection to counteract shrinkage alone will be

$$[0.002/0.017] \times [\text{maximum overall male-tool dimension (in.)}].$$

The overall tool dimension refers to width or length, whichever is the greater, since it is shrinkage in these directions which causes binding across the male tool.

Side of Ejection. In most mould designs, the female tool is positioned on the injection side, with the moulding remaining on the male when the mould opens. The moulding is thus ejected from the moving side. However, in some large or very large moulds, special provision is made to retain the moulding in the female and to eject it from the female, since, if the usual method were adopted, the total shrinkage of a large moulding might cause complete seizure on to the male, or the cracking of the component. The principal exception to the practice of the female being on the fixed side is where the component is to be fed on the inside, when the male tool is positioned on the injection side. Since such a moulding will normally remain on the male, this means that ejection will then take place from the fixed side. This procedure applies to both two- and three-plate tools. With the latter, however, if it is also desired to provide positive ejection of the runner system, this is initially retained on the fixed plate by moulded hooks or undercuts and then positively ejected in addition to the normal ejection of the component, usually by the same mechanism.

As a general rule, the ejectors are placed to operate on the underside or unseen face of the moulding, whenever possible. The appearance of ejector marks and print lines on the front face is thus avoided.

The principal methods of ejection are considered in the following sections. The exact type to be adopted will depend upon the shape of the component and the moulding material. Brittle materials, for instance, must be adequately supported during ejection, whereas flexible or semi-flexible materials such as low-density polythene must generally be eased off as a whole rather than forced off locally. Generally, too much ejection should be provided rather than too little, including if necessary, the ejection of runners and gates.

Stripper-plate Ejectors. In principle, the stripper-plate method consists in the provision of a closely-fitting plate surrounding the base of the male tool. During ejection, the plate moves along the male tool carrying the moulding before it, and hence stripping the moulding from the punch. This type of ejector is chiefly employed where the zone of ejection pressure is a thin edge, such as the wall of a box. For such conditions, a blade or pin ejector would have insufficient bearing area, and the pressure exerted on the small face to achieve ejection could be sufficient to crush the ejector into the moulding or to distort the moulding. On the other hand, a stripper plate may be used if no normal ejection marks are permissible, since an ejector pin or blade will leave a print line on the moulding where the plastics material penetrates the clearance between ejector and mould face.

The principle of stripper-plate ejection is shown in Fig. 6.16, where the provision of 0.01 in. minimum clearance to prevent scoring of the vertical wall of the male tool will be noted, together with a face angle of not less than 5 deg. (10 deg. in the Figure). View (*a*) shows a section of the complete stripper plate, and view (*b*) shows a stripper bar. This latter would be used where only one length of a particular moulding is to be ejected by this method. The use of a stripper on undercut mouldings is shown in Fig. 5.16, page 135. In this mode of ejection, it is most essential that the plate is provided with adequate guides during operation, and that

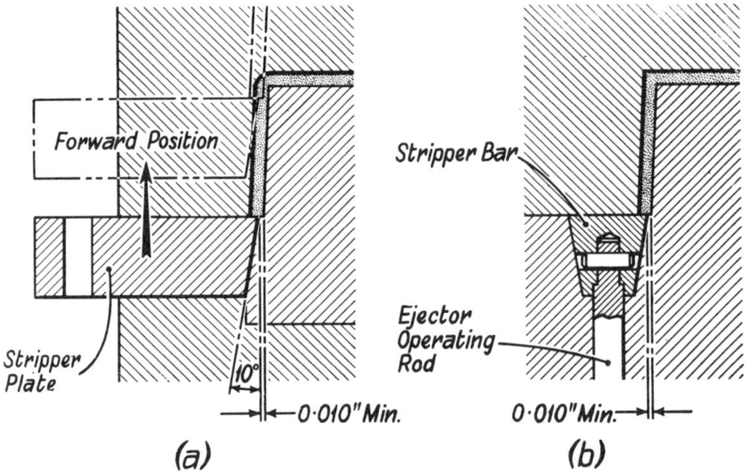

Fig. 6.16. Stripper-plate Ejection.

the dowels and bushes used as guide pins are of ample rigidity and bearing length; often the main tool guide pins are used.

Stripper plates provide an even and reliable form of ejection, but constitute a more expensive method in many applications.

Sleeve Ejection. Ejection by sleeve is substantially confined to circular mouldings or to circular portions of mouldings; e.g. the local ejection of bosses. The ejector is made in the form of a tubular member which surrounds the male core, the forward movement of the sleeve stripping off the component. Sleeve ejectors are seldom used for non-circular forms because of the expense of machining anything but a round sleeve. As with stripper plates, sleeves provide even and reliable ejection, free of marks or distortion. At (*a*) in Fig. 6.17 is shown diagrammatically a plain parallel sleeve, and it will be noted that for this type of ejection the

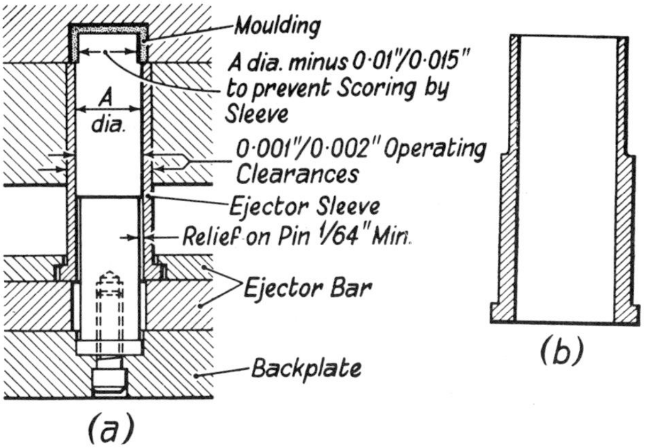

Fig. 6.17. Sleeve Ejection.

male core must be carried through to the backplate. The view (*b*) shows a stepped or reinforced sleeve. This is used where the actual width of sleeve at the point of ejection must be thin but would produce too weak a member if the width were carried the full length of the sleeve. One application of a sleeve ejector is shown in Fig. 5.2, page 116, and its use for stripping undercut mouldings is indicated in Fig. 5.16, page 135.

It is undesirable that the sleeve should be a close fit to the core pin for its whole length, since this may cause difficulties from lack of parallelism, and difficulty in reaming. To avoid this, the pin or core is relieved over a portion of its length as shown in the Figure. The relief should not be taken on the sleeve itself, since the effective bearing length would then become diminished as the ejector moved forward, and the relief step might run off the end of the pin itself on a long ejection stroke.

Ejector Pins. Ejection of mouldings by means of local pins is the most common and cheapest method. The chief types of pin employed are:

(i) Plain or parallel pins;
(ii) Stepped pins; and
(iii) D or flat-sided pins.

The first is employed wherever possible, since it is the simplest, and the proportions of pin and mounting are given in Fig. 6.18, in which the view (*a*) gives the proportions of the pin and (*b*) the details of the mounting recesses, etc. Relief is again provided as with the ejector sleeve, this being on the hole and not on the pin itself. Various other methods of securing the pin to the ejector bar are employed. One very common method is indicated at (*b*), in which a head is machined on the ejector pin and locked between the two plates of the ejector bar.

A cheaper method is shown at (*c*), in which a length of ground rod is used for the pin, this being peened over to form a head. Another method is illustrated at (*d*) where the ejector is retained by means of proprietary spring clips that fit in a groove turned in the parallel rod.

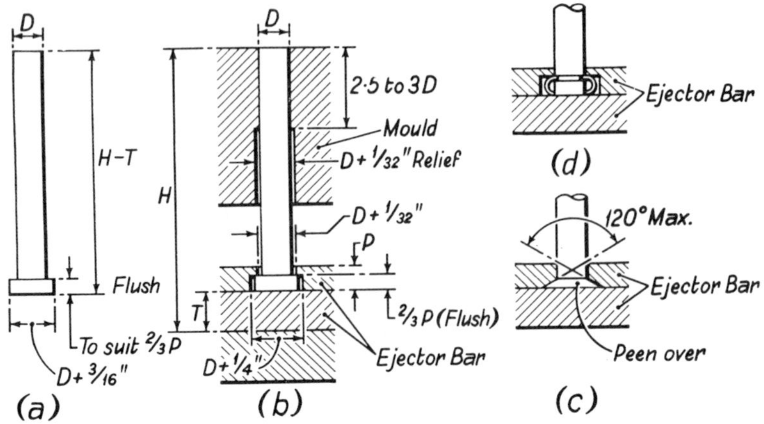

Fig. 6.18. Parallel Pin Ejectors.

Ejectors of the parallel-pin type are placed wherever required on the mould, care being taken in positioning that the ejector load is satisfactorily balanced and that sufficient are provided to prevent the moulding being locally deformed. Special attention must be paid to those points that present particular ejecting difficulty; e.g. deep local depressions. Also to be considered is the material being moulded. Care must be taken with brittle plastics such as polystyrene that the ejectors do not cause distortion and hence cracking. Thus, in the case of box-type mouldings, ejection should take place evenly around the rim, rather than by attempting a "pull-off" with ejectors placed at the bottom of the box.

Even when a particular layout of the ejection system has been decided, it is useful to consider at the design stage what further means could be adopted and added if necessary, without undue difficulty, when the tool is sampled and the means already provided are proved insufficient.

A special type of parallel pin ejector is the single large central ejector pin (see Fig. 5.16, page 135). This is of special use for circular mouldings of small size, since it is less expensive than several single ejectors in this application. However, as previously stated, the pin must be as large as possible in order to prevent bowing of the moulding across the top face.

Stepped pins are a special form of plain ejector pin in which the diameter of the pin at the ejection point is small. Such reduced diameter may be required, for instance, if the ejector has to pass up a slender core as shown in Fig. 5.2, page 116. It is not desirable, however, to continue this reduced diameter right back to the ejector bar since the ejector load may cause the pin to buckle. The pin is, therefore, increased in diameter behind the mould face, to provide a stronger construction. A common method is to braze a length of rod into one of larger diameter as shown in Fig. 6.19.

Ejector pins and blades are preferably made from hardened and tempered nickel-chromium steel, and although the use of silver steel is quite common, it is not to be recommended because of the danger of brittleness after hardening and consequent liability of severe tool damage.

Flat-sided or D-pins are again a special form of round ejector pin used for the ejection of mouldings by the edge or face of side walls; e.g. ejecting along the side edges of a box moulding. Round ejector pins are often used for this purpose, even in preference to rectangular blades. Fig. 6.20 illustrates ejection by a plain parallel pin, and also by means of a D-shaped pin. It will be seen that the area at the face of the ejector pin in contact with the moulding in the latter case is greater than that with the plain pin of the same diameter. This is due to the fact that the flat on the side of the pin allows the ejector to contact the moulding more closely

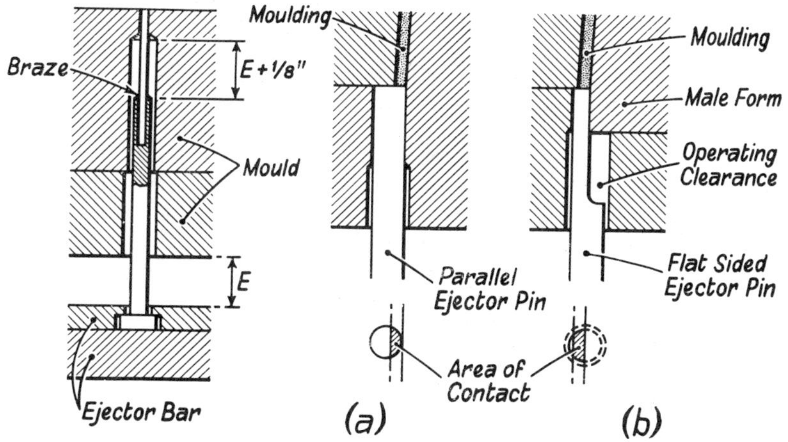

Fig. 6.19. [*Left*] Stepped Ejector Pin. Fig. 6.20. [*Right*] Comparison of Parallel and Flat-sided Ejector Pins.

to the pin centre and the point of maximum cross section. As shown in the figure, with r = radius of ejector pin, and t = thickness of moulding at point of ejection, then for the parallel pin at (*a*), with the inside edge of the moulding on the periphery of the pin:

$$\text{Area of moulding in contact} = (\pi r^2/180)\cos^{-1}[(r-t)/r] - (r-t)[t(2r-t)]^{\frac{1}{2}}.$$

With the flat-sided pin as at (*b*), with the inside edge of the moulding on the pin centre line:

$$\text{Area of moulding in contact} = \pi r^2/2 - (\pi r^2/180)\cos^{-1}(t/r) + t(r^2-t^2)^{\frac{1}{2}}.$$

As an example of the increased area of contact provided by the flat-sided pin, let the pin diameter in both cases be 0.5 in. and the thickness of the moulding 0.1 in. Then from the preceding formulae, the area of contact with the parallel pin is 0.0279 sq. in., while the area of contact with the flat-sided pin is 0.0491 sq. in.

With its greater bearing area, the flat-sided pin is much less likely to hob into the moulding. Also, this type is only slightly more expensive, since the flat is readily added to the basically round pin. It will be noted that the flat-sided hole in the mould to receive the ejector in the view (*b*) is facilitated by boring a round hole in the die plate and setting in the male form as a separate block to provide, *inter alia*, the flat side of the hole.

Ejector Blades. In general, these are employed:

(i) Where part of the moulding is trapped between opposite steel faces of the tool. This is of common occurrence with deep thin ribs, which may easily resist extraction unless the ribs themselves are ejected. Here the ejector is limited to the width of the rib since it has to operate between the side walls of the mould forming the rib. To provide sufficient contact area, a rectangular blade must be employed. This application is shown at (*a*) in Fig. 6.21. These ejectors are placed in the most effective positions, often close to cross ribs, and sufficient must be provided to eject the moulding without causing cracking across the unsupported lengths. For compartmented cross-ribbed mouldings of the "egg-crate" type, ejection is greatly facilitated by arranging that alternate blocks comprising the male cores are withdrawn before ejection, thus freeing one side of the ribs.

(ii) Blades may also be employed where the use of round pin ejectors would be impracticable, e.g. where the diameter would be too small to give effective contact area. This might be the case where the point of ejection is a very thin wall, so that with a blade, the contact area can be brought up to the desired amount by extending the length. Stepped blades are employed where the ejection edge carries form, but the use of the full edge is preferred to provide maximum contact area. This can occur on a box-type moulding where the edge is recessed to locate a lid.

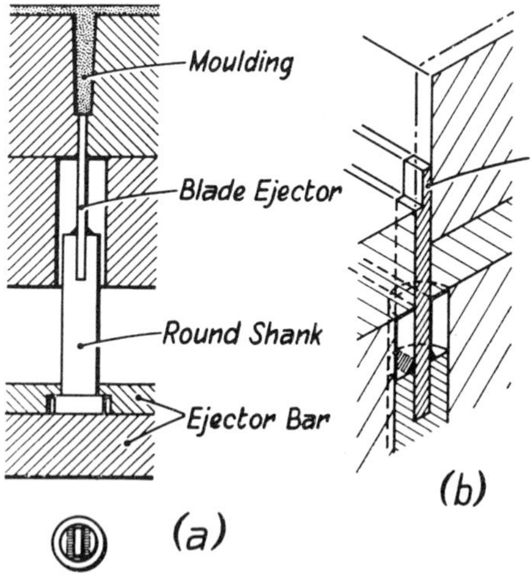

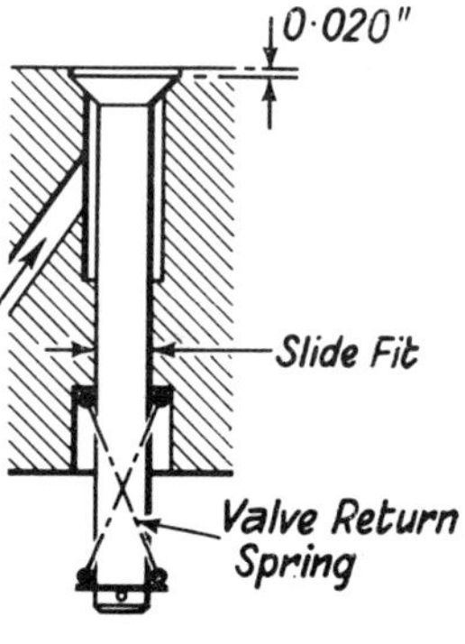

Fig. 6.21. [*Left*] Blade Ejector. Fig. 6.22. [*Right*] Air Ejector.

An instance is shown at (*b*) in Fig. 6.21, where the blade is stepped to bear locally, both on the edge of the moulding and on the bottom of the recess.

Ejector blades may be machined from the solid, but may also be fabricated by brazing the rectangular blade into the shank. The blade is kept to the minimum length to obviate buckling, and is then brought out to a full diameter.

Air Ejectors. Ejecting the moulding by introducing compressed air between the moulding and the mould face offers a cheap and effective method for suitable mouldings. Generally these are box- or container-type mouldings. The air must be introduced at the point farthest from the flash line, so that it must effectively separate the moulding from the mould face before it can escape. Thus, the method cannot be employed where any feature of the moulding offers any short-circuit escape route for the air. A common example in the use of air is the common washing-up bowl which is normally ejected from the male tool by means of an air ejector placed at the top of the male tool. If, however, a moulding of this type had a moulded hole in the bottom face, this method of ejection would not function, since the air would escape from the hole rather than take the longer path to the rim of the bowl. An illustration of an air ejector tool is shown in Fig. 5.10A, page 127.

Apart from providing a positive ejection, the introduction of compressed air also breaks the vacuum that is produced when a closed moulding is withdrawn from the male tool. This can easily cause distortion of the moulding, and many closed mouldings which are ejected by mechanical means have to be provided with air ejectors as well, to prevent such distortion.

The detail of a typical air ejector is shown in Fig. 6.22. The ejector is invariably a valve-seated nickel-chromium pin which is operated by the introduction of compressed air behind the head. The air is often controlled by the press operator through an externally-operated valve, but may also be linked to operate automatically from the press. Return of the valve is generally effected by means of a compression spring. In order to eliminate sharp edges, a parallel of 0.02 in. is provided on the valve and seat.

Ejection by Undercuts. Sometimes it may happen, that when the press opens, the moulding will fail to remain consistently on that side of the tool on which ejection has been provided. This is particularly liable to occur where the retaining form is present on both halves of the mould, when variation of press conditions or mould temperature can cause the moulding to remain, erratically, on either half. This difficulty can be overcome by filing or grinding slight grooves on that half of the mould on which the moulding is intended to remain. During moulding, these form undercuts will retain the component on that half of the tool. Such grooves must be placed in an inconspicuous position on the moulding and are not added to the tool until sampling has taken place.

A particular form of retainer is the moulded snatch hook illustrated in Fig. 6.23. Here a moulded undercut disc is connected to the main moulding by means of a stem. This often operates in a ferrule let into the die which forms one part of the moulded disc, and avoids boring a fine guide hole for the ejector pin in the main die plate itself.

In operation, the moulded snatch hook retains the moulding on the ejection side of the tool as the mould opens. When the ejectors are afterwards moved forwards, the snatch-hook ejector pin pierces the disc and ejects the snatch-hook stalk at the same time as the moulding. On the next moulding cycle, the incoming plastics material re-enters the stalk and again welds itself to the remaining portion of the previous disc. The snatch hooks are employed where mouldings have considerable form on both sides of the tool and where positive retention in the ejection side must be ensured; moulded plastics grilles are a common example. In

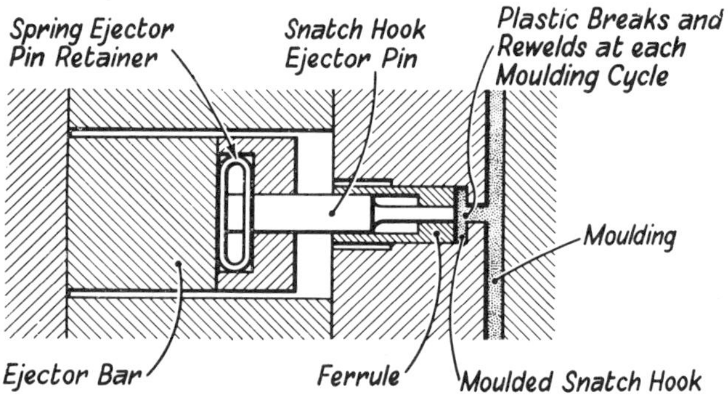

Fig. 6.23. Ejection by Moulded Snatch Hook.

many of these, the marks left by the use of push-off pins on the front face would not be acceptable. The snatch hooks are generally small in size, the length of the stalk being between $\frac{1}{8}$ in. and $\frac{3}{16}$ in., and the diameter of the stalk and ejector pin only from $\frac{1}{16}$ in. to $\frac{1}{8}$ in.

Ejector-operating Mechanism. The ejectors provided in the mould invariably derive their operation from the opening movement of the press. Where the ejectors are in the stationary half of the tool, they are normally operated by means of chains or links, one end of the chains or links being attached to the mould ejector bar and the other to the moving platen. When the ejectors are mounted in the moving tool, they are generally actuated by the fixed press ejector bar or pin, these latter being mounted on the press behind the moving platen.

Press ejectors are normally either of the centre-pin or outside type, according to the size and make of press. The former consists of a plate or pin which can make contact with the centre of the mould ejector bar through a hole in the moving platen, while the latter type consists of two pins which are positioned either side of the platen and which make contact at either end of the mould ejector bar. The central ejector should be used only if the ejection system on the mould is substantially balanced; otherwise canting about the centre pin can occur with consequent seizing of the ejectors and heavy damage to the tool.

In addition to mechanically-operated ejectors, hydraulic cylinders are sometimes employed in special cases. One instance arises where they are used to operate ejectors in the stationary tool and the use of chains is not convenient. The mould ejectors (pins, sleeves, blades, etc.), are brought through the back of the tool and secured to a common connecting member. This is the ejector bar, grid, or plate, and the construction is shown in Fig. 6.24.

This connecting member consists of two principal components: the bar or plate itself which transmits force to the individual ejectors, and the retaining plate. The ejector bar or plate is designed in the manner shown because:

(i) The method of assembly is easy. The ejector pins can be assembled

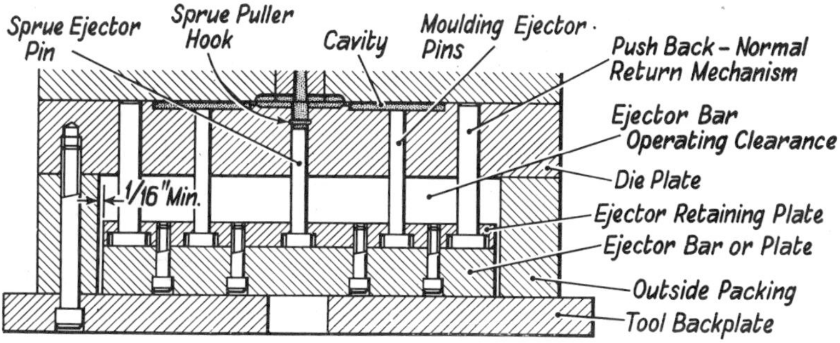

Fig. 6.24. General Arrangement of Ejector Mechanism

one-by-one in the die plate, to pass through the holes in the retaining plate, in which the ejector pins are a free fit; see (*b*) in Fig. 6.18. When all the ejectors are assembled, the ejector bar or plate can be screwed to the retaining plate. It is thus essential that the screws enter from the back of the tool as shown.

(ii) The method of securing the ejectors under a retaining plate makes for safety, since individual ejectors cannot work free during operation.

(iii) If all the ejectors have been made of correct length, this method of mounting ensures a constant ejector height on assembly.

The provision of undercut sprue puller hook and sprue ejector pin should be noted.

In the design shown, the ejector is operated by a central press ejector pin (not shown) passing through the hole in the tool backplate. The general details of design are, however, similar if the ejector bar or plate is operated by outside press ejectors, chains or rods, except that the length of the ejector bar has to be increased to suit and to come outside the tool. The forward motion of the ejector bar takes place in the clearance provided behind the die plate, this clearance being arranged through the packing blocks between die plate and backplate.

Ejector Bars. Where the ejectors are substantially in a straight line they can be secured to, and operated by, a bar. If there are several lines of ejectors these can be mounted in a number of bars, each bar being rigidly connected to the others by means of screwed or welded crosspieces. An assembly of this nature is frequently referred to as an ejector grid, and a two-bar grid is illustrated in Fig. 6.25. Here the moulding ejectors are provided in two parallel lines, these being mounted in the two ejector

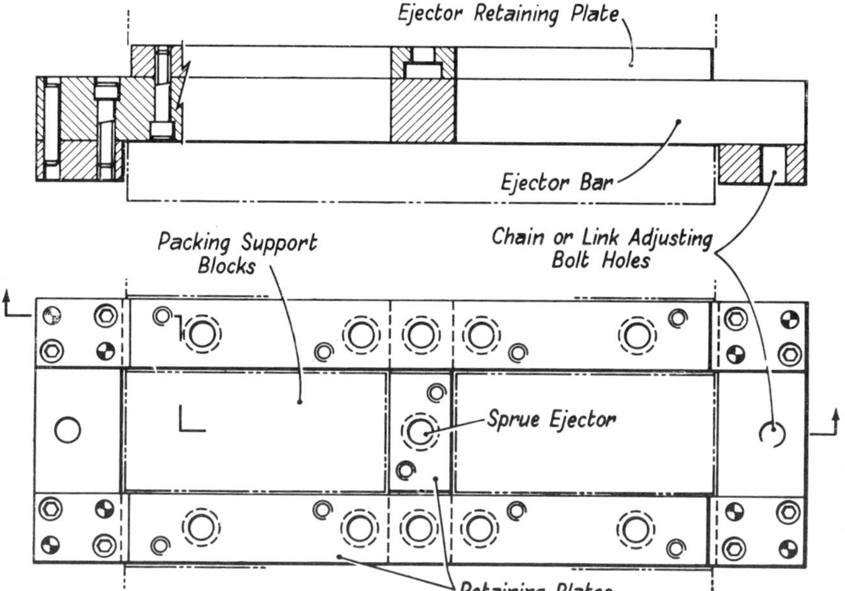

Fig. 6.25. Detail of Ejector Grid.

bars. The illustration shows the details of an ejector grid similar to that shown employed in Fig. 6.24, except that the latter is operated by a central press ejector, while the present detail shows the grid extended outside the tool and suitable for operation by chains or links. If this ejector were intended for operation by outside press ejectors, it would have to be extended outside the tool still further to suit the standard press ejector centres.

The two ejector bars are rigidly connected together by the two end crosspieces (to which are also connected the chains or links) and also by the centre crosspiece which, in addition, carries the sprue ejector. The crosspieces are shown screwed and dowelled to the ejector bars, but could be welded. The disposition of the crosspieces must be such that the load is carried by the solid member and not by the screws or welding. Thus in the case of the end crosspieces, these are placed under the ejection bar so that the applied load from the chains is applied direct to the ejector bar and not through the screws. In the case of the centre crosspiece, the sprue ejector applies a load on top of the ejector bar. Since it may not be convenient, for toolspace consideration, to mount this centre crosspiece directly on the top of the ejection bar, a half joint is provided as indicated in Fig. 6.26, care again being taken that the joint is provided on the side to support the live load directly on the members and not through the securing screws.

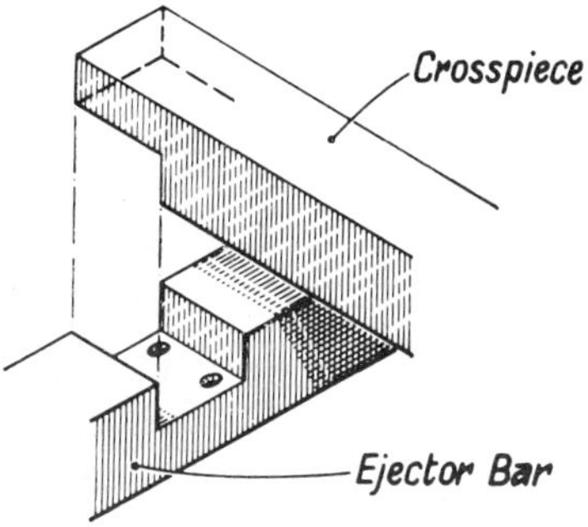

Fig. 6.26. Half Joint for Ejector-grid Crosspieces.

The ejector bars themselves must be of sufficient strength to withstand the highest likely ejection load, which varies from low values with flat-plate type mouldings to quite substantial loads on mouldings with deep thin fins, etc. The designed strength must achieve minimum deflection of the bar rather than a specific stress level, and can be calculated from standard beam theory. From this, it follows that improvement in the strength of the bar is more effectively achieved by increasing the thickness rather than the width.

Ejector Plates. Where there are large areas of ejection or the ejectors are irregularly spaced, ejector plates are used. In effect they replace the ejector grid with a single plate and may be operated by any of the normal means. Obviously, no crosspieces are required, and the ejectors are still secured by means of a retaining plate. Standard bolsters normally have to be provided with ejector plates in order to satisfy any ejector spacing requirements required on the ultimate mould.

Ejector Guides. The ejector bar or plate is normally guided in its travel by means of the ejectors themselves. If these are too slender or the ejector load is unbalanced, the provision of specific ejector guides is to be recommended. A typical design is shown in Fig. 6.27. Four such guides

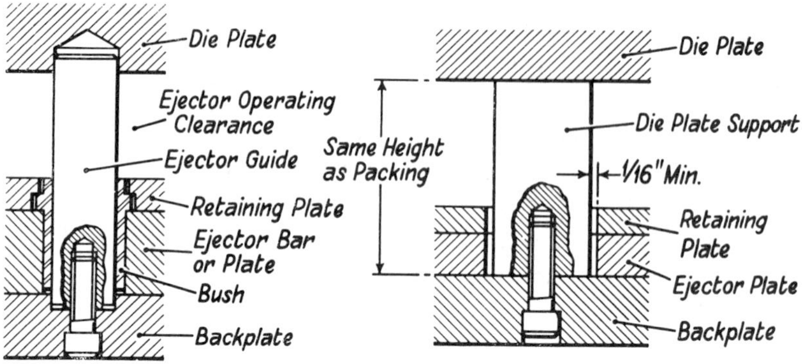

Fig. 6.27. [*Left*] Typical Ejector Guide. Fig. 6.28. [*Right*] Die Support for use with Ejector Plates.

are commonly provided on an ejector plate or grid, and two on a single ejector bar.

Retaining Plates. The plates to secure the heads of the ejectors to the ejector bars or plate are screwed and dowelled in position as shown in Fig. 6.24 and 6.25. The screws should be positioned as close to the ejectors as possible, since, if moulding material penetrates as flash down the sides of the ejector, the force to return the ejector may be high, and there will be a tendency to bend the retaining plate.

Die-Plate Supports. The provision of an operating space for the ejector grid or plate beneath the die plate can leave this plate unsupported and, hence, supports must be provided. In the case of a single ejector bar, the packing blocks can be brought up alongside and this is adequate. With an ejector grid, packing blocks can be added between the bars and crosspieces (see Fig. 6.25), and these, together with the outside packing, provide sufficient support. In the case of an ejector plate, special clearance holes have to be provided in the plate in order to allow the addition of die-support pillars, one of these being illustrated in Fig. 6.28.

An ample number of these supports should be included, since this type of support is inferior to that obtained from packing blocks with ejector bars or grids. As mentioned previously, standard bolsters* are often designed for ejector plates, and die supports have to be included when the ejectors are set out.

Centre Ejection. If the press is fitted with a centre ejector behind the moving platen, this will normally consist of either a plate or a pin of small or large diameter, according to the type of press. If the centre ejector is too large to enter the tool itself to operate the ejector bar, an intermediate pin is added to the tool, this pin protruding rearwards from

*B.S. 2012 "Mould die sets for injection moulding machines".

Fig. 6.29. Use of Ejector Plug with Press-centre Ejection.

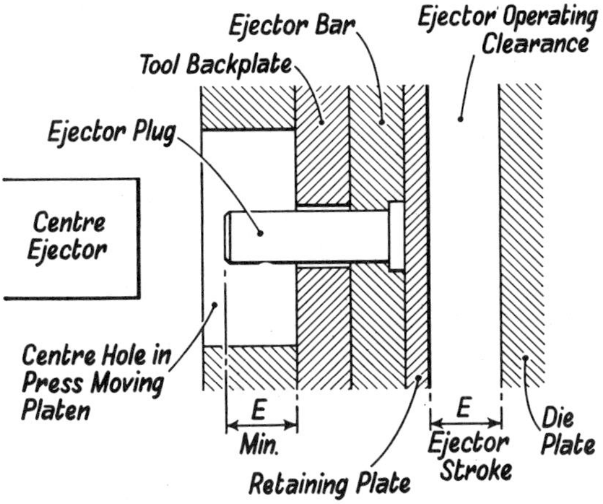

the tool into the centre hole of the platen and being brought into contact with the press centre ejector. This is illustrated in Fig. 6.29, from which it will be seen that the plug is retained in the ejector bar and extends through the backplate by an amount that is not less than the desired ejection stroke. Occasionally, a standard bolt is used in place of the plug but, while cheaper, it is not as good, since the thread is liable to damage.

Chain Ejection. Where ejectors are to be operated by chains, these are always positioned on the sides of the tool. If they were placed at the top and bottom, the chains would fall between the closing mould faces as the chains slacken on press closing. The length of chain allowed is sufficient to provide the required ejection stroke when the press is fully

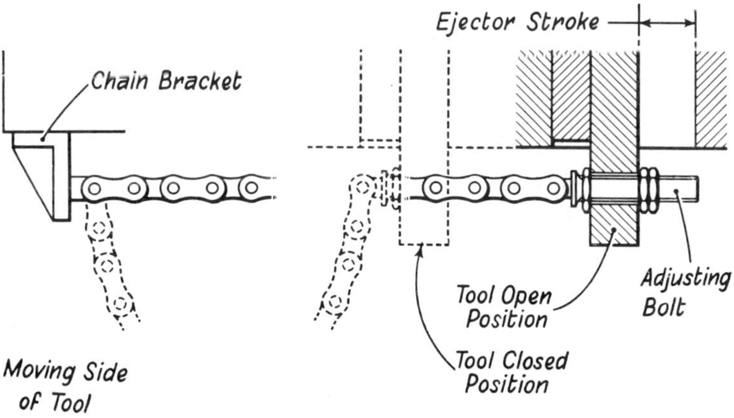

Fig. 6.30. Chain Ejection.

open, but means must also be included for adjusting the length of individual chains to at least one pitch so that the pull may be set equally on each—see Fig. 6.30.

The chains must be of adequate strength, and 4,000-lb. breaking-strain chain is often used. New chains should not be used in conjunction with old chains, since the difference in stretch over a period will unbalance the ejector. Two chains can be used with an ejector bar or narrow grid, but four should be used if there is a likelihood of tilting on the ejector. When four are used, the centre distance between top and bottom chains must be sufficient to allow the component to pass between them on extraction from the tool. Hence the bars to which the chains are secured are extended from the ejector sufficiently to allow this.

Rod Ejection. Rods are occasionally used in place of chains, and can be placed at the top and bottom of the tool if desired. In order to clear, they must normally be set outside the tool far enough to clear the platens when the press closes. The rods are generally of mild steel with adjustable stop collars to effect operation of the ejectors.

Ejector-return Pushbacks. This method of returning the ejectors to the back position is that most commonly used, since it is reliable, simple and economic. Typical return pins or pushbacks are shown in the mould depicted in Fig. 6.24.

When the mould is open and the ejectors are fully forward, these pins protrude above the surface of the mould. As the press closes, the return pins eventually make contact with the plain face of the opposite half of the tool, and further closing of the press causes the pins to move back, carrying the ejector bar and pins with them. When the ejector bar or plate is hard against the tool backplate, the ends of the ejectors and the return pins should be exactly flush with their relevant mould surfaces, final adjustment normally being effected by grinding during tool assembly.

Ejector-return Springs. It is often necessary to arrange that the ejectors are returned before the tool closes. This particularly applies to moulds with moving cores operated by the press, such as shown in Fig. 5.12, page 131. Here, as the tool commences to close, the ejector is still in the forward position, with the splits open. As the tool halves come together, the finger pins commence to close the splits. If the ejector were to be returned by conventional pushbacks, the ejector sleeve might be caught by the closing splits before it was fully back. This could occur in a number of tool designs, and hence early return of the ejector mechanism must be arranged.

One method is to use springs, and this type of return employing a press centre ejector is shown in Fig. 6.31. Here springs are provided both on the ejector plug and between ejector bar and die plate. It is difficult to estimate the return load required, especially when flash conditions are prevalent; hence the load should be overestimated five or six times. The springs should be of adequate section and of best commercial quality, not "home-made". Sufficient length of spring should be allowed so that it does not have to operate over its full range from free length to solid length; in no

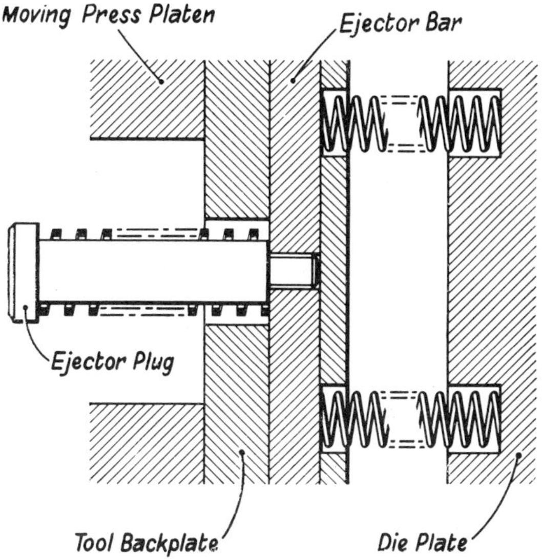

Fig. 6.31. Ejector Spring-return.

circumstances should the spring be closed solid when compressed.

Pockets are provided for the spring, with a total depth, where possible, of one-half of the maximum working depth. Since the return force available from the springs is limited and non-positive, it is very desirable that the ejector system should be provided with guides, if spring return is to be used. This minimizes return load friction caused by any tendency of the ejector bar to cant.

Finally, it is essential that the assembled tool should be fully tested in its working plane before being fitted to the press. Another example, employing ejector return pins and spring is shown in Fig. 5.2, page 116.

Ejector-return Cams. While spring-return mechanisms are frequently used, the non-positive action is a disadvantage, since excessive operating friction or heavy flashing down the ejector pins can cause faulty ejection. Another design giving positive operation is shown diagrammatically in Fig. 6.32. Two freely swinging cams are pivoted on either side of the moving die plate. When these are rotated towards the rear as shown at (*a*), the faces of the cams make contact with the ends of the ejector bar, or plate, thus forcing this back and returning the ejectors. Rotation of the cams is effected by means of finger cams that are fastened to each side of the stationary die plate. In the mould-open position indicated at (*b*), the ejectors are forward and the freely-pivoted cams hang downwards. (The design should provide that these cams drop down by gravity when free of the finger cams, but in any event, the cam form should be such that if the cams do not fall free, the forward action of the ejector bar will cause them to rotate anticlockwise.) As the tool closes, the fixed finger cams engage the pivoted cams, causing them to rotate and return

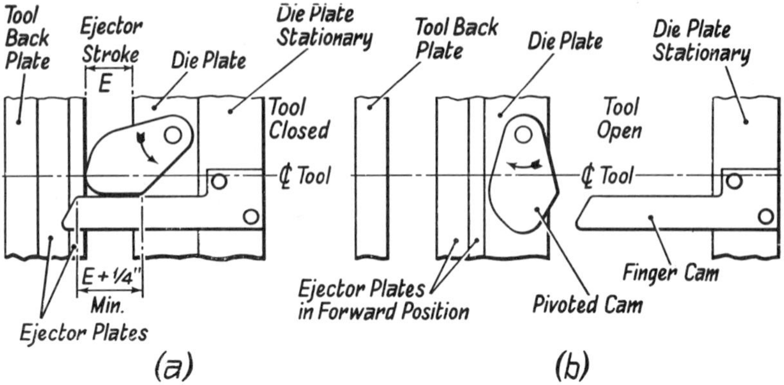

Fig. 6.32. Ejector-return Cam.

the ejectors before the tool fully closes, as described previously.

The pivots and the finger cams are of hardened and tempered nickel-chromium steels and are generally about 0.5 in. thick. The lead provided on the finger cams should be sufficient to ensure that the ejectors are

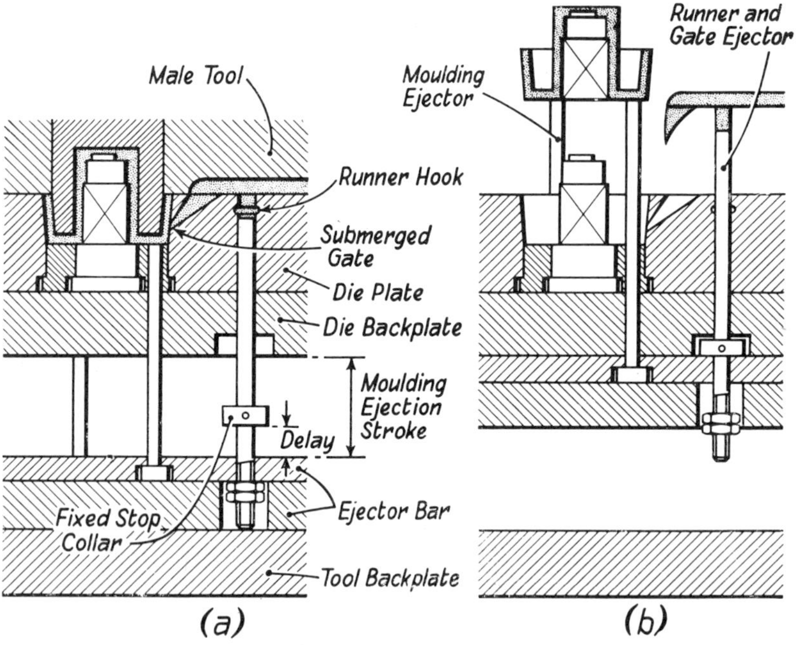

Fig. 6.33. Delayed-action Ejection.

fully back well in advance of any interference from other moving parts of the tool, such as splits or cores. This lead is shown in the illustration as $E+\frac{1}{4}$ in. minimum, but its actual value is dictated by the operational requirements of a given tool.

Delayed-action Ejection. In some moulds it is necessary to eject the runner system after the ejection of the moulding itself. This is required principally on automatic degating moulds where the ejection of the moulding shears off the gate prior to the ejection of the runner.

An example of a mould of this type is shown diagrammatically in Fig. 6.33. Here, the moulding ejector pins are secured in the normal manner to the ejector bar. The runner and gate ejector pin is not secured to the ejector bar but passes freely through it as indicated at (*a*). At the commencement of ejection, the moulding is moved out of the cavity by the moulding ejectors, at the same time shearing off the submerged gate. The runner ejector does not commence to move until the ejector bar strikes the stop collar, after which the runner ejector commences to move, ejecting the runner and extracting the feed from the submerged gate as indicated at (*b*).

Another application of delayed-action ejection is for the extraction of the runner in two-colour moulding; this is shown in Fig. 5.22, page 147.

Double Ejection. It is occasionally necessary to eject the moulding by two separate and different movements, and an example is shown in Fig. 6.34. Here the moulding has deep thin side walls trapped between opposite steel faces on the tool. Extraction of such sections as this is likely to be difficult, especially if the moulding material is brittle. Ejection can be considered effected in a number of ways:

(i) A thin sleeve ejector, or ejector blades, could be introduced around the male core to strip the moulding off the core and from the cavity. Owing to the large surface area of both steel faces, however, the initial adhesion of the moulding to the tool is likely to be high, and, since the moulding thickness presented to the ejector is thin, damage to the moulding is likely to result. If the moulding material is at all flexible, the action of the ejector would be to compress the material and cause it to adhere more tightly.

(ii) A large ejector might be provided in the top of the male core to pull the moulding off the core and from the cavity. Again, however, the moulding is still trapped between opposite steel faces of the tool and the resistance to extraction could result in cracking or distortion of the moulding.

The best approach to this problem is first to withdraw one of the entrapping steel faces and then eject the moulding. In the tool design illustrated, the female cavity is used as a stripper plate first to strip the moulding off the male core, after which its movement is arrested while the moulding is ejected from the cavity by the continued movement of the secondary ejectors.

The ejector bar comprises two units: the base unit (1) which operates the standard ejector pins (9)—three only are shown—and a secondary

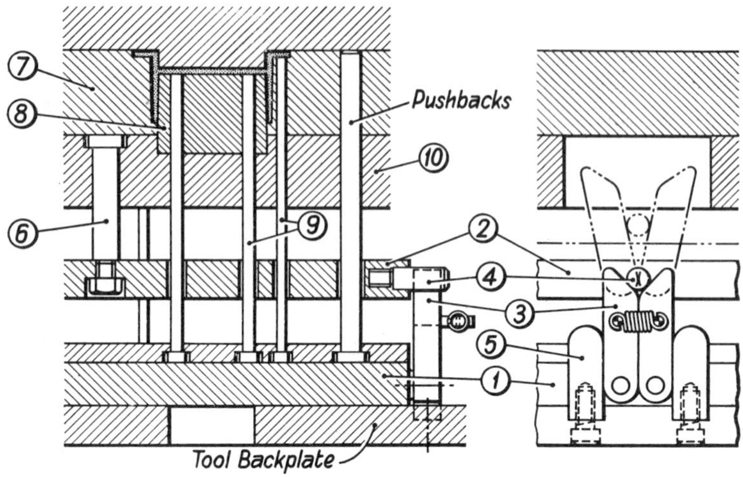

Fig. 6.34. Double Ejection.

bar (2) which operates the stripper plate (7) through the pins (6), of which only one is shown. The two ejector bars are connected by the pivoted links (3), these being held together by a tension spring and retained in this position between the cams (5) which are secured to the fixed-tool back-plate. The pivoted cams bear upon the pin (4) secured to the secondary ejection bar (2). This link system connecting the two ejector bars is provided at each end.

In operation, the press ejector commences the forward movement of the ejector bar (1). This, through the link system, moves the secondary ejector bar (2) at the same time. This action causes, through the pins (6) movement of the stripper plate (7) relative to the stationary male core (8). By this action, the moulding is stripped off the core without difficulty, since it is ejected from only one of the entrapping steel faces, and is securely supported on the other faces by the stripper plate. The movement of the ejector bars continues until the face of the secondary bar (2) makes contact with the back face of the die plate (10). At this point the moulding is fully stripped from the male core, while the connecting links (3) have moved out from between the fixed cams (5). Since further movement of the secondary ejector and stripper plate has been arrested, continued forward movement of the ejector bar (1) causes the connecting links (3) to open and ride over the pins (4).

This further movement of the main ejector bar induces the ejector pins (9) to continue their forward action and eject the moulding from the cavity and stripper plate. This again is achieved without difficulty, because the moulding, being clear of the male core, is free to shrink away from the cavity face.

It is possible that this moulding could be ejected off the core by the stripper plate in the manner just described, the moulding then being removed from the cavity manually. This would only require conventional single ejection, but would require a longer moulding cycle to allow the operator time to remove the moulding.

An alternative double-ejection design could be proposed in which the male core moves forward to strip the moulding from a fixed cavity, the ejection pins then continuing forward to eject the moulding off the core. This would be less satisfactory, however, since the moulding would have shrunk on the core, so that ejection off it, although not difficult, would not be as easy as ejecting it from the female as in the illustrated design.

Tool Support

BACKPLATE. The backplates are the base plates of each half of the mould; the remainder of the tool is built up on these plates and screwed back to them. When the tool is mounted in the press, the backplate is bolted or clipped to the platen, and thus supports the remainder of the tool. The backplate should therefore be substantial and, even on small tools, should never be less than 0.5 in. thick; on large tools it may be as thick as 3 in. or more. Backplates are nearly always made from mild steel, cleaned up but not necessarily ground, and with planed edges. If the tool is to be bolted to the press, the backplate is drilled suitably to match with some of the standard bolt holes in the platen.

If the size of backplate required does not correspond to any of the stock sizes of plate available, it can be made up from a number of plates or strips which are made up into a single rigid member by being screwed up to packing blocks that join the separate plates.

Some components of the tool which are mounted directly upon the backplate may be subjected to a high intensity of compressive stress. This can be produced if the loading is high and the cross-sectional area of the component is low, due to the design limitations on the given tool. Such components may be packing blocks, die blocks or core pins that are mounted directly on the backplate. Such excessive loading may cause the tool component to crush into the mild-steel backplate, but this may be avoided by the addition of an oversize pad of hardened nickel-chromium steel between the tool component and backplate to distribute the load.

PACKING. The spacing blocks positioned between the die plate and backplate to house the ejector gear and to provide ejector-operating clearance are normally made from mild steel unless high loadings are involved. If the height of the packing is not too great, or the tool not too heavy, the packing may be through-bolted as shown at (*a*) in Fig. 6.35. Since, however, long bolts are liable to stretch under load, double bolting as shown at (*b*) is preferable for tall packing or heavy tools. In assembling the tool, care must be taken that the height of all the packing blocks is the same, otherwise uneven loading or distortion of the die plate may result.

EYEBOLTS. In order to facilitate handling, both in the toolroom and in the press shop, all tools of more than 40 lb. weight should be fitted with eyebolts. In heavy tools, eyebolts should be provided on both halves of the

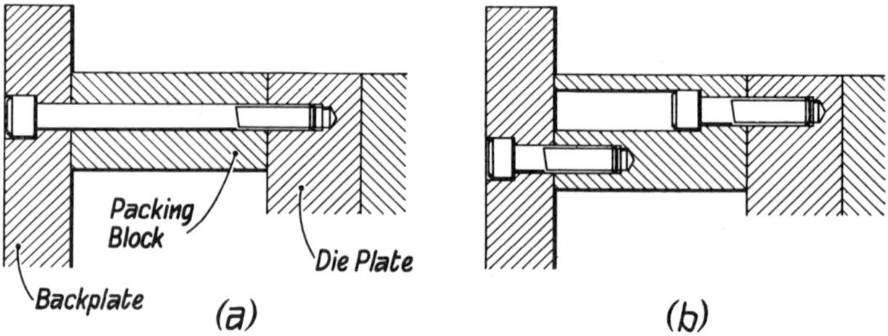

Fig. 6.35. Tool Packing.

tool, and they should always be positioned as nearly above the centre of gravity as possible. Standard load capacities of forged eyebolts are given below, and it should be noted that they are invariably screwed B.S.W.

Nominal Size (B.S.W.) *in.*	*Safe Load* *lb.*
½	630
¾	1500
1	2760
1¼	4450

Tool Location—Guide Pins. The majority of tools are provided with guide pins to locate the two halves, and these are preferably four in number. Tools are made with two guides, and even three, as dictated by the size of the tool and space available, but four pins provide superior control of location.

The size of pin is dependent upon the particular tool, but it should always be substantial. In practice the guide pins vary from ¾ in. diameter on small tools to 3 in. diameter on large moulds. One type of guide pin and bush is shown in Fig. 6.36. With this, the fitting shank of the pin and the outside of the bush are of the same diameter. This allows both halves of the tool to be aligned and both holes bored to finished size at the same time, thus ensuring accuracy and reducing the amount of work. The

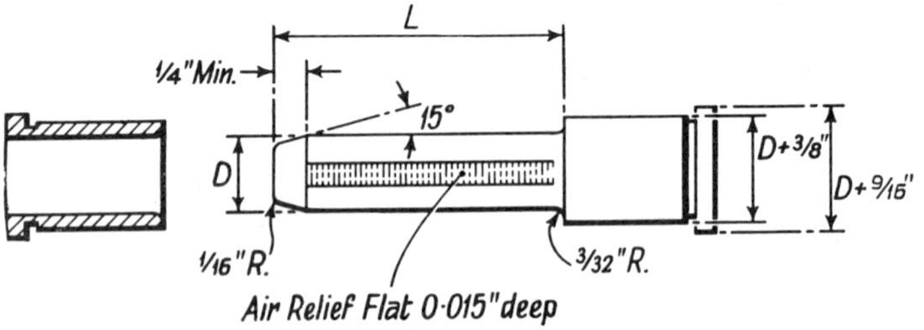

Fig. 6.36. Guide Pin and Bush.

length of the pin *L* should always be sufficient to ensure that the guide is adequately engaged in its bush before the mould form itself engages, while the inclusion of the taper at the end of the pin allows alignment of the tool halves on the press during operation, should any slight deviation have occurred through platen wear on the tie bars, or any bolt slipping, etc.

In Fig. 6.37 is shown an alternative parallel type of pin. Here, when the two tool halves have been aligned, they can only be bored to pin diameter or with a pilot hole; the hole for the bush must be subsequently opened out. The pin shown has a helical groove for lubrication and air relief.

Guide pins and bushes are normally made from hardened nickel-chromium steel; the use of case-hardened steel is not recommended since the case is liable to break up under bending loads.

The head of the guide pin is normally socketed into the die plate,

Fig. 6.37. Parallel Guide Pin.

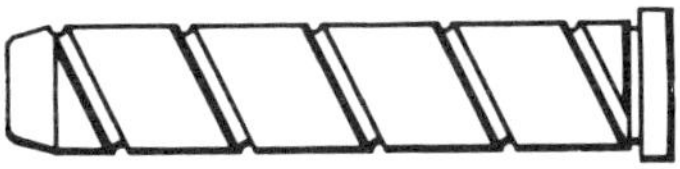

except in the case of three-plate tools when it is mounted in the backplate. To facilitate extraction of the moulding, the pins should be placed in the side opposite to the side of the mould from which the mould is ejected.

Tool Location—Taper Form. In some mouldings, close accuracy of location between the mould halves is necessary, often with the added necessity of ensuring that there will be no relative movement between male and female from the effect of cavity pressure. The moulding of thin-walled containers is a case in point. Here, the section thickness is so small that cavity pressures can rise to high values; and any eccentricity between the male and female will yield an uneven flow pattern, so that on the reduced side the flow may be retarded or totally stopped, while on the thicker side the material may tend to force the core sideways. To provide accurate location on cylindrical mouldings, a taper spigot is provided on one half of the mould to engage a similarly tapered seating in the other half: see Fig. 6.38. This positive form of location is superior to guide pins since the operating clearance between spigot and bush is obviated, the bearing surface of the taper spigot is greater than that between guide pin and bush, and, finally, the taper location is closer to the mould form than the spaced guides. However, it is common practice to include guide pins on taper location tools, but such pins are then only used to assist setting up on the press and are often fitted without bushes.

The angle of taper employed is usually between 10 and 30 deg., but not less than 10 deg. to avoid the danger of sticking, nor parallel to avoid scoring. The depth of taper provided depends on the size and nature of the tool, but it is seldom less than 0.5 in.

If it is essential that any movement between male and female be

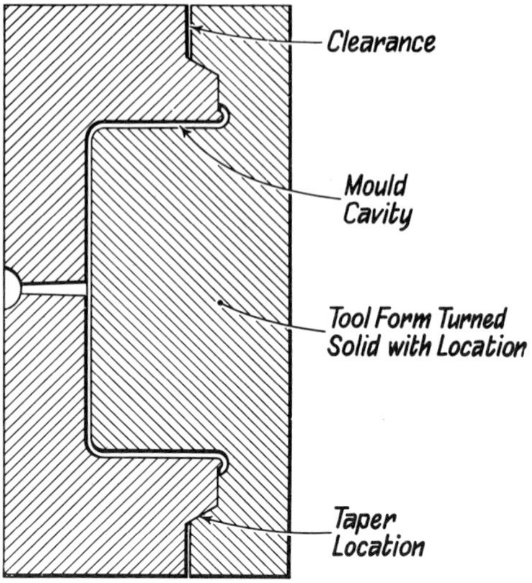

Fig. 6.38. Tool with Tapered Location.

eliminated, the core and the cavity are turned solid with taper location faces.

In the case of non-circular moulds, alignment is effected by straight taper locations on all sides, pads being inserted between the faces: see Fig. 5.10A, page 127. These pads, which are often of gauge plate, can be ground to thickness, thus allowing alignment of the tool halves to be adjusted. In addition, they can readily be replaced in the event of wear. The thickness of the pad must be sufficient to sink the screw head well below the surface.

Guide pins and bushes are fitted to this type of tool, not only to facilitate press setting, but to prevent rotation, such as might occur in operation by a slight slipping of the tool under the platen mounting bolts or clips. This latter point does not arise with completely circular tools, but if a circular mould does have a specific location feature, the same prevention of rotation will apply.

Similar tapered registers are sometimes employed in the construction of cavities from blocks, especially where one portion of the cavity is made up of a solid ring, which is then spigoted into another portion forming the remainder of the cavity and its back.

In all these instances, where registers are employed, care must be taken that the tapered face is large enough to prevent hobbing by the compressive forces involved. Similarly, the strength of the register must be adequate to resist the bending forces set up by the cavity pressure.

CHAPTER VII

TOOL CONSTRUCTION—Continued

Types of Insert. It is beyond the scope of this book to discuss the great variety of inserts* used in injection mouldings, but chiefly they fall into the following categories:

(i) Blind and open female threads;
(ii) Male threads;
(iii) Blind and open plain holes;
(iv) Special inserts such as pins, wires, metal pressings, etc.

Probably the largest usage is that of inserts to form female threads. The inserts themselves are of brass or aluminium, these metals being chosen both to resist corrosion and to minimize damage should the tool close on a displaced insert.

Female-thread Inserts. These are blind or open, and examples are shown in Fig. 7.1. The open type of insert is generally used where the thread must run right through the moulding. The insert is, therefore, clamped between the opposite faces of the tool and hence should be given just sufficient nip between the tool faces to prevent flash without distorting the insert. An interference of about 0.002 in. is usually sufficient.

Normally, blind holes are formed with blind inserts as shown at (*b*) in Fig. 7.1, the insert being moulded part way into the thickness or depth of the moulding or boss. This allows the carrier thread to be an easy fit without danger of flash penetration. The common domed head of the insert is preferably made more square, with only a small radius, since this decreases the thickness of moulding above the insert and hence the possibility of sinking. However, a completely sharp edge should be avoided, to prevent stress concentration.

The inserts themselves are inserted in the moulding tool, mounted upon carriers. The latter are special pins made of hardened nickel-chromium steel, and threaded at one end to receive the insert.

Two complete sets of carriers are made for each mould, so that a further set of inserts can be assembled to the spare set of carriers during a press cycle in preparation for the next cycle. When the press opens and the moulding is extracted, the carrier is ejected, the carriers being placed in the line of draw. The fresh set of carriers, together with their assembled

---

*See British Plastics Federation Publication No. 38, "Standardisation of Inserts. Code of Practice for the Use of Inserts".

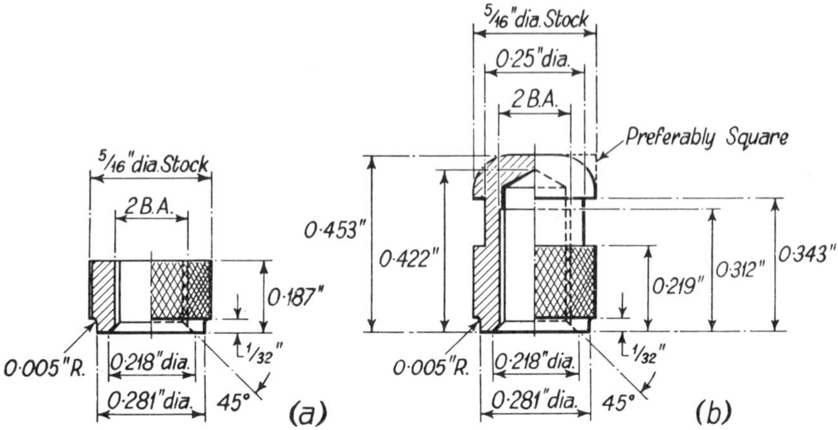

Fig. 7.1. Metal Inserts: (*a*) Through Insert; (*b*) Blind Insert.

inserts, is then rapidly placed in the tool and the next cycle commenced. The carriers still attached to the finished moulding are then unscrewed, usually by means of a fast unscrewing jig, using the square head provided on the shank of the carrier.

Moulded inserts not in the line of the draw are very infrequently used. It is possible to mould a plain insert by using a fixed carrier pin attached to a moving side core, but a full through insert and pin are inadvisable since they may interfere with the other half of the tool as the mould closes. The most commonly adopted method of providing inserts in side walls is to mould a hole by means of a side core and to cement the insert in after moulding.

Two principal types of carrier are employed: the parallel-shank type shown at (*a*) in Fig. 7.2, and the valve-seat type shown at (*b*). The first is the cheaper form of carrier, and the easiest to replace and maintain, but the latter, if carefully made, is probably the most efficient. With the parallel type, the only feature that prevents flash penetrating down the side of the carrier is the close fit of the shank in its mating hole. With the valve-seat type, the flash entry is positively prevented by the seal. The thread on insert carriers (but not on pins for moulded threads) may be run out into an undercut as indicated in the illustration. The shank slips into a closely fitting hole in the tool, the shoulder of the screw, and thus the face of the insert, being made flush with the mould face.

The position in the tool of the parallel type is controlled by the length of the carrier and the depth of its hole, whereas, with the valve-seat type, the carrier hole can be bored completely through, location being provided by the much shorter valve depth dimension.

In many cases, it is desirable and convenient to eject the moulding at the inserts; this is particularly the case where the insert is moulded into a boss which might otherwise tend to stick in the mould and break off. Here

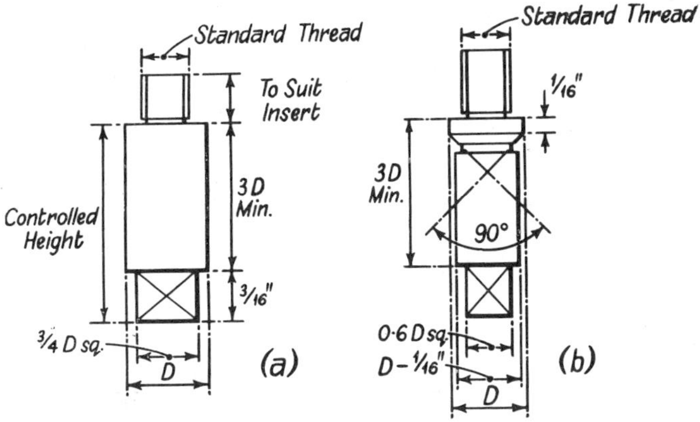

Fig. 7.2. Insert Carriers: (*a*) The Parallel-shank Type; (*b*) The Valve-seat Type.

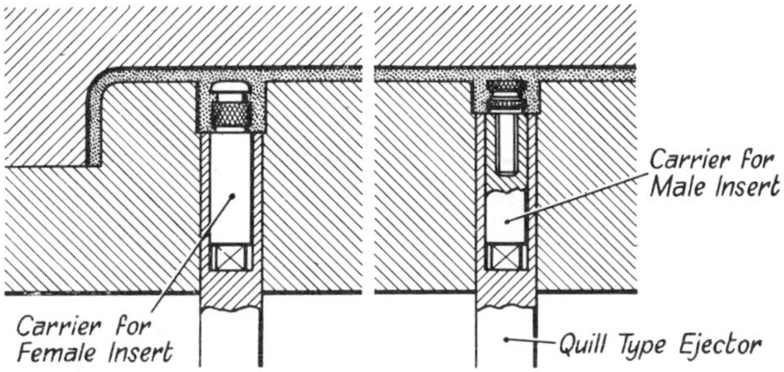

Fig. 7.3. Insert Carriers mounted in Ejectors.

the carrier is positioned in the ejector pin itself, as shown in Fig. 7.3. The ejector is bored to receive the carrier. This is essential, because, when the carriers are to be loaded into the mould in the press-open position, the ejectors will be forward. Parallel shank carriers are generally used with quill ejectors.

In Fig. 7.4 are illustrated two common methods of retaining the carrier in the tool, both employing spring clips. At (*a*), a length of spring wire is housed in a slot in the shank. This is compressed when the carrier is inserted into its hole and prevents the carrier from being shaken out of position by the movement of the tool. At (*b*), a similar result is achieved by employing a spring circlip around the shank.

While these can be employed on any mould, they are primarily of importance on carriers housed in ejectors. Here the return movement of the ejectors, as the mould closes, could easily displace the carrier unless it were retained.

The determination of the side of the tool where the carriers are to be placed may be decided by the design of the component, or may be decided

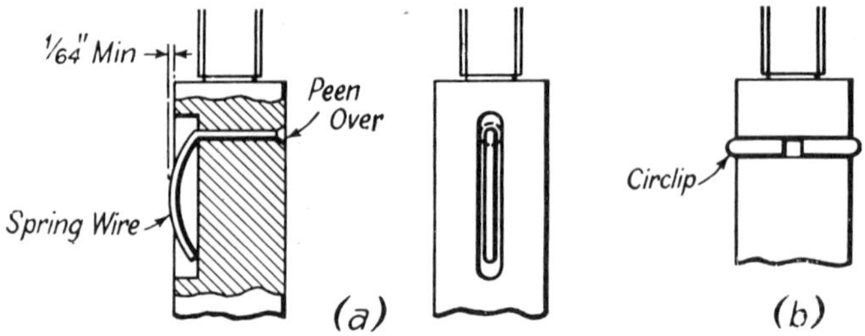

Fig. 7.4. (*a*) and (*b*) Carrier-retaining Methods.

by the requirement to eject off the inserts. For full through inserts, the choice may arise at to whether the carriers should be placed in the male or the female tool. Normally they are placed on the ejection side, but the main point in making a choice is ease of loading the carriers. Difficulty may be encountered if the carriers are to be loaded behind a large male punch on the side remote from the operator, but these conditions are occasionally inescapable.

Female Plain-hole Inserts. These inserts may be open or blind, and both types are basically the same as the threaded variety except that a plain hole replaces the threaded hole. If the insert is large and is to be placed in an accessible position on the male tool or in a shallow cavity, no carrier need be employed since it has only to be placed on the pin without any screwing. If, however, the inserts are small or the cavity deep, loading of the inserts into the tool is facilitated if they are first assembled to carriers, since the carriers are easier to handle and position than the inserts only. The carriers are identical with those employed for threaded inserts, except that the thread is replaced by a plain diameter. This can be made

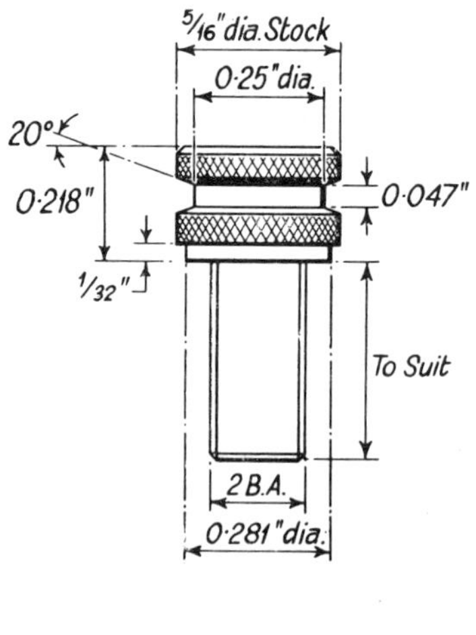

Fig. 7.5. Standard 2 B.A. Male Threaded Insert.

a push fit to hold the insert, and if desired a spring retainer may also be included.

Caution should be exercised that the flow pattern of the plastics material within the tool does not tend to lift the insert off its carrier.

Male Inserts. These are much less commonly employed than female inserts, but are occasionally required. A typical threaded male insert is shown in Fig. 7.5, and similar plain inserts are also used. Carriers are used under the same conditions as for female inserts. However, in place of the threaded or plain pin on the carrier, the male carrier has a threaded or plain hole to take the insert shank. For many threaded male inserts, the hole in the carrier is made plain, the diameter being that of the outside dimension of the thread, i.e. the major diameter. The close fit thus produced and the fact that the shoulder of the insert seals off against the head of the carrier substantially prevent the entry of flash. If full preven-

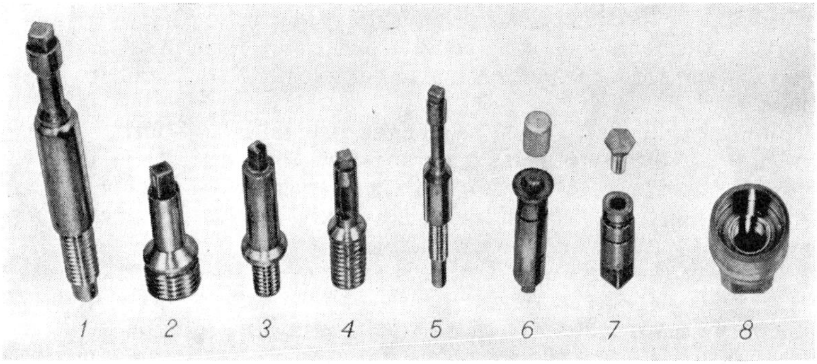

Fig. 7.6. Carrier Pins for Inserts and Moulded Thread Pins: 1. Pin to produce Male Moulded Thread and Boss, with Parallel Seating; 2. Carrier for Male Insert with Circlip Location and Parallel Seating (Only the Bottom of the Thread Carrier Hole is threaded to minimize Time in screwing Insert in and out); 3. Carrier for Blind-threaded Female Insert with Circlip Location and Valve Seating (Note the Short Length of Screw Engagement to minimize Time in screwing Insert on and off); 4. Valve-seated Threaded Pin to produce Female Moulded Threads and Two Parallel Bores, with Spring-wire Location (Note the Turned-down Portion on Locating Length to provide Relief on this Length); 5. Valve-seated Threaded Pin to produce Female Moulded Thread, with Spring-wire Location; 6. Valve-seated Threaded Pin to Produce Female Moulded Thread, without Spring Location; 7. Carrier for Full Through-threaded Female Insert, with Valve Seating; 8. Threaded Pin with Valve Seat and Spring-wire Location to produce Female Moulded Threads and Two Parallel Bores (Note the Turned-down Portion on the Locating Length to provide Relief).

tion of flash is necessary, then the hole in the carrier is also threaded to engage the insert thread.

A male carrier is illustrated at 7 in Fig. 7.6.

Wall Section around Inserts. The amount of moulding material to be left above inserts is generally determined by the adjacent wall thickness of the moulding. The thickness should not be noticeably less than that of the surrounding area, since it will retard flow across the top of the insert; nor should it be appreciably greater because of the danger of sinking above the insert. The amount of material to be provided around the insert is dependent upon the size of insert and the nature of the plastics material, but is seldom less than $\frac{1}{16}$ in. This is necessary to remove the possibility of the shrinking material cracking around the insert. Again, however, too great a thickness may lead to sinking. This necessity to keep wall sections around inserts to a minimum to prevent sinkage often leads to cracked bosses as the moulding cools. This is particularly so with brittle materials and larger types of insert. For this reason, it is frequent practice to mould only a plain hole and to cement in the insert after moulding; generally however, this means a lower insert pull-out strength.

Plate-type Inserts. It is occasionally necessary to mould in metal tabs, plates or strips as inserts. An example is given in Fig. 7.7. This indicates the moulding-in of a metal tag into a knob. Such an insert generally requires the use of splits for loading, either in a completely split tool, or a locally split portion as illustrated.

Metal parts intended as inserts must be produced to close tolerances, as otherwise the moulding material will penetrate down the sides, or the insert will not enter the tool. A maximum tolerance of ± 0.002 in. is quite usual.

Similarly, the inserts must be straight, flat and free of burrs. They are generally produced as metal stampings, so that apart from die wear, accuracy is reasonably assured. If there is any doubt about the final size of the insert, it is preferable to delay cutting the locating recesses in the mould until actual samples are available.

Whatever the form of insert, it is essential that the form of location in the splits will prevent movement in any direction, when under the influence of plastics flow and pressure. This positive location can be seen in the illustration, and the use of the dowel to prevent forward motion of the insert, in preference to attempting to locate on the radius of the insert neck, should be noted.

The recess to receive the insert is invariably cut into one face only,

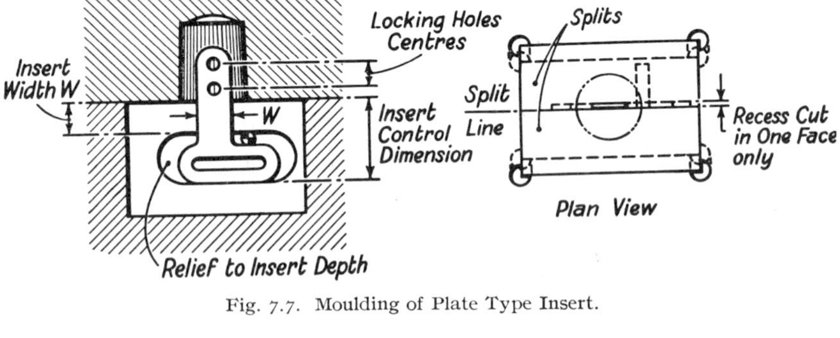

Fig. 7.7. Moulding of Plate Type Insert.

and the width and depth at the moulding face must be a close fit to minimize flash. The length of the closely fitting portion of the recess should if possible be at least as long as the insert is wide. However, if it is made too long, imperfections in the insert may prevent it from lying flush.

Means of locking the insert into the moulding must be provided, since shrinkage alone may not be sufficient. One common method is shown in the illustration, where holes are stamped through the metal. The plastics material flows through these holes and thus locks the insert in position. These holes should be set as far apart as possible, to achieve maximum rigidity.

Wire-type Inserts. In many electrical mouldings, small wires have to be moulded in as inserts. These may be single straight wires for use as terminals, or may be bent forms. For the former, the length of unsupported wire left projecting into the mould must be kept small, since the pressure of incoming plastics material will readily bend any excessive

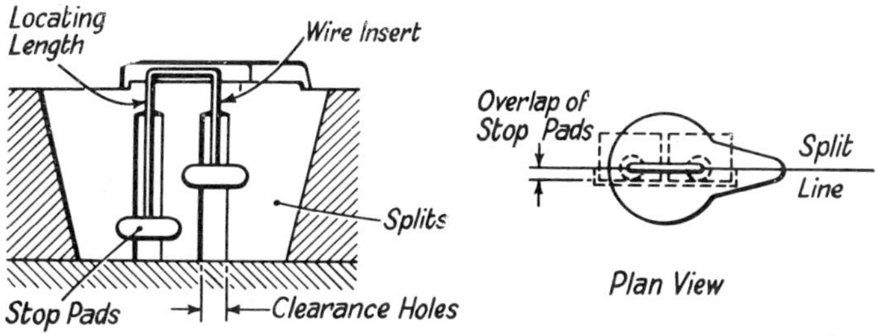

Fig. 7.8. Moulding of Wire Type Insert.

length. Such wires are generally keyed into the moulding by the provision of a flattened spade end on the wire.

In the case of bent wires, it is essential that both ends of the wire are held in the tool. This is necessary to maintain control of the position and direction of the wire; if one end is not required, it can be cut off after moulding. The moulding of a bent-wire insert is indicated in Fig. 7.8.

If the wires are produced free of flattening or burrs at the ends, they may be inserted direct into locating holes drilled in the tool; however, the possibility of flash entry and blockage should be considered. This is seldom the case, however, and to obviate the difficulty, they are generally loaded into splits. Since the wire diameter is generally of close tolerance, the locating hole can be held to between +0.002 in. and −0.000 in., but the length of the locating portion should not exceed 0.25 in., since otherwise frictional resistance set up against ejection, from flash, may be excessive.

The wires are located for depth by means of stop pads. In the case of splits, these must overlap the split line to ensure that the wires locate on a firm face. If the pads were divided at the split line, the wires might pass over the edge and become trapped between them, and out of position.

Preform Inserts. In some instances it is convenient to produce an insert in a plastics material and then to include this insert in the main moulding. Such inserts may be plain or threaded, and either produced in a separate mould or in a separate cavity in the main mould, the inserts so produced being moulded into the main moulding during the next cycle.

Preform inserts are used:

(a) To reduce the bulk of the moulding. In the case of thick sections, the resulting excessive shrinkage can cause objectionable sink marks, such sinking occurring either on the face of the moulding or in the bore of a hole. By using a preform, the section thickness to be produced on the final moulding can be reduced.

(b) In the case of threaded holes, it may be desirable to produce the thread in a tougher material than that being used for the main moulding. These preforms may either be moulded already threaded, or produced with a plain hole for use with a self-tapping screw on final assembly.

Preforms are normally used where bosses are to be included on the main moulding for the direct attachment of screws. Their use resolves the difficulty of bosses bursting due to insufficient wall thickness yet causing sinking if their thickness is increased. It is not uncommon to design the tool with a boss of adequate strength and to adopt a preform only if this is proved necessary on sampling. This does mean, however, that the design and estimating of the job is carried out with this eventually in mind. The preforms are mounted in the main tool either on fixed pins (plain hole preforms), or on carrier pins as other inserts.

With preforms for use with standard screws or self-tapping screws, these can be moulded in various lengths on stock multi-impression tools. For self-tapping screws, proved hole sizes must be used to ensure that maximum thread is engaged without the danger of bursting the boss.

Preforms are normally produced with a plain exterior without serrations or knurl. They are bonded on the second moulding by the melting of the outer skin, which thus welds the preform to the main moulding. It therefore follows that, in general, the material used for the preform

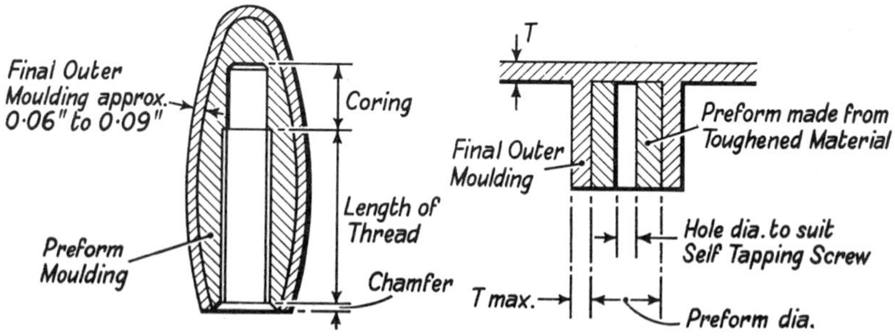

Fig. 7.9. [*Left*] Use of Preform in Handle Moulding. Fig. 7.10. [*Right*] Use of Preform in Moulded Boss.

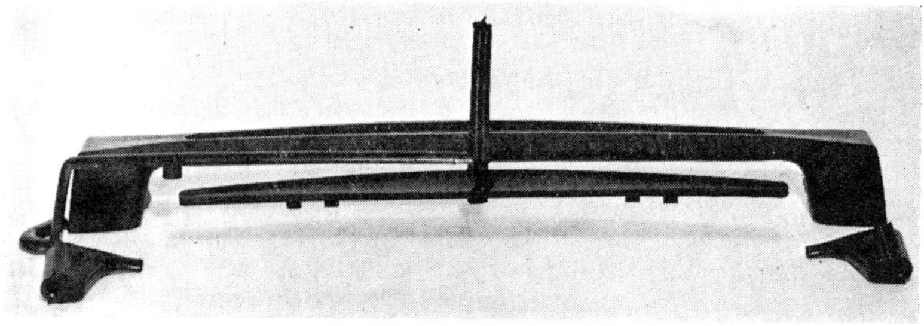

Fig. 7.10A. Hand and Cover Moulding in ABS Material produced on a Four-ounce Press. The Tool used has One Impression of the Handle gated at one End, and one Impression of Cover edge-gated at the Centre. Due to the Heavy Section of the Handle Ends, the Component is moulded with Preforms at these Points. The Preforms used are shown lying in the Foreground and are Moulded separately in Another Tool. [*Courtesy: Ekco Plastics Ltd.*]

should have a lower melting point than that used for the main moulding, but should be compatible with it. However, in practice, the same material used for the main moulding is often used for the preform.

The use of preform to reduce the bulk of two handle mouldings is indicated in Fig. 7.9 and 7.10A. The application of a preform to reduce the final moulding thickness of a boss (and thus to eliminate sinking in the front face of the moulding) is illustrated in Fig. 7.10. In such a case, the preform might be produced in toughened polystyrene while the main moulding might be in unmodified styrene.

Moulded Internal Threads. The moulding of female threads requires a male-thread tool form, cut to the dimensions of a nut plus shrinkage; i.e. the final moulding required is a nut form, not a bolt, and with the standard operating clearance, this is larger than the corresponding screw. Small moulded female threads up to about 0.5 in. diameter are normally intended to mate with a metal screw, whereas larger sizes

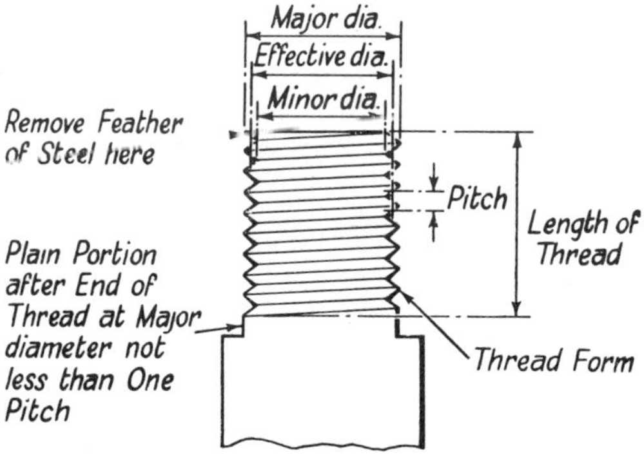

Fig. 7.11. The Required Thread Dimensions and Particulars.

frequently engage with a moulded male thread on another moulded component.

The thread form on the tool is generally cut by thread milling, since this method permits a clean runout of the thread at the shoulder of the pin, see Fig. 7.11.

The precise dimensions to be given for the thread form on the tool are dependent upon component requirements and shrinkage allowances. The initial requirement is to establish the type of fit required between the moulded thread and its engaging screw. If normal B.S.W. or B.S.F. threads are quoted, then a medium fit should be assumed, unless the precise fit is stated. American and Unified thread standards generally quote the class of fit in the drawing specification.

These comments are applicable to cases where the moulded thread engages a metal screw. If two moulded threads are to engage, a much freer fit should always be provided.

Example : If the component drawing states

2 B.A. . . . . . . assume a medium fit;
$\frac{1}{2}$ in. B.S.W. . . . . assume a medium fit;
$\frac{1}{2}$ in. B.S.W., close fit . . assume a close fit as stated.

An example of how a Unified Standard Thread is specified is as follows:

$\frac{1}{4}$—20—U.N.C.—1A

where: "$\frac{1}{4}$" = nominal size (in.);
"20" = number of threads per in.;
"U.N.C." = Unified Standard Coarse Series; others being
U.N.F.—Fine Series;
U.N.S.—Supplementary Selected Series;
U.N.—Special.
"1" = class of fit; others being
2, Medium fit;
3, Close fit.
"A" = Male thread; (B = Female thread).

The American Coarse/Fine Thread Series is specified in generally the same manner. For details of thread dimensions, pitch etc., reference should be made to the appropriate Thread Standards.

The dimensions to be stated on the drawing of the tool thread form will usually include:

(1) Pitch; (2) Effective diameter; (3) Major diameter; (4) Minor diameter.

These dimensions will be controlled by the thread size and type of fit, together with the appropriate shrinkage allowances. This latter can be added to the diameters without difficulty, but allowance for shrinkage on pitch is more difficult, since the pitch plus shrinkage becomes non-standard and difficult to cut on standard equipment. However, many mould-making concerns now possess thread-turning lathes or milling machines specially equipped to provide shrinkage allowance, and no difficulty is encountered. But if this equipment is not available, other means have to be adopted, utilizing standard equipment. These include:

1. The use of special combinations of change wheels on lathes;
2. The use of taper-turning attachments;

3. The use of the sine-bar attachment;
4. Variation of thread diameter to compensate for pitch shrinkage.

These toolmaking methods are fully described in the British Plastics Federation publication, "Tolerances for Moulded Screw Threads in Thermosetting Plastics", to which the reader is referred. Although the publication is concerned with compression-moulded materials, the section on pitch compensation and many other points apply equally well to injection-moulded materials. Compensation for pitch shrinkage by design changes is discussed on page 203.

Threads less than 1-in. diameter. In setting the dimensions of a pin for a moulded thread, three conditions have to be observed:

1. The provision of shrinkage allowances;
2. The maintenance of maximum metal conditions;
3. Ensuring that the thread dimensions given maintain the standard thread form and allow the use of standard cutters.

In order to satisfy these conditions, a variety of methods can be adopted, but one good approach is based upon the maximum effective diameter permitted by the thread standard and determining the major and minor diameters from the basic depth of thread. In particular, this method satisfies items 2 and 3 of the requirements. The maintenance of maximum metal conditions and maximum moulded thread size are of especial importance when the large shrinkage effects and variations which occur on moulding are realized.

The maximum metal conditions on the threaded pin allow subsequent adjustment, if necessary, but also, by ensuring the largest possible thread diameter, minimizes the possibility of interference with the mating thread should the pitch shrinkage allowance be in error.

The method adopted is applicable to all normal lengths of engagement, but the use of short lengths of engagement reduces the difficulty arising from pitch error. An engagement of not more than one-and-a-half times the diameter is frequently adopted.

The procedure for threaded pins for the production of moulded female threads having a depth of thread that extends equally above and below the effective diameter (e.g. Whitworth form such as B.S.W., B.S.F., and B.S.P.) is as follows:

p = pitch of threaded pin = [Pitch of moulded component thread required plus shrinkage allowance on pitch dimension]

D_e = [Effective diameter of threaded pin] = [(Maximum effective diameter of moulded thread allowed by thread standard minus toolmaking tolerance of 0.0005 in.) + shrinkage allowance]

$D_{maj.}$ = [Major diameter of threaded pin] = [D_e + (basic depth of thread on moulded component as specified by thread standard) + (shrinkage allowance on depth of thread dimension)]

$$D_{\text{min.}} = \begin{bmatrix}\text{minor diameter of}\\ \text{threaded pin}\end{bmatrix} = \begin{bmatrix} D_e - (\text{basic depth of thread on moulded component as specified by thread standard} + \text{shrinkage allowance on depth of thread dimension}) \end{bmatrix}$$

It should again be stated that the thread dimensions to be adopted from the relevant thread standard in the case of female moulded threads, using a male threaded pin in the tool, are those pertaining to a nut or internal thread.

EXAMPLE. A threaded pin is required to produce a moulded female thread to $\frac{7}{16}$ in. B.S.W., with a length of thread equal to $\frac{5}{8}$ in.
Calculate on standard dimensions of a medium-fit nut.
From B.S. 84, pitch = 0.0714 in.; maximum effective diameter = 0.3966 in.; basic depth of thread = 0.0457 in.; take the shrinkage allowance of the moulding material as 0.004 in. per inch.

Then,

$$D_e = (0.3966 - 0.0005) \times 1.004 = 0.3976 \text{ in.}$$
$$D_{\text{maj.}} = 0.3976 + (1.004 \times 0.0457) = 0.4435 \text{ in.}$$
$$D_{\text{min.}} = 0.3976 - (1.004 \times 0.0457) = 0.3517 \text{ in.}$$
$$p = 0.0714 \times 1.004 = 0.0717 \text{ in.}$$

These results may be checked against the limits set by the standard. The major diameter is specified as only 0.4375 in. minimum, hence with shrinkage, this is equal to $0.4375 \times 1.004 = 0.4393$ in. minimum. Since this is exceeded by the previously calculated figure of 0.4435 in., the major diameter of the threaded pins is satisfactory.

The major diameter is specified as 0.3674 in. maximum and 0.3461 in. minimum. With shrinkage, these figures become $0.3674 \times 1.004 = 0.3689$ in., and $0.3461 \times 1.004 = 0.3475$ in. Since the previously calculated figure falls between these limits, the major diameter of the threaded pin is satisfactory.

For those threads where the depth of thread above and below the effective diameter differ (e.g. Unified and American Series), the formulae for the major and minor diameters are modified accordingly.

$$D_{\text{maj.}} = \begin{bmatrix}\text{major diameter of}\\ \text{threaded pin}\end{bmatrix} = \begin{bmatrix} D_e + 2\ (\text{height of thread above effective diameter as specified by the thread standard} + \text{shrinkage allowance on height of thread dimension})^{(1)} \end{bmatrix}$$

$$D_{\text{min.}} = \begin{bmatrix}\text{minor diameter of}\\ \text{threaded pin}\end{bmatrix} = \begin{bmatrix} D_e - 2\ (\text{depth of thread below basic effective diameter as specified by the thread standard} + \text{shrinkage allowance on height of thread dimension})^{(2)} \end{bmatrix}$$

---

(1)Alternatively, this may be expressed as $D_{\text{maj.}} = D_e +$ (difference between basic major diameter and basic effective diameter as specified by the thread standard + shrinkage allowance on this difference dimension).

(2)Alternatively, this may be expressed as $D_{\text{min.}} = D_e -$ (difference between basic effective diameter and basic minor diameter as specified by the thread standard + shrinkage allowance on this difference dimension).

EXAMPLE. A threaded pin is required to produce a moulded female thread to the standard designation, No. 10–24 U.N.C.—2B.
From B.S. 1580, pitch = 0.04167 in.; maximum effective diameter = 0.1672 in.; basic height of thread above basic effective diameter = 0.0136 in.; basic depth of thread below basic effective diameter = 0.009 in.
Taking the shrinkage allowance of the plastics material as 0.004 in. per inch: then

$D_e = (0.1672 - 0.0005) \times 1.004 = 0.1673$ in.;
$D_{maj.} = 0.1673 + 2(0.0136 \times 1.004) = 0.1946$ in.;
$D_{min.} = 0.1673 - 2(0.009 \times 1.004) = 0.1492$ in.;
$p = 0.04167 \times 1.004 = 0.04183$ in.

Checking these results against the limits set by B.S. 1580.

Major diameter is specified only as 0.1900 in. minimum; hence with shrinkage this becomes $0.1900 \times 1.004 = 0.1907$ in. minimum. Since, therefore, the previously calculated value of 0.1946 in. exceeds this figure, the major diameter of the threaded pin is satisfactory.

Minor diameter is specified as 0.1555 in. maximum, and 0.1449 in. minimum. With shrinkage, these figures become
$0.1555 \times 1.004 = 0.1561$ in. and $0.1449 \times 1.004 = 0.1455$ in.
Since, therefore the previously calculated value of 0.1492 in. falls between these limits, the minor diameter of the threaded pin is satisfactory.

Compensation of Diameter for Pitch Shrinkage. Adjustment of the pitch on threaded pins to compensate for shrinkage should be included whenever possible. However, if the length of thread engagement is short, or if pitch-compensation equipment is not available, adjustment can be made by varying the diameter of the threaded pin but leaving the pitch unaltered at the standard value. By this means, the pitch error can be absorbed in the reduced depth of engagement of the thread between the standard metal screw and the moulded thread. This method is not recommended for large screws, nor for screws carrying heavy loads, because the stress is not evenly distributed over the threads.

It will be seen from Fig. 7.12, that if A is the pitch error between the standard thread and the moulded thread, then the radial displacement required is given by

$$B = A \cot \theta,$$

where θ = the flank angle of the thread form. Thus, for Whitworth

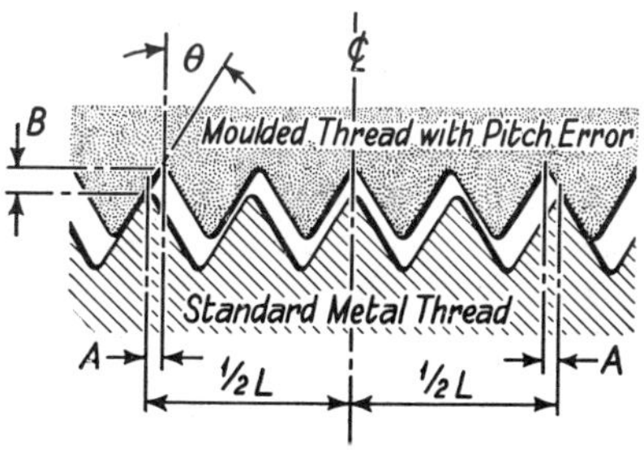

Fig. 7.12. The Pitch Error produced by Shrinking.

form $\theta=27$ deg. 30 min.; for B.A. form $\theta=23$ deg. 45 min.; for Unified and American form, $\theta=30$ deg.

This value of B represents the radial displacement necessary to accommodate the pitch error A that occurs over the length of engagement on either side of the centre line, i.e. $L/2$. The full shrinkage is an amount $+A$ at one end of the thread, and $-A$ at the other, and both are accommodated by the radial displacement B. The diametric displacement is twice the radial value,

$$\text{i.e. } 2A \cot \theta.$$

In terms of material shrinkage, the value of A is $S\times L/2$,
where $S=$ plastics material shrinkage (in. per inch);
$L=$ length of engagement of the thread (in.);
so that the diametric compensation is given by

$$T=2(S\times L/2) \cot \theta=SL \cot \theta.$$

where $T=$ the amount of diametric compensation to be added to the calculated pin diameter in order to engage with a standard screw. This value is subtracted in the case of a moulded male thread that engages a standard metal nut.

EXAMPLE. Let the above be applied to the previous example of a female moulded thread $\frac{7}{16}$ in. B.S.W. and with a $\frac{5}{8}$ in. length of engagement. In the present example, however, the threaded pin is to have a standard pitch, without pitch compensation.

Therefore, the diametric allowance necessary to accept the pitch error is

$$T=0.004\times 0.625\times 1.921=0.0048 \text{ in.}$$

This amount must be added to the diameters previously calculated; i.e.

$D_e=0.3976+0.0048=0.4024$ in.;
$D_{maj.}=0.4435+0.0048=0.4483$ in.;
$D_{min.}=0.3517+0.0048=0.3565$ in.

The pitch is now specified as standard, without shrinkage allowance, i.e. 0.0714 (or 14.t.p.i.).
Also, the length will include normal shrinkage allowance, and will be equal to $0.625\times 1.004=0.627$ in.

Moulded Threads of Larger Size. Usually, large male moulded threads engage with corresponding female plastics threads. Here, it is common practice to add shrinkage allowance to only the effective diameter, leaving the depth of thread and pitch as standard. Since the shrinkage will be approximately the same on both male and female threads, no further adjustment is generally necessary. Close fits are seldom attempted on this type of thread.

Extraction of Male-threaded Steel Pins. The largest number of moulded threads are of small diameter, generally for the attachment of screws. Such threads are most commonly produced by means of loose threaded pins, similar in many ways to insert carrier pins. These are placed in appropriate holes bored in the body of the tool and, after moulding, are extracted complete with the component. They are subsequently removed from the moulding outside the press by being screwed in a jig,

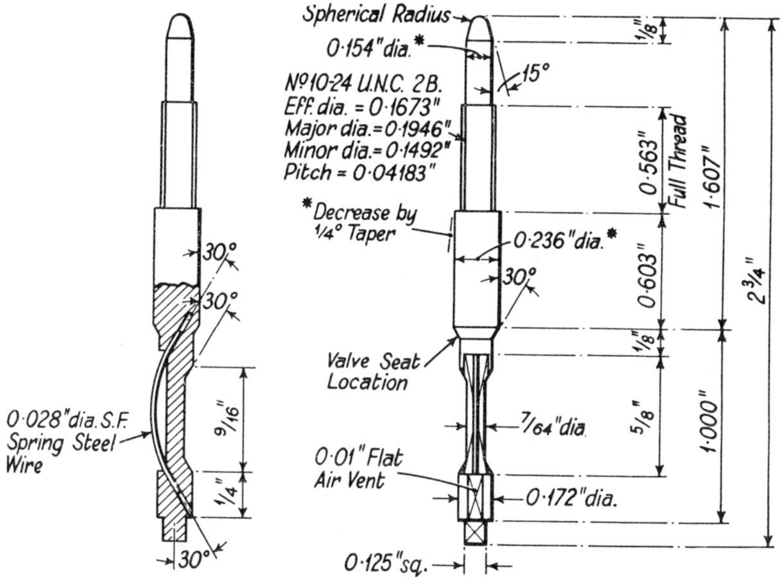

Fig. 7.13. Typical Carrier Pin to produce Moulded Female Thread and Two Plain Bores.

the square provided being used. Two sets of threaded pins are therefore required, one in the mould and one being removed from the component last produced. As with insert carriers, the threaded pins may locate in a plain bore or on a valve seat, being located for depth in the same manner, and may be spring-retained.

In Fig. 7.13 is shown an example of a fully dimensioned threaded pin, and a number of others are illustrated in Fig. 7.6, the latter also showing insert carriers and a loose pin to produce a moulded male thread.

In the case of a through-moulded screw, the feather edge of steel occurring at the end of the thread can easily be broken off or damaged

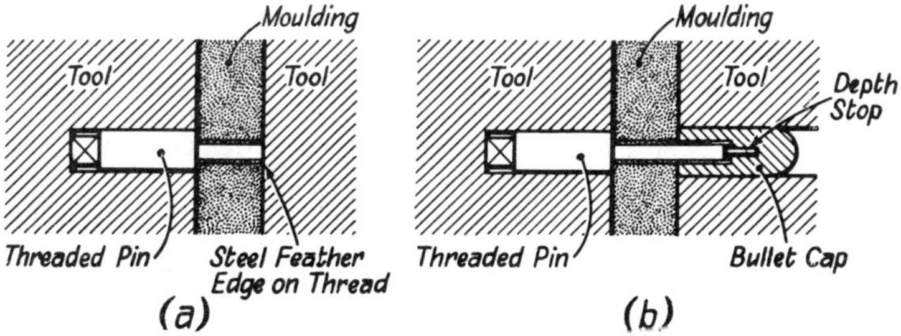

Fig. 7.14. Moulding of Full Through Threads.

where it meets the face of the opposite tool, see Fig. 7.14. This, in turn, will produce a moulded thread of distorted form at the end. This difficulty can be avoided by producing the threaded pin overlength and fitting a bullet cap before inserting the pin in the tool. This cap is screwed down until it makes contact with the depth stop (at the end of the thread is a clearance), and engages in a corresponding hole in the opposite half of the tool. The cap is removed from the completed moulding, before the threaded pin is extracted. A similar method can be used for full through inserts.

In Fig. 7.11 is demonstrated the need to leave a plain portion between the end of a thread and the shoulder of the pin. This is to provide clearance for the cutter and must not be less than the major diameter of the screw, since anything less would prevent extraction of the pin after moulding. In the case of carrier inserts, however, where the insert abuts on the shoulder, an undercut may be used to run-out the thread.

Rotating Cores. Where moulds are required for fast production or large quantities, the thread may be unscrewed automatically, either during or after the opening of the press. Most of the methods used operate by rotating the threaded pin or core, and the modes of operation may be any of the following:

(i) *Rack and Pinion :* Here the male core is secured to the pinion which is rotated by a rack that traverses across the tool. Only a limited number of threads can be unscrewed, due to the practical limitation on rack length. The rack itself can be operated by pneumatic or hydraulic cylinder.

(ii) *Helical Gears :* The male core is secured to a spiral gear or to a pinion at the end of a reduction train. This is driven through a second spiral gear from a rotating shaft across the tool, the shaft being usually driven by a geared electric motor.

(iii) *Worm and Wheel :* Here again the core is attached to the worm wheel, that is driven by a worm operated by an electric motor on the outside of the tool.

The actual power required to rotate the cores is difficult to assess, but the initial load due to friction and the starting of any flash is heavy, and all parts must be designed to take this loading. The threads of the moulding core must be of nickel-chromium steel, hardened and ground.

The method of unscrewing the core to be adopted will depend on the following factors:

(a) *The length and number of threads to be unscrewed.* For a great number of threads, a rotational helical or worm wheel can be used, the gearing being driven by an electric motor or an external handwheel. With short threads or a small number of threads, a rack-and-pinion drive can be utilized, the linear motion of the rack being derived from external cylinders, from rotation of a handwheel, or from the operation of cam pins within the tool.

(b) *The diameter of the thread.* For small threads of, say, below half-inch diameter, and limited length, a pneumatic cylinder or handwheel can be used to power the core, but for larger threads a more positive means such as hydraulic cylinders or geared electric motors must be employed. The power requirements are obviously dependent upon the area of the moulding in contact with the core.

(c) *Whether unscrewing must take place: (i) before the tool opens; (ii) during tool opening; (iii) after tool opening.* If the means of holding the component stationary against the rotary action of the core exist only in the half of the tool opposite to that from which the core will be unscrewed, then unscrewing must take place either (1) before the mould is opened, with the core pin rotating and moving down the thread, or (2) with the core pin rotating as the initial opening of the tool takes place, the component moving forward relative to the core pin which itself remains stationary in the axial plane. In this case the component is finally left in the female half of the mould, whence it is subsequently separately ejected. In some instances, especially where hand operation of the unscrewing gear is performed, the tool must be open before unscrewing can commence.

An example of a rack-and-pinion unscrewing gear is shown in Fig. 7.15. This may be used on a single-impression mould or on a multi-impression tool, but in the latter case, the cores must obviously be placed along the line of rack action. In this tool, unscrewing takes place while the tool

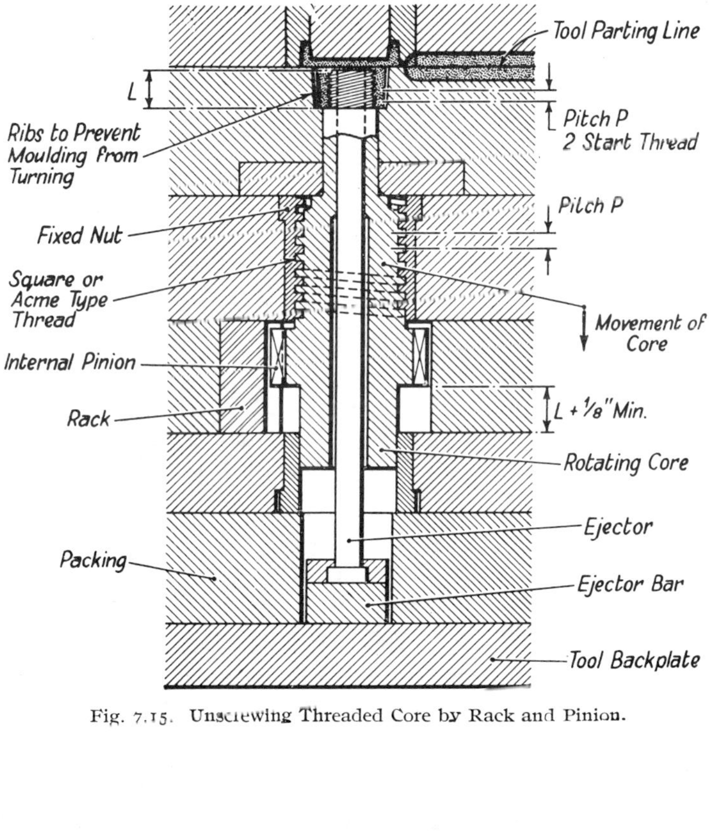

Fig. 7.15. Unscrewing Threaded Core by Rack and Pinion.

remains closed, the core rotating in its own thread and thus moving downwards into the tool, at the same rate as the threaded end unscrews from the component. The pitch of these two threads must clearly be the same. After the core has completely unscrewed, the component is left in the same side of the mould, whence it is ejected by the central ejector pin passing up through the core, after the tool has opened. In tools of this type, the unscrewing thread must be of ample strength to withstand the cavity pressure that will be transmitted down the core on injection, and this thread is preferably of square or acme form. The core must rotate

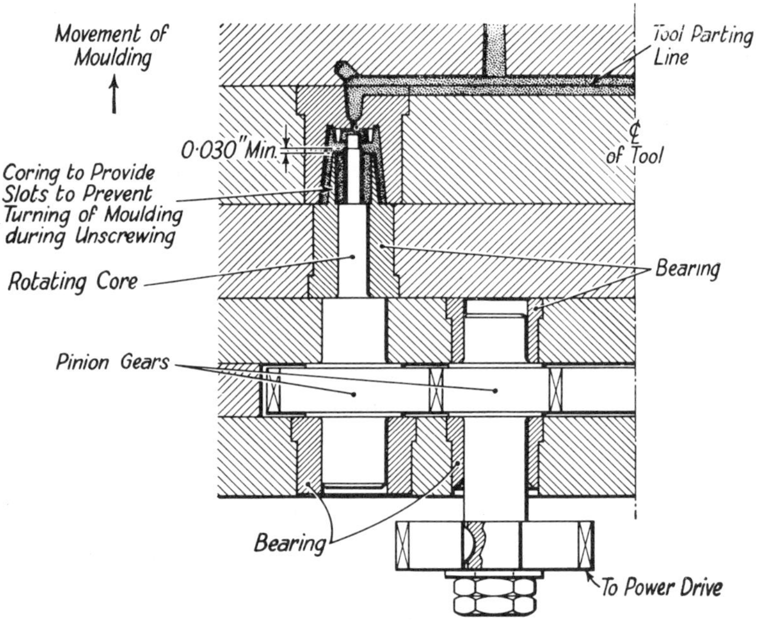

Fig. 7.16. Multiple-impression Unscrewing Tool operated by Gear.

in hardened bearings of adequate size. The provision of features on the moulding to prevent rotation on unscrewing is essential. The rack itself is connected to a cylinder mounted externally on the tool, and adjustable positive stops should be provided on the stroke of the rack in order to control the amount of rotation on the core.

Shown in Fig. 7.16 is an unscrewing mould operated by gearing. This form is well suited to multiple-impression tools in which the cavities are arranged about the pitch-circle diameter of the central gear, which then rotates all the cores simultaneously. This unscrewing method operates during or after tool opening. The core does not move axially, but causes the moulding to ride up the rotating thread and thus to be ejected. Fixed internal cores are shown which serve to prevent the component rotating during unscrewing, but it will be noticed that the length of these cores is made slightly less than that of the thread, this being done to ensure that the moulding is ejected completely. It is again essential that the rotating core is mounted in bearings of adequate strength, and especially that thrust bearings are provided to resist the axial load arising from the cavity pressure. It should be noted that positioning the start of the thread in the same place each time prior to moulding is more difficult with a motor-driven tool than with a cylinder-operated rack. Often, however, this is not required.

These examples illustrate two typical basic methods of operating unscrewing tools, but numerous other details and alternative mechanisms are used for this purpose.

Both examples are applicable to mouldings fed on the side remote from the thread. On occasion, however, mouldings must be fed on the inside surface (e.g. bottle caps in which no feed pip must appear on the top surface). In this case the threaded core is on the injection side.

Stripped Threads. The usual form of thread used for stripping is the roll-type thread generally similar to that shown in Fig. 7.17, where $R = 0.288 \times \text{pitch}$. With flexible or semi-flexible plastics materials, these

Fig. 7.17. Stripping of Roll-type Thread.

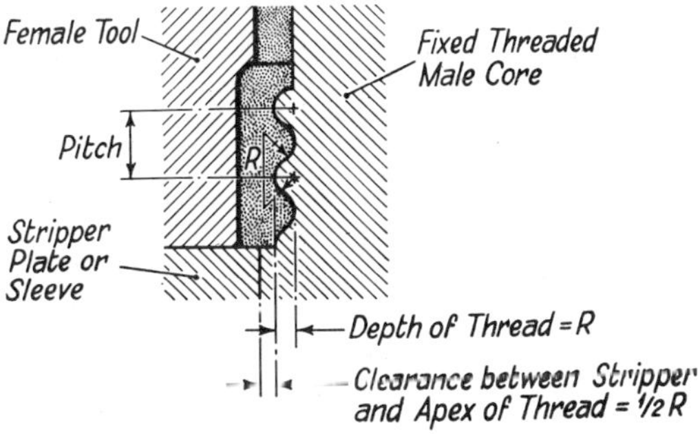

can generally be stripped off the core in a manner similar to that employed in stripping undercuts; and the relevant section (Chapter V, page 113) may be consulted for methods of operation, procedure and limitations, since the essential features are the same. Stripped threads are generally applied to those of a larger size, i.e. those having a diameter greater than 20 times the wall thickness in semi-flexible materials, but considerably smaller in flexible materials. Stripping cannot usually be recommended for brittle materials such as unmodified polystyrene.

Moulded Male Threads. The determination of thread sizes for moulded male threads is carried out in exactly the same manner as was indicated for threaded pins, except that:

(i) the basic sizes upon which the thread is calculated are based upon the *bolt* dimensions of the thread standard;

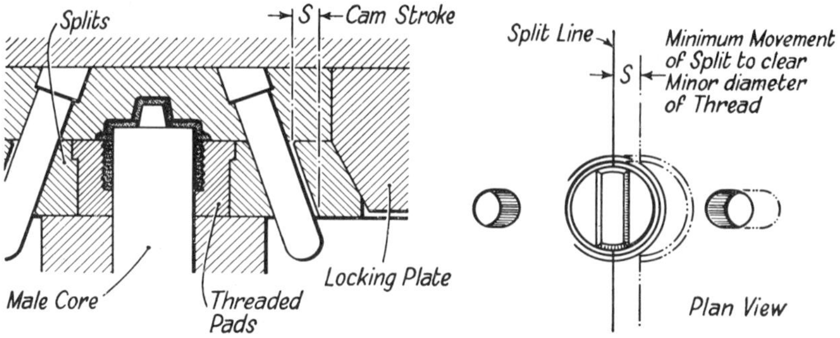

Fig. 7.18. Split Tool for moulding Male Thread.

(ii) to achieve maximum metal conditions, the *minimum* effective diameter permitted by the thread standard is used *plus* the toolroom tolerance of 0.0005 in.;

(iii) if diametric compensation is used for pitch error, this value is *subtracted* from the calculated diameters.

As before, these calculations are chiefly necessary in the case of small threads which have to engage with standard metal nuts, but in the case of large threads engaging with a moulded female thread, it is again common practice to add shrinkage allowances only to the effective diameter, leaving the depth of thread and pitch as standard.

Nearly all male threads are moulded in split-type tools: see Fig. 7.18. Many methods of operating split tools are used, but the most common is the employment of cams or finger pins similar to those described on split tools for undercuts. The minimum cam movement necessary is that required to ensure that the minor diameter of the thread is cleared, including clearance at the split line.

The minimum stroke or movement of the cam in inches is given by

$$S_{\text{min.}} = \sqrt{[(D^2 - d^2)/4]} + 0.125,$$

where D = major diameter of the thread (in.); and

d = minor diameter of the thread (in.).

These tools may be completely split about a component in which the thread is a major item, but where the screw is only a small feature, small local splits may be used.

In the production of solid screws, a split tool may also be used as shown at (*a*) in Fig. 7.19, but they may be moulded in a conventional tool as indicated at (*b*).

The problem of split-line flash must always be considered with moulded threads, since flash can cause interference on assembly with the mating thread. To eliminate flash, extreme accuracy of the matching threads in the splits is essential, together with close shut-off on the opposing pads. This is particularly true when a penetrative material such as nylon is employed.

The difficulty can sometimes be obviated by providing flats on the

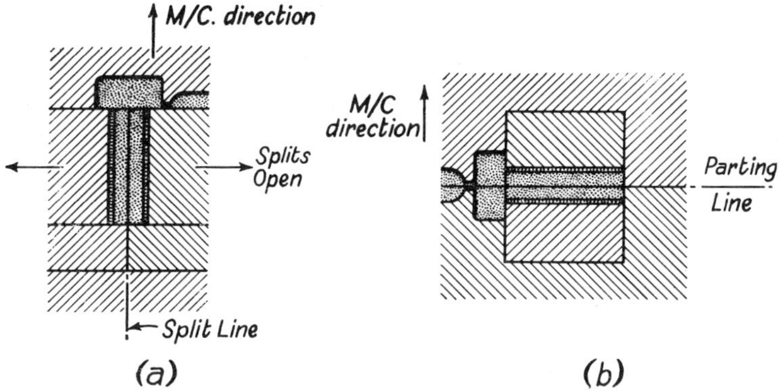

Fig. 7.19. Methods of moulding Solid Male Threads.

thread at the splits or parting line. Such flats are set just below the minor diameter, and thus any slight flash will not interfere on assembly with the mating thread. This is indicated in Fig. 7.20. The flats are formed by means of subsidiary pads set into the threaded splits.

If full threads free from flash are essential, then these can be moulded in a threaded cavity cut into the solid, but this method may increase the length of the press cycle because of the necessity to unscrew. Sometimes such threads can be produced in loose pins having a nut thread, and one of these is shown in Fig. 7.6. Such pins have a design similar to those used for female moulded threads.

If threads are to be moulded in a split tool, the design of the thread form, pitch and lead must avoid the formation of undercuts as shown in Fig. 7.21. This type of thread can, however, be unwound from a solid threaded cavity.

Threads cut into a blind hole with a final parallel portion should preferably run out as shown in Fig. 7.22. These threads will normally be produced by screw cutting, bringing the tool out at the end of

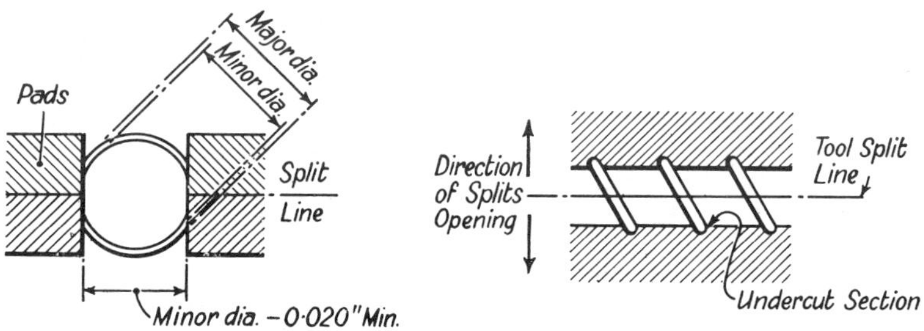

Fig. 7.20. [*Left*] Flats on Thread Split Lines to reduce Moulding Flash. Fig. 7.21. [*Right*] Undercuts on Threaded Split Tool.

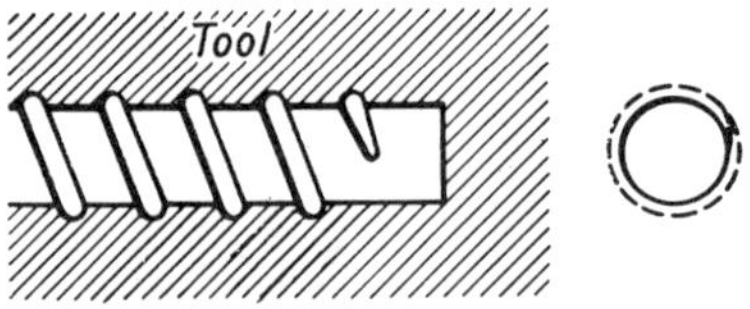

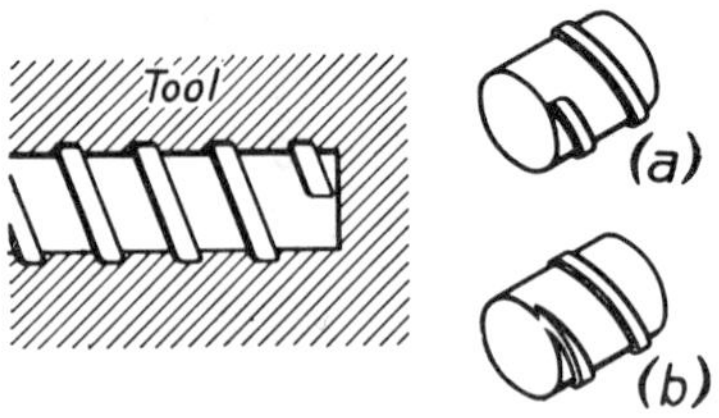

Fig. 7.22. [*Left*] Runout of Thread in Blind Hole. Fig. 7.23. [*Right*] Finish of Thread at End of Blind Cavity.

the thread; or by milling. For deep small threads, i.e. $\frac{5}{16}$ in. and smaller, a convenient method is to produce the thread by electro-deposition. Full through threads may, of course, be tapped.

Where threads run to the end of a blind threaded cavity, they should run out and stop short as shown in Fig. 7.23, producing a moulded thread as indicated at (*a*). If the thread in the cavity were taken completely to the end, the moulding would have as indicated at (*b*) a feather edge which could easily become broken.

For a similar reason, the screw entry of a male thread cavity is desirably chamfered as shown in Fig. 7.24, otherwise the feather edge of steel at the commencement of the thread is liable to break off. With solid threaded cavities, it is also clearly necessary not to allow any undercut features at the thread entry which could prevent unscrewing of the moulding.

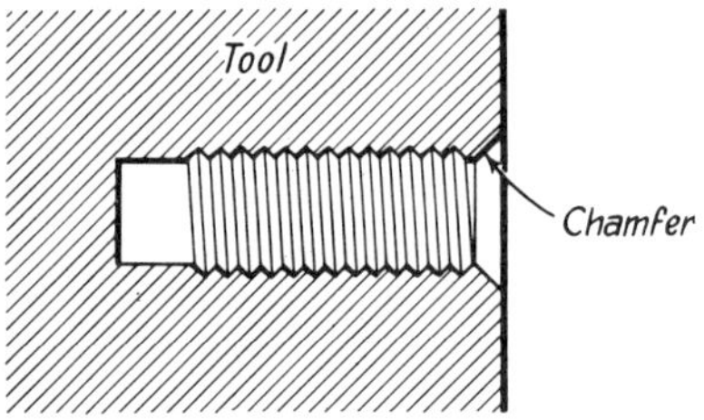

Fig. 7.24. Chamfer at Thread Entry.

Moving Cores. The provision of holes in mouldings which occur in the line of draw are readily taken care of by fixed cores having faces which shut off against the opposite tool. Where the hole is at right-angles to the tool centre line, e.g. the base of a moulding, the core is straightforward. Sometimes, however, apertures are arranged in the side walls of components, using fixed cores that shut off on the line of draw against the side faces of the opposite tool. Various methods of achieving this are possible, and one is illustrated in Fig. 7.25. Here, it will be noticed that the face forming the aperture can pass through the open end of the moulding, on ejection, to permit extraction of the component. This method of operation means that the shut-off face of the

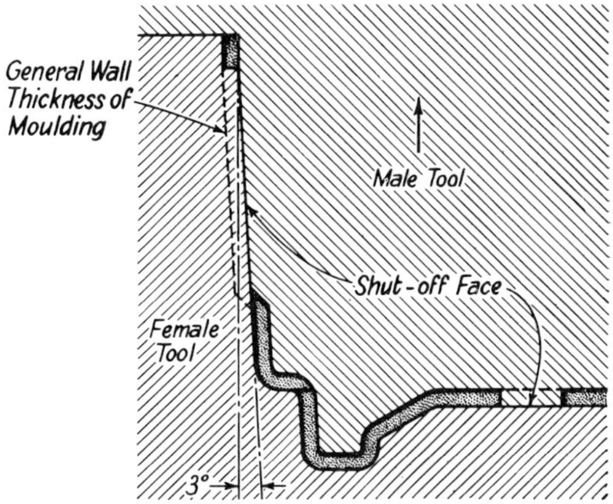

Fig. 7.25. Moulded Apertures in Line of Draw.

female tool slides against the male tool as the mould opens and closes. To avoid excessive scoring, the shut-off faces are made tapered with the maximum taper possible, which must not be less than 3 deg.

If the fixed shut-off faces can be employed in this manner to produce side holes, the tool is simpler, more reliable and cheaper. The face forming the aperture may be on the male or female tool, depending upon the design of the component and the ease of machining the tool. Also shown in Fig. 7.25 is a fixed core directly in the line of draw, to produce an aperture in the base of the moulding.

Where the holes are to be produced off the line of draw, a moving core must be used, this again shutting off against the opposite face of the tool to form the desired aperture.

Often, the moving core is mounted wholly within, and shuts off against, one half of the tool only, and scoring problems do not arise, since the core is moved in before the tool closes and is moved out before the moulding is ejected, unless the core is left in at tool opening to retain the moulding on the ejection side. If it is essential that the moving core is in one half of the tool and shuts off against the opposite half, the liability of interference and scoring must be considered, and, if necessary, the core must be moved in after the tool has closed and withdrawn before the tool is opened.

The operation of cores by means of pneumatic and hydraulic cylinders was considered in principle in Chapter II, page 46, but typical details of a moving core are shown in Fig. 7.26.

This forms a round hole in the side of the moulding, and is operated by a double-acting air cylinder. At (*a*) is shown a side section of the tool with the core fully in. A plan of the moving-core assembly is shown at (*b*), while at (*c*) is given an end view of the core and its guides. The moving core is mounted upon, and shuts off against, the male tool, the female tool being cut away locally to shut off around the face *Y* of the core.

The air cylinder is mounted to the male tool by the bridge piece,

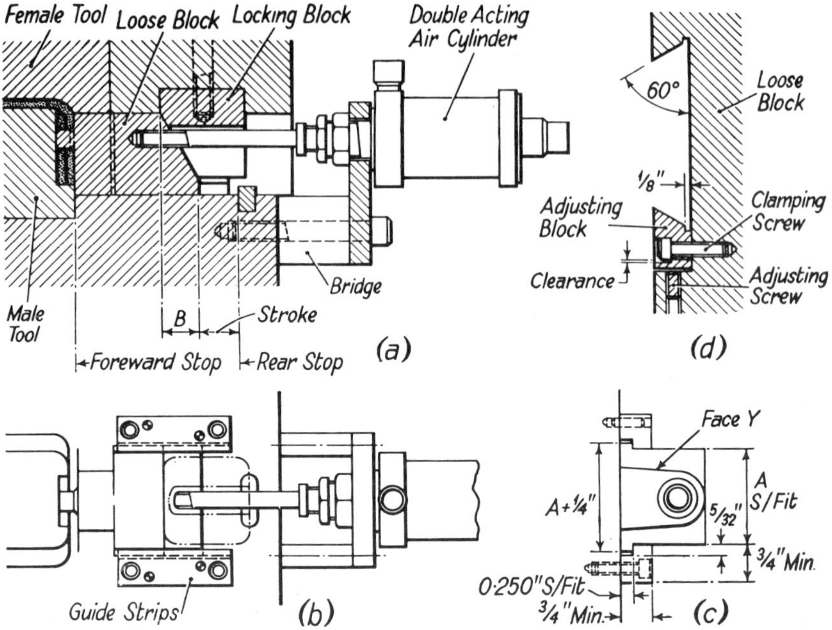

Fig. 7.26. Operation of Moving Core by Air Cylinder.

thus allowing easy access to the cylinder lock nuts, since it may be necessary to remove the cylinder to permit the tool being mounted in the press. The cylinder stroke should be in excess of the stroke required, to avoid the end plates of the cylinder being used as stops. Independent stops should be provided. In the example illustrated, the rear stop is formed by a steel lozenge let into the tool, while the forward stop is effected by contact of the lower portion of the front face of the moving core against the male tool. If small shut-off faces are used as the forward stop, there is danger that the high unit pressure may either cause the core pin to hob into the opposite face or lead to a peening over of the end of the core pin. Such metal distortion can lead eventually to difficulty in extracting the core from the moulding; hence a separate forward stop should be provided as shown.

In operation, the core is moved in before the tool closes. When the tool is closed, the locking block mounted on the female tool is positioned behind the core block, thus preventing any tendency of the core to move back under cavity pressure. The amount of engagement provided is just sufficient to hold the core shut-off face firmly against the male tool without excessive pressure. If the amount of core stroke necessary is small, the locking block can be arranged to overlap the rear taper of the core block. This will ensure that, on closing of the tool, the locking face will move the core into position, even if the actuating cylinder fails to operate.

A common method of guiding the core block is shown at (*c*). The moving core block should always be of sufficient length relative to its width for easy sliding. For the same reason, the depth and proportions of the actual guide surfaces should be adequate, and sufficient width provided on the guides to accommodate the mounting screws and dowels.

In view (*d*) is illustrated a dovetail guide which, although not widely used, does provide ready means of adjustment to take up wear and to facilitate assembly fitting. The adjusting block is preferably made of phosphor-bronze. In many cases, the core will pass through the wall of the tool, particularly with cores in the female tool, instead of being mounted on the flash line. Here the core is guided by the aperture in the tool in which it is made a sliding fit, without additional guides. All sliding surfaces of blocks, locating blocks and guides should be of hardened nickel-chromium steel with a ground finish.

In the example illustrated, the moving core can be mounted upon the male tool because the shut-off face of the female around the surface *Y* is reasonably effective. If, however, this face were of difficult profile, shut-off might be poor or impractical. Under these conditions, it would perhaps be necessary to mount the core in the female tool, either at split line level, or in the depth of the cavity. If this were done, however, the cylinder would have to operate before the tool opens, presuming ejection to take place off the male tool. Then it would not be possible to use a locking block on the opposite half of the tool, and a hydraulic cylinder would have to be used in place of the pneumatic cylinder. This will generally apply in any tool where the moving core cannot be secured by a locking block.

Air cylinders normally operate on a supply pressure of 80 to 100 lb./sq. in. and hence on a 2-in. diameter cylinder the maximum total pressure exerted is only 314 lb. At a cavity pressure of 9,000 lb./sq. in. this would only hold a face of 0.035 sq. in. Hence a hydraulic cylinder is used which can apply a solid locking pressure at a much higher value.

A single-impression mould employing four moving core blocks is illustrated in Figs. 7.27A, 7.27B, 7.27C, three of these being operated by finger cams, and the fourth by means of a double-acting hydraulic cylinder.

The cam-operated cores are mounted on the ejection side of the tool, the opposite half of the tool shutting off around the core down to the split line, on the principle shown in Fig. 7.26 at the face *Y* in the view (*c*). In the case of the two holes to be produced by the core block attached to the hydraulic cylinder, this is not convenient. Moreover, these two holes are in the fixed half of the tool and hence the cores must be withdrawn before the tool opens, since the moulding will go away on the moving half of the tool, held by the finger cam cores. Employing a hydraulic cylinder mounted on the fixed tool allows these core pins to be withdrawn prior to the press opening. In section *BB*, Fig. 7.27C, it can be seen that both core blocks form part of the actual cavity wall, the right-hand block forming two under-cut features and the left-hand block forming an under-cut feature and a moulded hole. These blocks are water-cooled, connections being made with flexible pipes.

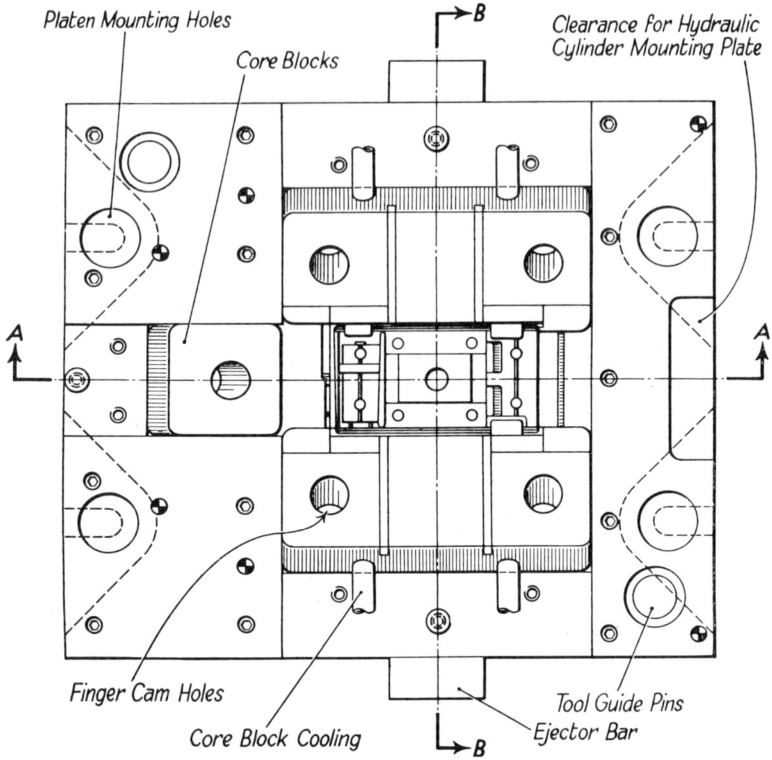

Fig. 7.27A. Mould with Hydraulic and Cam-operated Cores: Plan View of Tool.

In Fig. 7.28 is illustrated a typical single-impression split tool operated by dog-leg cams, to produce a small transformer bobbin moulding. It is seen that the splits are in the form of a split circle with a conical rim. This form is very convenient and widely used, since the splits can be positively located and locked in the conical recess provided in the fixed half of the tool. The splits themselves are secured to, and guided by, blocks (view *c*) sliding in tee-slot slides cut into the solid body of the moving tool. The tool is operated by two cam pins, view (*X*), so designed that opening of the splits is delayed until the tool opening has withdrawn the moulding off the central core pin to within approximately $\frac{1}{8}$ in. of its end. The splits then open, leaving the moulding held centrally on the end of the core pin, from which it is easily removed by hand. These cranked finger cams are made rectangular in section to facilitate machining.

For moving cores or splits operated mechanically through finger cams, measures must be taken to prevent the core or split displacing itself once the finger cam disengages on completion of press opening. On the follow-

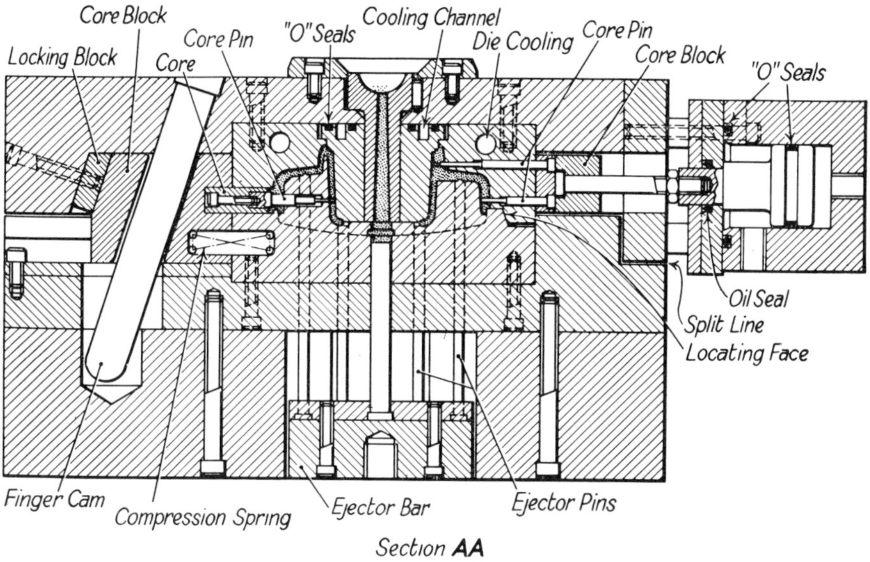

Fig. 7.27B. Section *AA* of Fig. 7.27A.

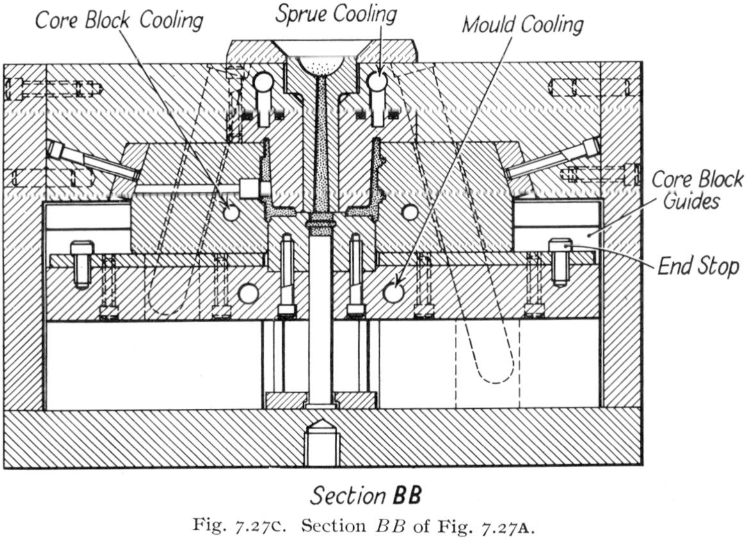

Fig. 7.27C. Section *BB* of Fig. 7.27A.

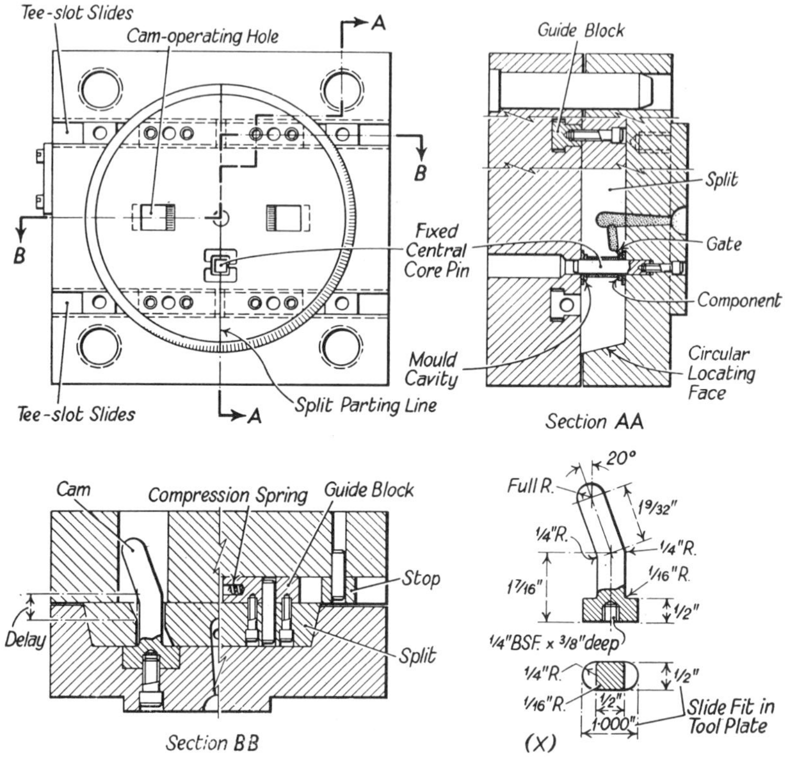

Fig. 7.28. Delayed-action Split Tool.

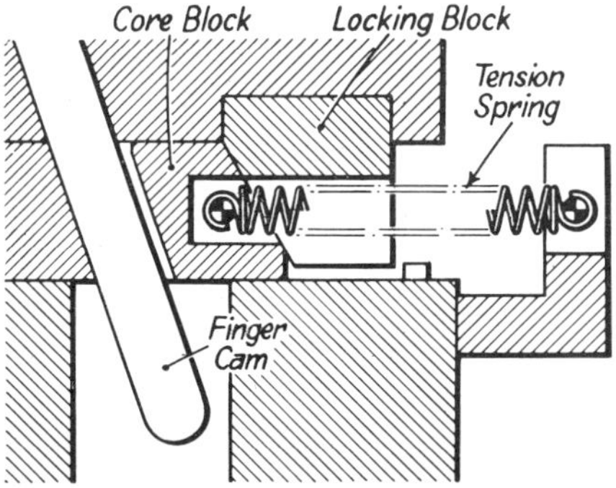

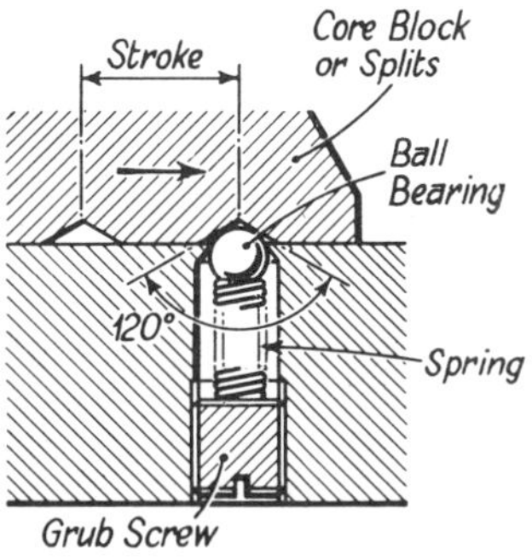

Fig. 7.29. [*Left*] Position Control of Moving Core or Splits by Tension Spring. Fig. 7.30. [*Right*] Position Control of Moving Core or Splits by Ball Catch.

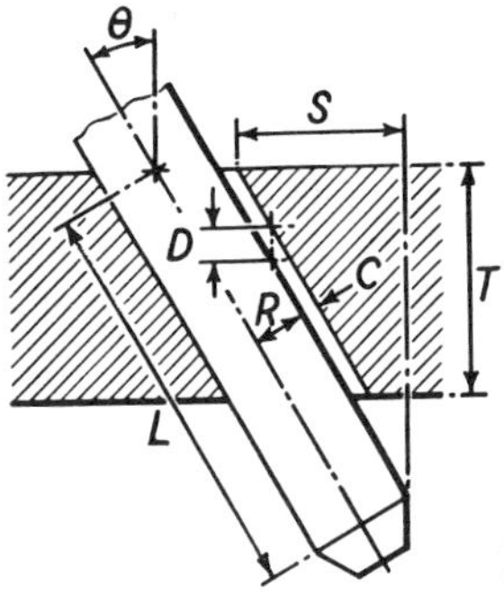

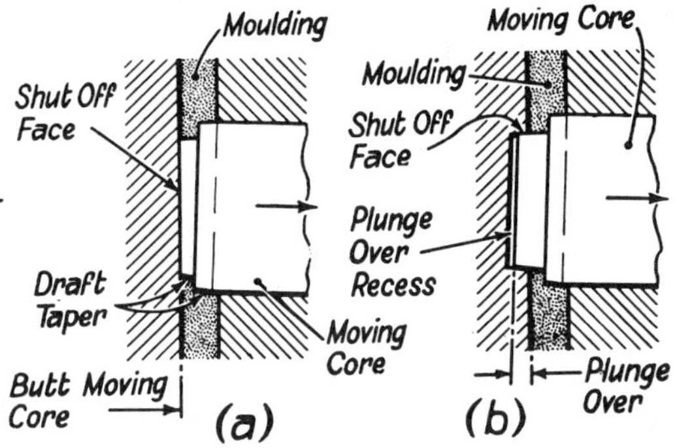

Fig. 7.31. [*Left*] Finger Cam. Fig. 7.32. [*Right*] Shut Off on Moving Cores.

ing cycle, as the press closes, the finger cam must again pick up its locating hole in the split or core, and if this has become displaced, damage to the tool can result. Two common methods of locking the core to obviate this danger are shown. Fig. 7.29 shows the use of a tension spring mounted on an outrigger, to retain the core in the open position when the press is open. The use of a locking block and rear stop similar to that shown on the cylinder-operated core illustrated in Fig. 7.26 will be noted. An alternative method is indicated in Fig. 7.30. This comprises a ball bearing that is urged upwards against the underside of the core or split and locates in a recess at the closed and open positions. A further type, using a compression spring, is employed on the mould shown in Fig. 7.27, and also in Fig. 7.28.

The length of finger cam, Fig. 7.31, can be determined from the formula

$$L=(C \sec. \theta+S) \operatorname{cosec} \theta+R \tan \theta,$$

where, L = effective length;
R = radius of cam pin;
C = clearance (usually from $\frac{1}{16}$ in. to $\frac{1}{8}$ in.);
T = thickness of core block or splits;
S = stroke;
θ = angle of cam pin;
$C \operatorname{cosec} \theta = D$ = delay.

Finger pins are not normally made less than 0.5 in. diameter, but in any event, should always be made as substantial as possible in order to minimize deflection, and of hardened and ground nickel-chromium steel. The operating angle of the pin should not exceed 30 deg. since otherwise the lifting component of the opening force causes the pressure on the guide faces to increase excessively, and hence makes the core-opening-force requirements greater than necessary. Even at this angle and below, the guides must be adequately screwed down to resist the lifting force.

The finger pin shown in Fig. 7.27 operates in an oversize hole in the core block, thus providing both delay and operating clearance. If the design is such that no clearance is allowed, it is essential that the angle of the locking block and its mating face at the back of the core block is greater than the angle of the finger cam. If this is not done and small angular errors exist, there is danger that, on tool opening, the finger pin will attempt to bring out the core block faster than the locking block is cleared out of the way, with seizure of the tool as a result. The locking block angle, therefore, is generally made 5 deg. greater than the finger-cam angle, this practice being extended for the sake of safety to all tools, even if the pin has a clearance.

Occasionally, moving cores are operated by means of screws or by rack and pinion. Generally, however, this leads to difficulties in providing a locking block to sustain the end pressure. However, the widespread use of hydraulic cylinders has made this method of operation largely obsolete.

Where cores are provided to produce full-through apertures in the moulding, it is clearly essential that the face of the core should shut off cleanly against the opposite face of the tool; therefore the end of the core must include a matching draft taper.

Normally, the only cavity pressure attempting to force back the core is that exerted directly on the core face. For example, in view (*a*) of Fig. 7.32, this is the end pressure exerted on the shouldered recess. However, if any mismatching of the shut-off face occurs, the moulding will flash across this face, and the cavity pressure, tending to push back the core, will thereby be considerably increased. For this reason, it is safe practice to design the core-locking device so that it can resist a cavity pressure of about 9,000 lb./sq. in. acting over the whole of the core end face.

If possible, the shut-off face of the core should be relieved, thereby increasing the seal-off pressure and minimizing flashing. Unfortunately, if this seal-off pressure is made too high, the core may hob into the opposite face, resulting in scoring of the tool and moulding. The difficulty is often avoided by carrying the core beyond the moulding face into a recess as shown at (*b*) in Fig. 7.32. The core must have as close a fit as possible in the recess to prevent entry of horizontal flash, and hence difficulties of ejection, but is provided with a clearance at the bottom of the recess. The amount of "plunge-over" is generally about $\frac{1}{8}$ in. This method is very effective, but requires that the core block be withdrawn *before* the tool opens.

On some mould designs, the position of the cored holes usually requires that the core pin is mounted on one half of the mould and that it plunges over into the other half. An example of this is illustrated in Fig. 7.33, where the normal split-line shut-off face would be *X*. This, however, means that the core would be operating in the female tool but would plunge over into the male tool. To contrive this would require extreme accuracy in toolmaking to align the holes on both halves of the tool. More serious, however, is the fact that the entry of any flash in the tool split-line *X* would cause misalignment of the core pin and plunge-over recess, causing difficulties in operation, and possible damage to the tool. The design would be improved by making the split line locally at the face *Y*, and the resulting print line on the moulding would have to be accepted.

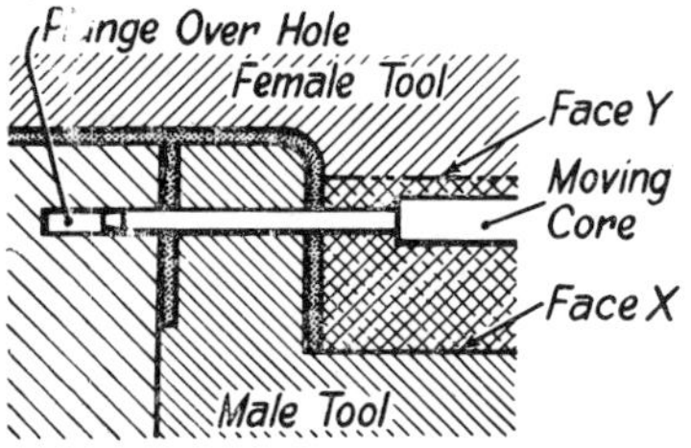

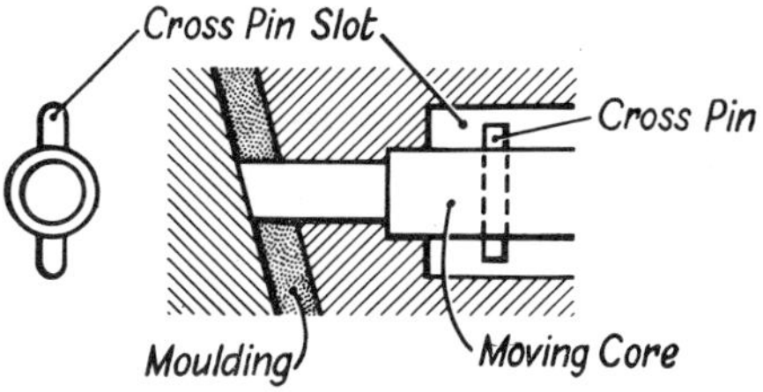

Fig. 7.33. [*Left*] Plunge-over Cores between Male and Female Tools. Fig. 7.34. [*Right*] Preventing Rotation on Circular Core Pins.

This would allow both the core and plunge-over hole to be on the same side of the tool (for other illustrations of this problem, see Figs. 7.26 and 7.27). If this method cannot be adopted, it is generally safer to leave the moulded hole and drill the moulding separately.

With non-circular moving cores, rotation of the core is not a problem, since it is prevented by the fit of the core in the tool. With circular cores, however, any rotation must be prevented if it is essential that the front face of the core must match the opposite face of the tool. One convenient way of achieving this is shown in Fig. 7.34, where the cross pin is driven through the shank of the core pin, and operates in a corresponding slot in the tool body.

Deflection of Slender Core Pins. Where core pins are secured at the mounting only, the calculations for deflection are based on considering the core as a cantilever, ignoring the non-uniformity of cross-section due to taper. This requires, however, that the fixed end is held rigidly for at least $3D$ in length, where D is the diameter of the pin, as indicated in Fig. 7.35; the normal maximum length for an unsupported pin is $5D$, although much more slender pins are sometimes used successfully.

To minimize deflection of the core, an even filling around the pin is necessary, and this is often conveniently achieved by means of a pin gate at the apex as shown in the illustration. Unfortunately, it is almost impossible to ensure absolute evenness in filling, and material will invariably fill down one side of the cavity faster than the rest. As soon as this happens, side pressure is developed on the core pin, resulting in deflection of the pin, which in turn increases the wall thickness at the side of maximum flow. This further facilitates flow on this side. On reaching the bottom of the cavity, the material will commence to fill back up the remainder of the cavity, finally causing bad weld lines and gas traps. The ultimate deflection of the core pin caused by such uneven flow will depend upon the amount and position of the leading material.

To improve the evenness of flow in top-fed cavities, a decreasing wall thickness can be adopted as shown in Fig. 7.36, the wall-thickness variation being effected by a taper on the core pin. With this method, the pressure required to feed the material down the taper rises steadily. Thus, if one side tends to feed faster than another, it is resisted by the rising cavity-pressure requirement, so that the faster flowing material tends to slow up while the feed is continued in the unfilled cavity portion of greater wall thickness. When this, in turn, has filled, the pressure is equalised and even progressive flow continues.

If the proportion of the length of the core pin rises to 8*D*, it is better to feed the cavity at the fixed end of the pin, thereby minimizing the deflection due to uneven feeding. One method is to employ a ring feed as indicated in Fig. 7.37. While this is often used successfully, a ring feed may, nevertheless, tend to feed unevenly at the point adjacent to the runner. An alternative is to feed by means of several gates, two or four, placed around the base of the core. This reduces the tendency of the injection pressure to apply an off-balance load on the core, and the separate gates can more easily be adjusted to provide even filling.

With base feeds of these types, a vent pin at the apex is essential to allow the escape of air or gas. Sometimes, this vent pin plunges over into the core pin and thus tends to hold the core pin against the effects of small initial differences of pressure.

Above a length-to-diameter ratio of 8*D*, the feed is, of course, still placed at the base, but core deflection becomes increasingly serious; unless side supports can be introduced. At lengths of this proportion, it is much preferable to employ a top support.

A core pin supported at the free end is shown in Fig. 7.38. Here, deflection is greatly less, since the core is effectively much stiffer than the unsupported pin.

Long slender cores similar to those illustrated present cooling

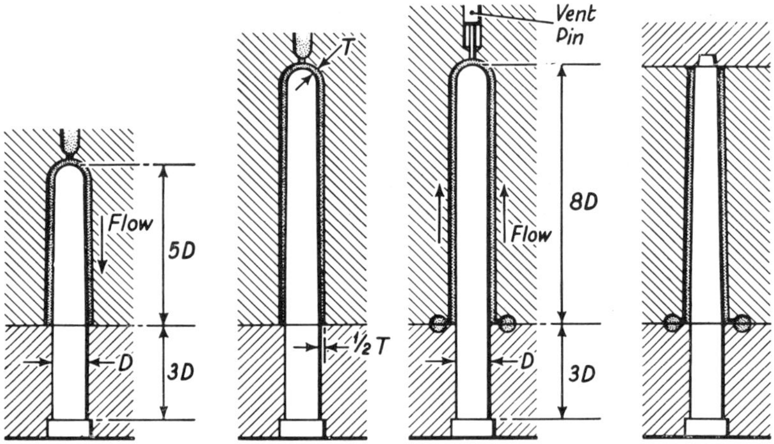

FIG.35 FIG.36 FIG.37 FIG.38

Fig. 7.35. Unsupported Core Pin fed from Top. Fig. 7.36. Tapered Wall Thickness to stabilize Flow. Fig. 7.37. Unsupported Core Pin with Ring-gate Feed to Base. Fig. 7.38. Supported Core Pin with Ring-gate Feed fed at Base.

difficulties if it is not possible to bore the core for direct air or water cooling. One method is to continue the core back into the body of the tool, and into a chamber containing circulating water. Conduction of heat to the water is facilitated by adding cooling fins around that portion of the core which is within the chamber. Copper inserts in the core are occasionally used for the same purpose.

In Fig. 7.39 is indicated a method of top feeding a supported core pin. The feed is taken to four runners and gates that are spaced around the location support. The latter locates on the recess surfaces between the runners. This method gives four weld lines on the moulding, but enables a rigidly supported core to be used, and is often successfully employed.

The actual deflection of the core can be estimated from standard beam formulae, although varying flow, changes in viscosity and dynamic rate of loading make any close calculation impossible. The load is that set up by the leading material on one side, and this final lead condition can be assumed as 0.1 of the length of the core. This will result in a load due to

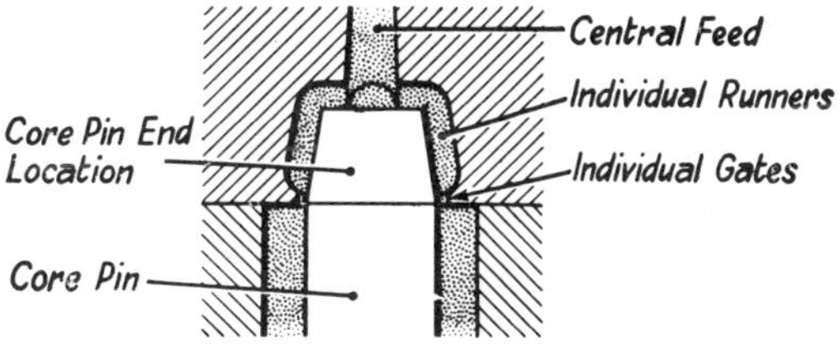

Fig. 7.39. Supported Core Pin with Top Feed.

the cavity pressure exerted over the projected area, the latter being equal to 0.1 times the length of the pin multiplied by the diameter, i.e. $0.1Ld$. Assuming a cavity pressure of 9,000 lb./sq. in., then the load on the pin is given by

$$W = 0.1 \times 9{,}000 \times Ld$$

where L = free length of core (in.); and d = diameter or width of core pin (in.).

In the case of the unsupported top-fed core pin of the type shown in Fig. 7.35, the maximum deflection will occur, if it be assumed that the uneven load W due to the leading material is exerted at the free end. Then

$$D = WL^3/3EI$$

where I = moment of inertia, which for round pins $= \pi d^4/64$,

so that $$D = 64WL^3/3E\pi d^4$$

where D = deflection of the core pin (in.);

d = diameter of core pin (in.);

E = modulus of elasticity($= 30 \times 10^6$, for steel).

When an unsupported core pin is fed at the base, as in Fig. 7.37, the actual deflection may be estimated by considering the core as cantilever with a uniform load, which is again considered as being set up by the

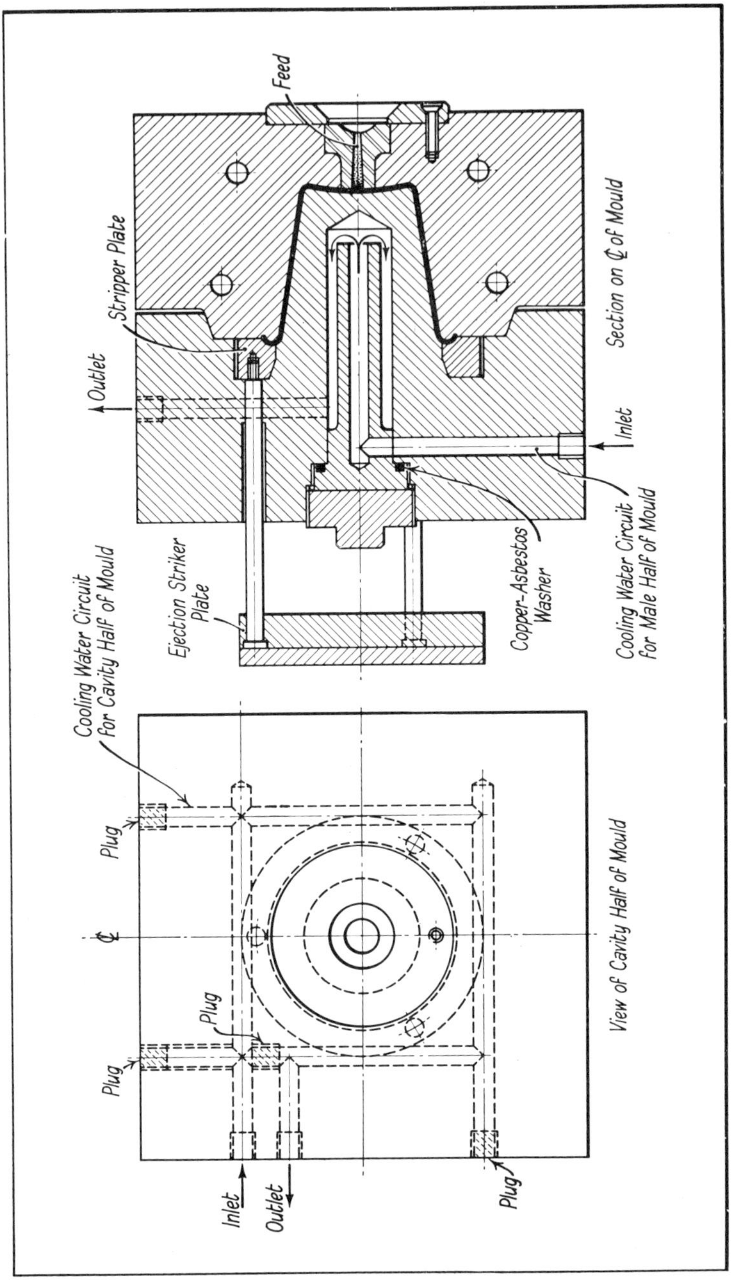

Fig. 7.40. Thin-walled Moulding.

leading material. Then the maximum deflection at the end of the core is given by

$$D = WL^3/8EI.$$

If the core pin is round

$$D = 64WL^3/8E\pi d^4$$

in which the value of W is equal to $900Ld$.

With a supported core pin similar to those of Figs. 7.38 and 7.39, the deflection can be approximately estimated by considering the core as a beam fixed at one end and supported at the other, with a uniform load due to the leading material. The deflection set up when the feed is at the fixed end of the core, as shown in Fig. 7.38, will normally be somewhat less than that where the feed is at the supported end, Fig. 7.39, but for the purpose of estimating, both are treated alike.

The maximum deflection is given by

$D = WL^3/185EI$, and if the core pin is round $I = \pi d^4/64$, so that $D = 64WL^3/185E\pi d^4$.

Fig. 7.40 illustrates the rigid mounting of a core in a mould for a thin-walled moulding. With this type of component, any slight errors in the thin-wall section (often between 0.015 and 0.04 in.) can cause failure to fill the cavity. To minimize any danger of core deflection or misalignment, it is turned solid with the backplate, and is located relative to the female by means of an integral conical location. The use of a tapered-wall section as shown in Fig. 7.36 is also desirable on thin-wall mouldings of this type. An accuracy in concentricity between the male and female of 0.001 in. is essential for success in this type of moulding.

If the core has an asymmetrical form such as shown in Fig. 7.41, the correct location for the gate to ensure even filling of the cavity is difficult to determine, and it is almost impossible to prevent deflection of the core.

The use of blades in the mould as shown in Fig. 7.42 again raises problems of core deflection. Here, however, the flow of moulding material filling the cavity often occurs first on one face of the core, which is therefore subjected to heavy side pressure. In such cases, the height of the core is often confined to three times the thickness.

Mould Cooling. The general aspects of mould cooling were discussed in Chapter II, page 46, including the theoretical calculations for the

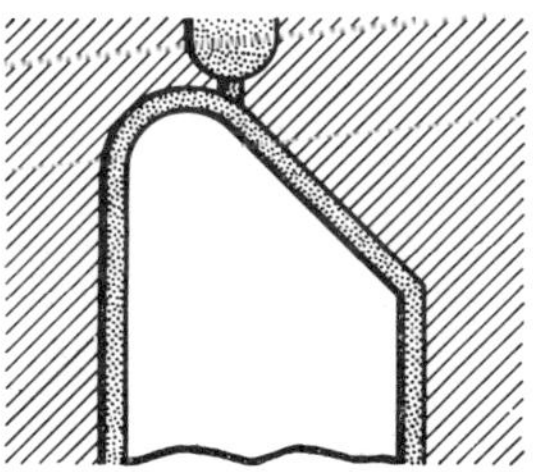

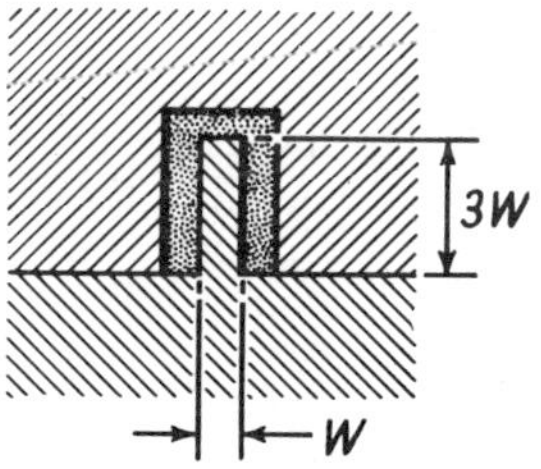

Fig. 7.41. [*Left*] Asymmetrical Core Pin. Fig. 7.42. [*Right*] Blade Proportions.

quantity of water required to remove the heat from the mould. In general, the types of water-cooling arrangements within the mould include the following.

1. Open Channel: (a) Bore type with cooling holes drilled through the mould and backplates;

(b) open recesses with grooves cut into the face of mould components and closed at the top by the fit of the other mould components.

2. Pipe: (a) Pipe used within the tool where it is required to cross the mould joint lines;

(b) pipe set in open channel and recesses and secured to the walls by setting in a low-melting-point alloy.

A typical water connection to a mould is illustrated in Fig. 7.43. Here the tool is cooled by means of bored holes, the connection to the mould being made by means of a length of copper tube. The flexible rubber hose from the press cooling valves is joined to this extension when the tool is mounted in the machine. Occasionally, due to restrictions on space, a bore smaller than $\frac{3}{8}$ in. diameter may be necessary in the tool. This should

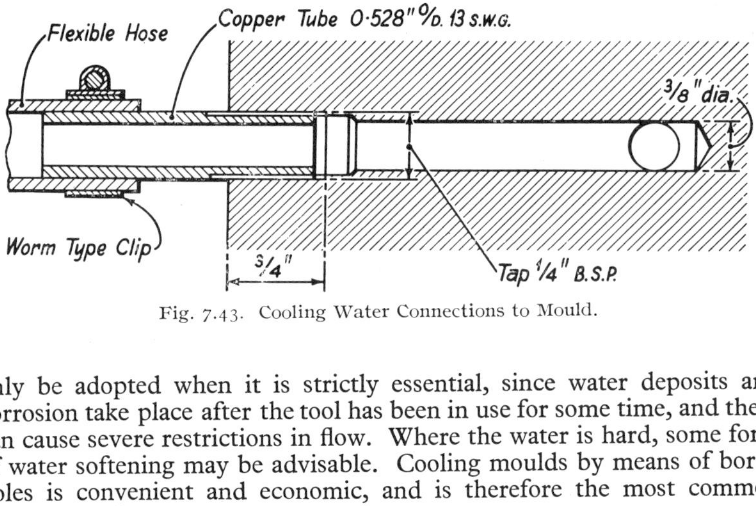

Fig. 7.43. Cooling Water Connections to Mould.

only be adopted when it is strictly essential, since water deposits and corrosion take place after the tool has been in use for some time, and these can cause severe restrictions in flow. Where the water is hard, some form of water softening may be advisable. Cooling moulds by means of bored holes is convenient and economic, and is therefore the most common method employed.

Where possible cooling holes should not be bored closer than 1 in. to the moulding face. Severe local cooling takes place around the hole, and if the hole is close to the moulding face, local cold spots will be produced. This can cause restrictions in the flow of moulding material, and hence bad surface marking. This problem is very common in female tools of round shape where drilled cooling holes are placed around the periphery in a square configuration. This leads to a series of four cold spots where the cooling is closest to the cavity wall. In such serious cases of unevenness of cooling, the minimum distance should be increased to 1.5 in.

Another consideration in keeping the cooling a sufficient distance from the mould face is the danger of tool collapse under cavity pressure.

Fig. 7.44 illustrates the use of open-channel cooling. This method gives a greater area of steel in contact with the water than is possible with bored holes. In many cases, also, it is possible to design the recesses so that they follow the contour of the cavity, thus improving the effectiveness and uniformity of cooling. This technique increases the cost of the tool, but this is more than compensated for by the better mould-temperature control, and the quality of the mouldings obtained. It is essential to ensure that the cooling water is completely sealed, particularly from the cavity. Leakage would cause serious stoppage of production, since the whole tool may have to be taken off the machine and dismantled to repair the leak. Because of this, cheap methods of sealing should not be used. Rubber seals, i.e. O-rings, provide the best method, but the seal recesses

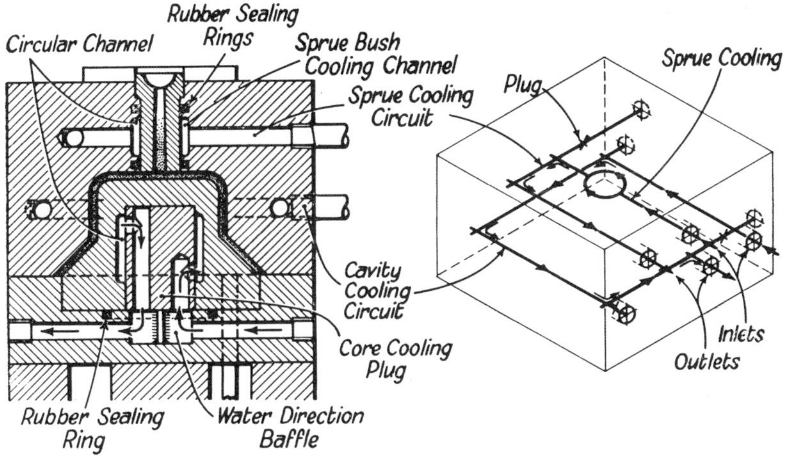

Fig. 7.44. Mould Cooling: Water Circuit for Female Tool.

must be accurately machined to the manufacturer's recommendations, and the opposite faces should be free of machining marks. Open-channel cooling is also shown in Fig. 7.46.

To obviate the necessity to seal all joint faces, copper tubes may be employed, the connection between the mould block and tube being brazed.

In Figs. 2.11 and 2.12, pages 57 and 58, examples are shown of coil-type copper tubing used for cooling, this being sealed into position by means of low-melting-point alloy. While this method is widely used, its cooling efficiency is not always very high. It is difficult to get a complete bond between the low-melting-point alloy and the tube, and the mould steel, hence there may be poor heat transfer at these faces.

In many applications, it is often convenient to cool only the backplates, especially if the mould form is comparatively shallow. However, it should be borne in mind that any small air gap existing between the tool and back-plate will act as an effective insulator; hence either very intimate contact must be ensured across these faces, or else a high temperature gradient must be maintained by the use of ample cooling water at a fairly low temperature.

General points to be observed in all cooling circuits are:

(1) The aim must be to provide a uniform temperature gradient across the mould faces without local hot or cold spots.

(2) The initial water inlet should be as near as possible to the sprue vicinity, since this is an area of maximum heat, and the water circulated away in stages from this point. This sprue cooling is illustrated in Fig. 7.44.

(3) The water connections on most presses are located on the rear side, and it is important that the designed mould inlet and outlet of the water circuits are located adjacent to this part of the machine. This prevents the water connections from obstructing the operator.

(4) Multiple cooling circuits should be connected in parallel and not in series. Similarly, independent control of temperature between male and female tools is imperative.

As has been mentioned, the sprue is a point of high heat and requires special attention in cooling. This is due to the constant contact of the hot injection cylinder nozzle, to the passage of plasticised material, and also to the presence of the hotter melt at the end of the injection stroke. This latter is important with large-diameter sprues and materials of high heat content such as polythene. In such cases, the direct cooling of the sprue

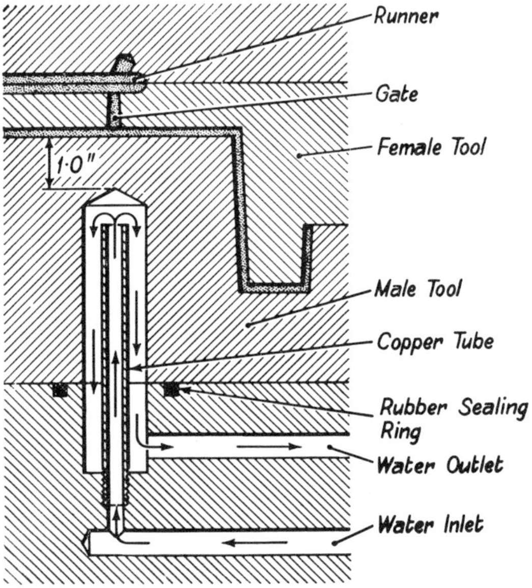

Fig. 7.45. Water Circulator in Male Tool.

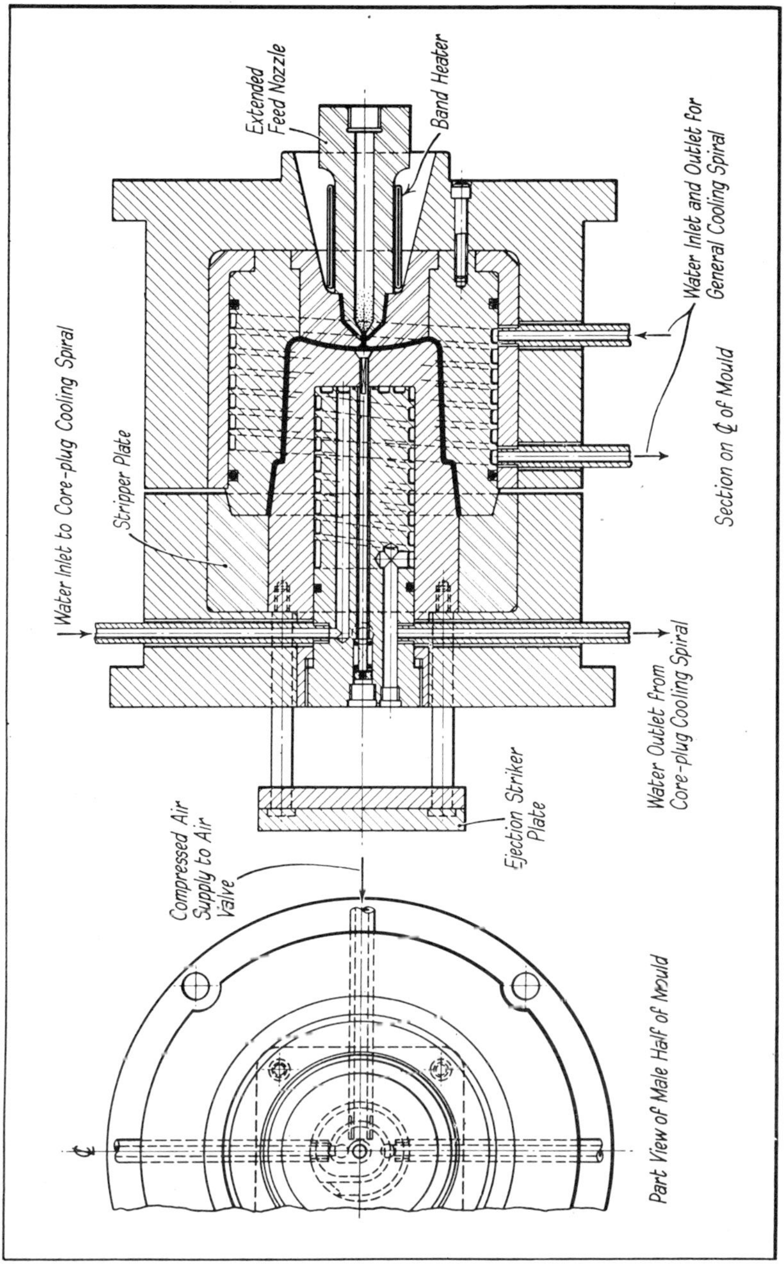

Fig. 7.46. Helical Cooling Arrangement on Core.

bush, as indicated in Fig. 7.44, is desirable. The male tool also is an area that receives maximum heat. This is due to the following facts:

(i) The end of the male tool may be directly opposite the sprue and hence receives the hottest melt at the end of the injection stroke.

(ii) Where restricted gates are used, the energy expended to force the material through is dissipated in heat which is transferred to the area around the gate.

(iii) The moulding remains in intimate contact with the male tool due to shrinkage, whilst the female tool is insulated by the air gap formed as the moulding shrinks away. Hence the male tool receives the major portion of the heat of the moulding, and will thus have to be provided with a large amount of cooling.

(iv) The male tool adjacent to the moulding is often of smaller mass than the female. As a result, its temperature will rise more quickly.

(v) High-efficiency cooling is essential on moulds intended for fast cycling.

To offset the excess heat at the gate, it is desirable to provide cooling in the male immediately opposite this position, in the form of free water, since this gives the maximum efficiency. This type of cooling is shown in Fig. 7.45. Another type, suitable for deep cores, is illustrated in Fig. 7.46. This comprises a plug with grooves cut across its face in a spiral. The water enters the centre of the spiral, from the bore up the centre of the plug, and is then led down the side of the plug through the helical groove cut on its periphery. The use of a central air ejector through the cooling core will be noted. A further design of a cooled core is shown in Fig. 7.40.

With materials liable to give distortion on moulding such as polythenes, one of the measures necessary to minimize this effect is adequate cooling arrangements. Hence the bubbler or spiral cooling methods shown in Figs. 7.45 and 7.46* are necessary to extract the considerable heat content of these materials and to ensure a temperature gradient which is constant across the tool, with cold water entering at the sprue and being directed to the outer regions, whilst it is being warmed. The application of water cooling directly opposite the sprue also lessens the sinking of the moulding face opposite the sprue, the sinking occurring more within the sprue itself. This reduction in sinking at the front face reduces, in turn, the need for packing pressure, and hence improves the strength of the moulding.

In the case of moulds with deep cavities, arrangements for cooling the core can be as illustrated with channel or bore cooling in the main portions of the tool. In the case of a shallow moulding, a spiral can be cut across the face of the backplate to correspond with the mould area, the spiral being sealed between dieplate and backplate with an O-ring.

---

*Various special detail features of component and tool design are advised by the manufacturers of each plastic material and these should be studied. An example of a domed base to the component, as recommended for materials such as high-density polythene, is shown, for instance, in Fig. 7.46.

CHAPTER VIII

NOZZLES, RUNNERS AND GATES

Standard Nozzles. There are two general types of standard nozzle:

(a) Spherical nose—as shown in Fig. 8.1;

(b) Flat nose—as shown in Fig. 8.2.

The overall form and dimensions of the nozzle may vary with the size and type of injection machine, but the style of seating is either spherical or flat, the former being the most common. The spherical nose itself must be accurately machined to the specified dimension, in order that a satisfactory seal will be obtained on all the moulds. Similarly also, the spherical recess in the sprue bush must be accurately machined, but the radius of this spherical recess is usually made $\frac{1}{32}$ in. greater than the radius of the nozzle nose. This allows positive seating of the nozzle in the sprue bush around the orifice, and prevents leakage.

The nozzle orifice size varies with the size of the mould and the material to be moulded, but is generally in the range of $\frac{1}{8}$ in. to $\frac{5}{16}$ in. Too small a nozzle and sprue-bush orifice should not be employed, since these tend to increase the length of cycle. A small nozzle orifice in conjunction with a

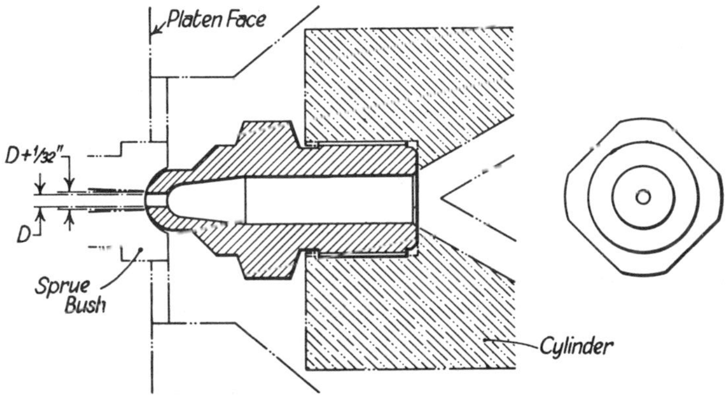

Fig. 8.1. Standard Spherical-nosed Nozzle.

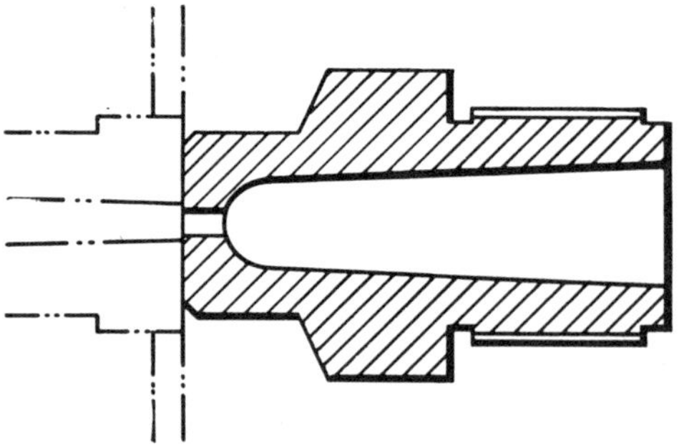

Fig. 8.2. Standard Flat-nosed Nozzle.

much larger sprue bush orifice can lead to entrapped air, because of jetting and folding of the material.

Machines having a screw pre-plasticiser that is also used as the injection ram are often provided with nozzles having sliding valves or needle valves. Here, the nozzles used are generally of the spherical-nosed type.

The nozzle selected to feed a given mould normally has a slightly smaller orifice than the sprue bush. One disadvantage of the spherical type of nozzle is that the pressure maintaining the nozzle against the sprue bush can cause hobbing with possible partial closing of the orifice. To minimize this, nozzle and sprue bush should be of hardened and tempered steel.

The use of the flat nozzle, which seats on a corresponding plane surface on the sprue bush, avoids the possibility of hobbing. However, this type allows a greater unwanted heat transfer from nozzle to mould, and also can cause greater leakage at the nozzle because of the lower sealing pressure and lack of a positive sealing line.

Many present-day nozzles have a bore which tapers down to the orifice, as shown in Fig. 8.2, or a parallel bore which tapers to the opening as indicated in Fig. 8.1. However, the use of a parallel bore changing with a sharp radius at the orifice and with a short land length on the orifice itself can effect some reduction in pressure loss

Extension Nozzles. The standard nozzle is designed to operate with a sprue bush positioned at the fixed platen location hole as shown in the previous illustrations, and Fig. 8.3, and is provided with limited axial adjustment at this point. In some tools, however, the gate or runner is some distance forward of the front platen face; a common instance is on moulds with the ejectors on the fixed side and where the housing of the ejection gear requires the sprue side of the mould to be deeper than normal. With a standard nozzle, this would require a sprue bush that is abnormally long, with consequent material wastage, loss of machine plasticising capacity, and material temperature and pressure losses along the sprue. To overcome this difficulty, extension nozzles are employed, similar to that illustrated in Fig. 8.3. Here the sprue bush is made of normal length, but is set back in the body of the tool, the nozzle being

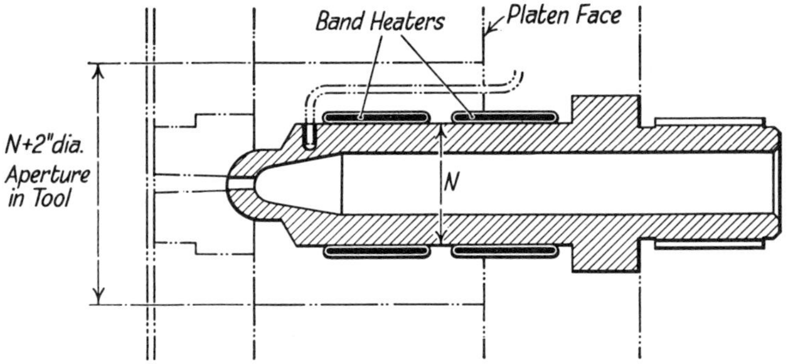

Fig. 8.3. Extension Nozzle.

extended correspondingly. With this type of nozzle, the provision of external heater bands is essential, since otherwise, the material is liable to freeze. Means of controlling the heating applied to the extension nozzle is necessary, this often taking the form of some type of energy regulator, but the use of a thermocouple and temperature controller is preferred.

To provide space for the nozzle heaters and their leads, the hole in the tool must be of adequate size—2 in. on nozzle diameter if possible. This air gap surrounding the heaters is normally sufficient to prevent excessive transfer of heat to the tool, but if not, asbestos insulation around the nozzle can be used.

Nozzles for Nylon. The very low viscosity of nylon at moulding temperature presents problems, since it is difficult to prevent the material dribbling from the nozzle between moulding cycles.

Numerous mechanically shut-off nozzles have been designed to prevent this, but many of them have had only limited success because of the adverse effect of the high moulding temperature on the mechanisms and the ease with which nylon is able to penetrate even small clearances.

One type of nozzle currently employed, shown in Fig. 8.4, uses a fine orifice with a reverse taper. Prevention of dribble is achieved by accurate temperature control of the nozzle heater, the temperature being maintained at a point just below that at which the material would dribble through the

Fig. 8.4 Nozzle for Nylon.

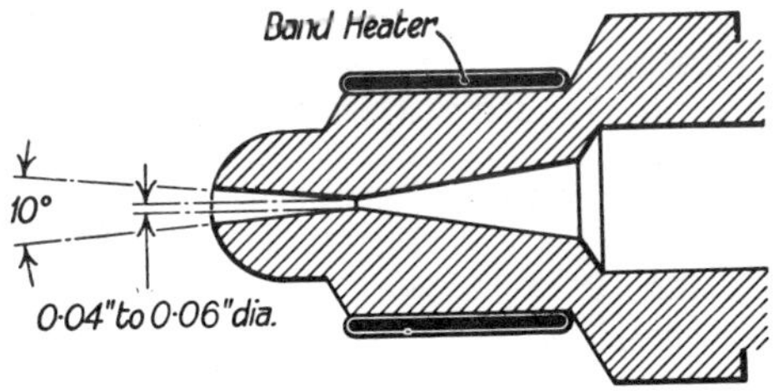

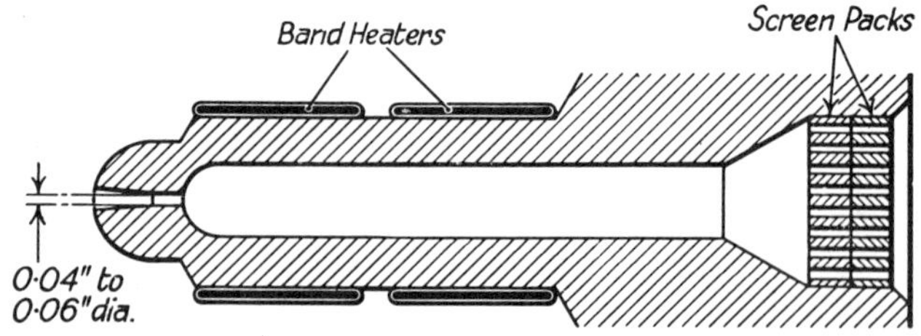

Fig. 8.5. Nylon Extension Nozzle with Screen Packs.

choke restriction. Should the material at the tip of the nozzle freeze, then its extraction is facilitated by the reverse taper.

Another approach is shown by the nylon extension nozzle depicted in Fig. 8.5. Here, in addition to the reverse-taper orifice, a series of perforated plates is incorporated, these plates being from $\frac{3}{4}$ in. to $\frac{3}{8}$ in. thick and drilled with a large number of holes of about $\frac{1}{32}$ in. diameter. These screen packs not only prevent the passage of unplasticised material, but also, by providing an internal restriction, cause sufficient pressure drop across themselves to reduce the tendency of the material to feed forward and dribble when the ram is in the rear position and the material is consequently under the influence of the weight of the material in the heating cylinder.

Various modifications of these types and alternative designs are also suggested by the different manufacturers of nylon material.

Italian Nozzle. This is a special type of nozzle, chiefly used for pin feeding components and is illustrated in Fig. 8.6. The nose of the nozzle seats in a conical sprue bush to give a line-contact seal. On injection, the plastic material first fills the space between the nozzle end and the narrow conical end of the sprue bush. Continued injection pressure then forces the material through the centre of this mass (which insulates this following material from the cold tool) to the gate, and then fills the cavity. At the end of the cooling cycle, the sprue-break mechanism is operated and the nozzle moves slightly away from the tool. The sprue bush maintains contact with the nozzle, being slidable in the tool, and moving back at the same time under the influence of the compression springs. This separation reduces excessive heat losses from sprue bush to cavity face. The plastic mass remains in the sprue bush, subsequent injections taking place through the molten interior. If the cycle is interrupted for any period this mass will, of course, solidify. In this case the frozen slug must be removed before recommencing moulding. Care must be taken that the maximum diameter of the conical bush pre-chamber in contact with the nozzle is not too great. If excessive, the pressure of the molten material at injection pressure acting on the face of the nozzle, may force back the injection carriage and cause leakage. On presses with hydraulic injection carriage

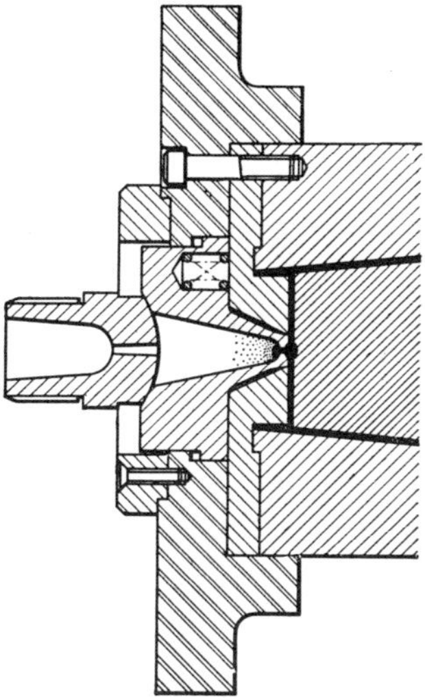

Fig. 8.6. Direct Injection Sprue—Italian Type.

movement, the maximum pre-chamber diameter will be controlled by the ratio:

$$\left[\begin{array}{c}\text{Projected area of}\\ \text{nozzle face in contact}\\ \text{with pre-chamber}\end{array}\right]\times\left[\begin{array}{c}\text{Material}\\ \text{pressure}\end{array}\right]=\left[\begin{array}{c}\text{Area of carri-}\\ \text{age hydraulic}\\ \text{cylinder}\end{array}\right]\times\left[\begin{array}{c}\text{Hydraulic}\\ \text{line}\\ \text{pressure}\end{array}\right]$$

Sprue Bush. As previously intimated, the hardened sprue bush provides the seating, either flat or spherical, for the injection-cylinder

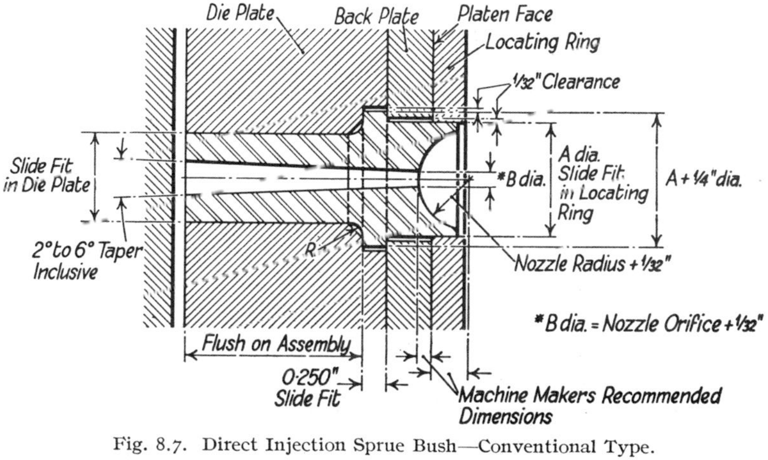

Fig. 8.7. Direct Injection Sprue Bush—Conventional Type.

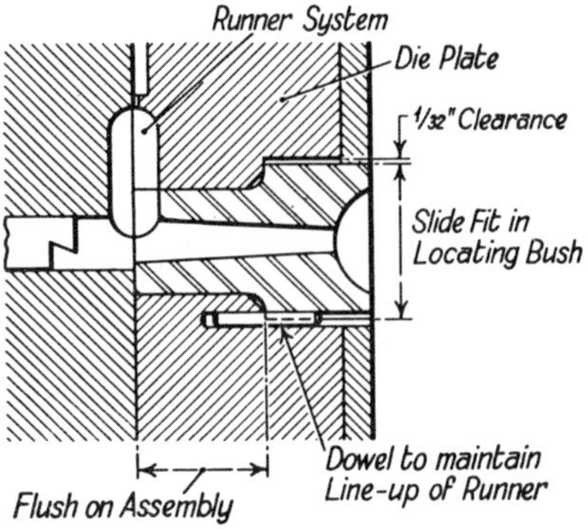

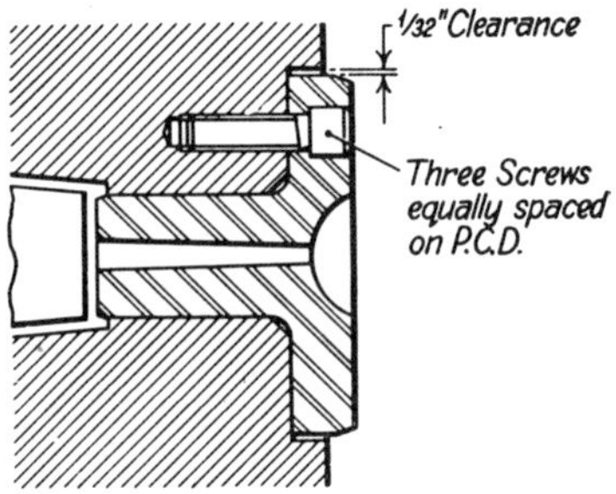

Fig. 8.8. [*Left*] Non-retained Sprue Bush. Fig. 8.9. [*Right*] Screw-retained Sprue Bush.

nozzle and conducts the hot moulding material from the injection cylinder to the mould cavity, either direct or through runners.

The bush is provided with a shoulder to prevent its being displaced into the tool when it sustains the end pressure from the injection cylinder. Conversely, the sprue bush must be restrained from movement in the opposite direction due to reaction from the cavity pressure over the exposed face. See Figs. 8.7 and 8.9. The orifice of the bush that forms the sprue stalk is nearly always round, polished and tapered, to facilitate extraction of the sprue, the amount of taper varying between the inclusive angles of 2 and 6 deg. The diameter of the small end varies according to the size of moulding and press, usually from $\frac{1}{8}$ in. diameter on the smallest to $\frac{9}{16}$ in. on the largest. In all cases, the diameter should be made slightly larger than the nozzle orifice diameter. Within the basic functional requirements of the sprue bush, its actual dimensions can vary considerably to suit the requirements of specific mould design, and a number of typical examples are shown in Figs. 8.7, 8.8 and 8.9.

Almost all injection moulds include a sprue bush in their construction. This provides the locally hardened seating for the nozzle, facilitates machining of the seating and tapered bore, and allows changes in their form or dimensions with ease. Sprue water-cooling arrangements are also facilitated.

With direct injection feed, the sprue bush should be provided with a collar to prevent its being pushed back by cavity pressure over its exposed end.

Cold Well and Sprue Hook. At those periods during the press cycle when material is not being injected, the material in the extreme end of the nozzle is becoming cooler. It is necessary to prevent this cold material from entering the runner system or cavity at the next injection stroke, since in the former it will restrict flow, and in the latter it might cause scab marks on the surface of the moulding. To prevent such occurrences, a cold well is provided at the mould parting face end of the sprue.

Positive ejection of the sprue from its bush is desirable, especially when a runner system is used, since any machining marks left in the orifice, which are not polished out, will tend to hold the sprue in the bush, leaving the ejection of the sprue to the runner system. This is very uncertain and can cause cracking or distortion of the moulding.

The difficulty is avoided by providing positive means of extracting the sprue by means of a sprue-puller hook placed in the tool opposite the end. This portion fills at each cycle, with the first plastics material to be injected, and thus serves as a cold well for the initial cold material in the nozzle as well as a sprue hook. For this reason, the diameter of the cold-well sprue hook must not be less than the diameter at the large end of the sprue bush, in order to take this cold charge. After the tool has opened and the moulding is ejected, the sprue and sprue hook are ejected at the same time by means of the sprue ejector pin.

Various types of combined cold well and sprue hook are used and the three principal types are

1. Undercut ring, as illustrated in Fig. 8.10;
2. Z-hook, as shown in Fig. 8.11; and
3. Stripper-plate hook, as depicted in Fig. 8.12.

In the illustrations mentioned, typical dimensions for all but the very small or very large tools are shown.

The first type usually works well with most materials except nylon or high-density polythene. On ejection of the sprue, the undercut ring is sheared off and remains in the tool, the fresh material welding to it during the next cycle. On some more flexible materials, the undercut ring may be sprung out with the hook. Typical proportions are shown in Fig. 8.10.

The Z-hook is very positive in action, and can be used with high-density polythene and nylon, since no shearing action or rewelding action is involved. The main disadvantage is a tendency, sometimes, for the runner to continue to hang on to the hook after it is finally ejected, instead of dropping away. This hook can also be seen in the illustration of the two-impression tool, Fig. 8.29. A further variety of arrangement is sometimes used with non-rigid materials. This employs a reverse taper on the cold-well bore with a plain end in place of the Z-hook, the sprue being pulled out by means of the undercut so formed.

With the stripper-plate hook, a stationary pin is employed, passing up through the stripper plate. This pin has an undercut button head around which the material fills and acts as a key to extract the sprue. On final ejection, the stripper plates move forward relative to the stationary pin, shearing off, or springing out, the material in the undercut. It is important that the pin is placed in the cold well with the head below the runner level, in order not to offer any obstruction to the flow of material.

Another kind of sprue hook, used where ejection is not necessary, is shown in Fig. 8.13. This is produced by plunging an angle milling cutter to produce the undercut hook shown. After the tool has opened and the moulding has been extracted, the sprue is removed from the tool

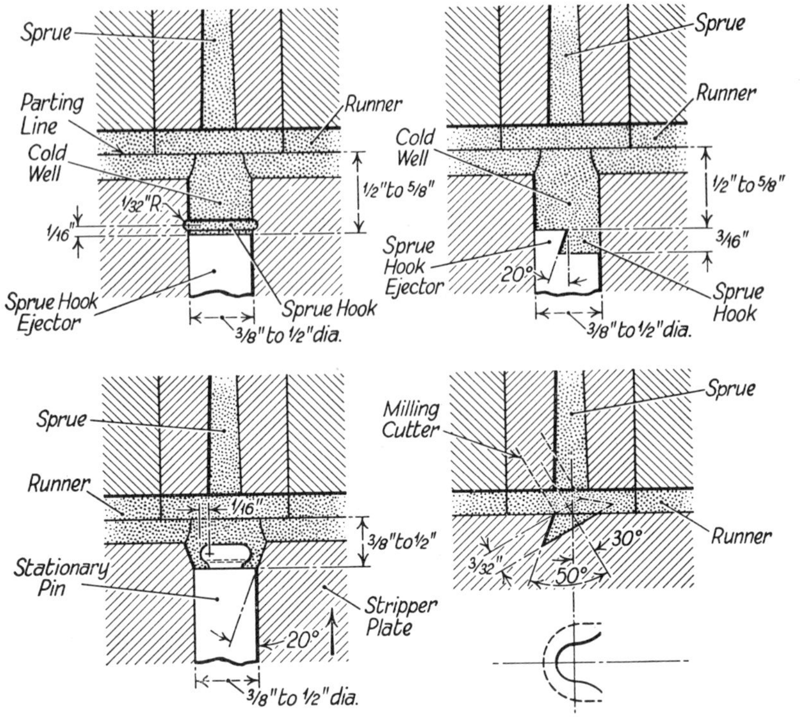

Fig. 8.10. [*Top Left*] Undercut Ring Sprue Hook. Fig. 8.11. [*Top Right*] Z-type Sprue Hook. Fig. 8.12. [*Bottom Left*] Strip-plate Sprue Hook. Fig. 8.13. [*Bottom Right*] Jag Sprue Hook.

by hand, moving it sideways to free the hook. This method is often used on disc- or diaphragm-fed flat mouldings where the component is left free on the head of the male tool.

Another common use for this type is to extract the sprue on three-plate tools which utilise a runner system to multiple impressions or multiple gates, the runner system being extracted manually. This is illustrated in Fig. 5.10A, page 127.

Where the sprue gates directly into the moulding, it is occasional but not general practice to include a cold well or sprue hook to remain on the moulding, but these are very seldom acceptable visually.

Hooks for Pin Feeds. For centre-fed restricted-gate mouldings on three-plate tools, means must be provided to retain the runner in the fixed plate on the injection side when the third plate opens, in order to extract the sprue to the pin gate. In many moulds, this is achieved by using an undercut-ring or Z-hook sprue puller, placed in the fixed gate opposite the sprue with normal ejection or stripper plate (Fig. 8.14B). However, since these types normally require positive ejection, it does mean that

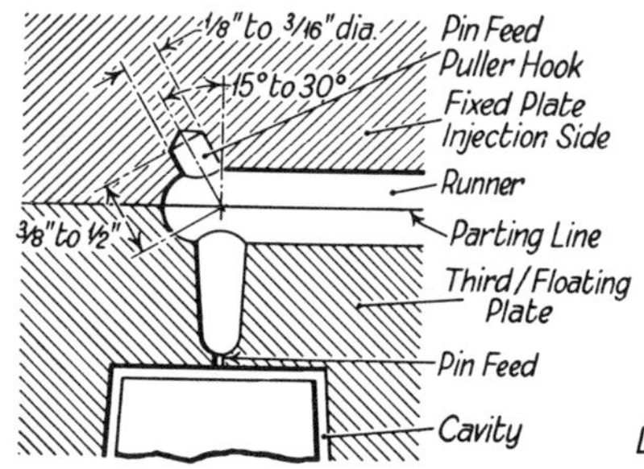

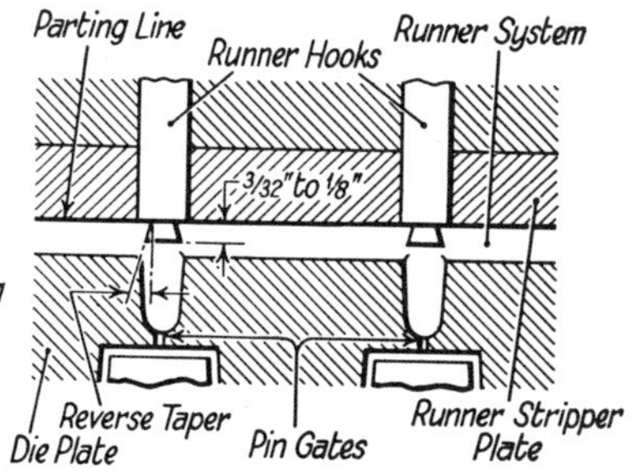

Fig. 8.14A. [*Left*] Puller Hook for Offset Pin Feed. Fig. 8.14B. [*Right*] Stripper-plate Ejection of Runner System for Three-plate Tool.

ejectors have to be placed on both sides of the tool, one set to eject the sprue hooks, and the other to eject the moulding. An alternative is shown in Fig. 8.14A, where the hook consists of a short blind hole drilled at an angle to the runner. This method is simple and effective, and the runner system requires only a small manual effort to pull it free from between the floating plate and the fixed plate, when the tool is open. It is restricted, however, to cases where the pin feed is offset to the main sprue, e.g. multiple-impression or multi-gate moulds, and is generally applied to low-density polythene. Its application to a three-plate tool is shown in Fig. 5.10A, page 127.

With single-impression tools where the feed is in line with the main sprue, the method shown in Fig. 8.15 can be used. This involves the use of a further small plate positioned between the floating plate and the

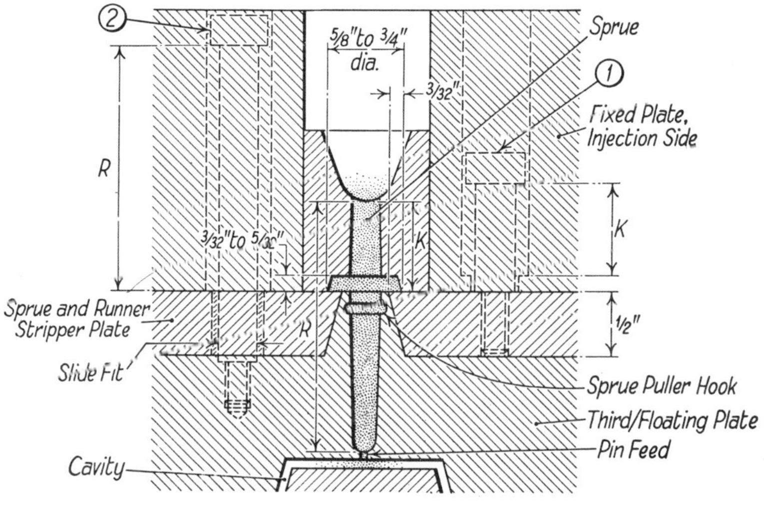

Fig. 8.15. Ejection of Sprue and Runner on Direct Pin Feed.

injection-side fixed plate. It is held and guided by stripper bolts the lengths of which are adjusted to provide sufficient movement to eject their respective feeds.

The operating length *K* of the bolt (1) must be greater than the length of the main sprue. This is often difficult to achieve due to limitations on tool-opening space, unless a short sprue and extension nozzle are employed.

The opening length *R* of the bolt (2) must be greater than that of the feed to the pin gate, and its full length is related to the total opening stroke of the mould.

In operation, as the tool commences to open, the main sprue is pulled from the sprue bush by the undercut-ring puller hook. As the mould continues to open, the stripper plate moves away with the third plate until it is prevented when the head of the bolt (1) strikes the end of its recess, this occurring when the main sprue has been extracted. Further opening movement of the tool then separates the stripper plate and third plate, causing the pin feed to be extracted from the floating plate. The tool continues to open until the head of the bolt (2) is just in contact with the stripper plate. The aperture in the stripper plate through which the sprue and runner pass is conveniently made a slot, so that, when ejection is complete, they can drop freely downwards out of the tool.

A similar device to effect sprue-and-runner ejection on a two-impression three-plate mould is shown in Fig. 5.8, page 124.

Runner Hooks. With runners cut into one half of the tool only, they should be placed on the ejection side, if at all possible, thereby assisting the moulding to remain on that side of the tool from which it is to be ejected. However, in certain cases, the runner recess is cut into the fixed (injection) side of the tool and extends for some length from the sprue. If it is intended that the feed system is to move away with the moulding on tool opening, it is often desirable to provide additional hooks to extract the runner itself from its recess. These hooks are similar to the sprue hook, and are generally placed adjacent to the gate, in order to prevent the pull-off of the runner shattering the gate or distorting the moulding. The undercut-ring type is often used as shown in Fig. 5.22, page 147. This type is employed also on the moulding shown in Fig. 8.36, although the undercut, having been sheared off on ejection, cannot be seen. A similar set of conditions can arise with full-round runners cut into both plates.

Runner Profiles. The purpose of the runner system is to convey the hot moulding material from the sprue to the gate with minimum loss in pressure. At the same time, the weight of the material in the runners must be kept to a minimum, since this material is eventually cut off from the final moulding, except in hot-runner tools, and represents a waste value.

In order to reduce pressure loss, it is necessary to ensure the lowest possible value of the ratio

(surface area) : (cross-sectional area).

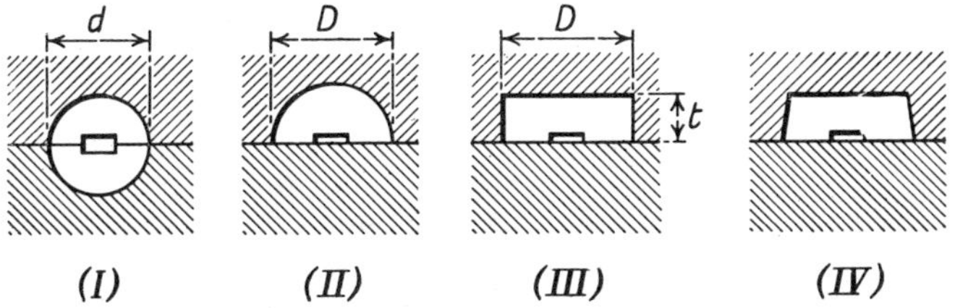

Fig. 8.16. Types of Runner Profile: (*I*) Full Round; (*II*) Half Round; (*III*) Rectangular; (*IV*) Trapezoidal.

The surface area is a direct function of the runner circumference or perimeter, and it is this surface which causes loss by heat transfer from the moulding material, this leading, in turn, to pressure losses, since the viscosity of the melt is increased by loss of temperature. In Fig. 8.16 are illustrated four basic runner profiles, for which the surface area/sectional area ratios may be obtained as follows:

(I) *Full-round Runner*

Surface area $=\pi d$; Sectional area $=\pi d/4$;

Ratio $=4/d$ .. .. (1)

(II) *Half-round Runner*

Surface area $=D(\pi/2+1)$; Sectional area $=\pi D^2/8$;

Ratio $=6.54/D$ .. .. (2)

If the cross-sectional area is to be maintained the same as that in the full-round runner, then $D=\sqrt{2}\times d$, and

Ratio $=4.62/d$ .. .. (3)

(III) *Rectangular Runner*

In the rectangular runner, assume the width D to be the same as the diameter D of the half-round runner (II), and the cross-sectional area to be the same, then

Sectional area $=Dt=\pi D^2/8$; i.e. $t=\pi D/8$.

Hence, Surface area $=2D(1+\pi/8)$, and

Ratio $=7.09/D$ .. .. (4)

But since $D=\sqrt{2}\times d$,

Ratio $=5.01/d$ .. .. (5)

From the comparison of the ratios in formulae (1), (2) and (3), based on identical cross-sectional areas, the superiority of the full-round runner is evident. In practice, however, due to the reduced amount of machining effected by cutting into one plate only, half-round, rectangular or trapezoidal runners are often used in spite of their reduced efficiency. Of these, trapezoidal are preferred to rectangular ones, because of their greater ease of ejection. Proportions commonly adopted for rectangular or trapezoidal runners correspond to

Depth $=(\frac{2}{3}$ to $\frac{4}{5})\times$ Width.

The slope of the sides is usually from 5 to 15 deg. per side.

In general, the full diameter of the runner system should not be less than three times the thickness of the thickest section of the component being moulded. In practice, this varies between

$\frac{1}{8}$ in. (minimum), and $\frac{3}{8}$ in. (maximum).

Runners should not be made below this minimum, but on thick-section moulding, it may be necessary to increase the runner section above the maximum stated, in order to maintain sufficient pressure in the cavity to obviate sinking.

EXAMPLE. Thickest portion of moulding $=\frac{1}{16}$ in. Adopt equal cross-sectional area in each runner.

(I) Full-round runner diameter $=d=3\times\frac{1}{16}=\frac{3}{16}$ in.

(II) Half-round runner diameter $=D$
based on formula (2), $D=\sqrt{2}\times\frac{3}{16}=0.265$ in., say $\frac{1}{4}$ in.

(III) Rectangular runner:
Width D based on formula (2) $=\sqrt{2}\times\frac{3}{16}=0.265$ in., say $\frac{1}{4}$ in.
Depth t based on formula (5) $=\pi/8\times\frac{1}{4}=0.098$ in., say $\frac{3}{32}$ in.

The design criterion of all runners is that they must always be of sufficient size to ensure that they do not solidify before the cavity is full and packing pressure has been applied to compensate for the shrinkage in the moulding. Nor must they be so small that the incoming material chills before it reaches the gate and causes surface marks on the moulding. On the other hand, it must not be too large so that the cooling cycle has to be increased to allow the runner to set sufficiently before it can be ejected.

An exception to this latter restriction occurs on mouldings where optical clarity is of prime importance, especially with acrylic materials. Here the runner system may be large, and a long cycle is consequently essential. High-viscosity plastics such as this require large runners to compensate for the increased flow resistance, and this infers the shortest possible length to reduce material wastage. For this reason, multi-impression tools of small components with their long runner systems are often uneconomical in such materials.

Runner Length. In order to reduce heat and pressure losses in the system, it is essential to keep the runners as short as possible. This conservation of available pressure assists in lowering the temperature required in the material, and therefore reduces the stresses in machine and tool.

On multiple-impression tools with main and side runners, the primary runners should carry on beyond the junction of the last side runner, in order to provide a cold well, for trapping the cooler front portion of the incoming moulding material—see Fig. 8.17.

Where several secondary runners branch off from the primary runner, the cross-sectional area of the latter should not be less than the total of the areas of the secondary runners that it feeds. Runners on multiple-impression tools should be kept as short and direct as possible. It is frequent practice to lead off to individual cavities from the main runner as shown in Fig. 5.4, but there are disadvantages, since a more even filling may be obtained by arranging the runner pattern so that the length of runner from each cavity to the sprue is the same, as indicated in Fig. 8.38.

Fig. 8.17. Cold Moulding Material Trap at Intersection of Runner.

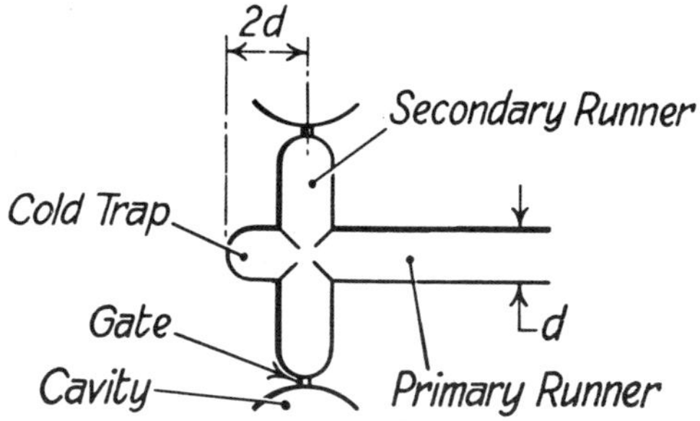

Pressure Drop in Runner Systems. The equation for the quantity of viscous fluid flowing though a circular channel is given by

$Q = \pi d^4 \Delta p / 2^7 \mu L$ (for round runners) .. .. (6)

$Q = wt^3 \Delta p / 2^5 \mu L$ (for rectangular runners) .. .. .. (7)

where: d = diameter of runner;
Δp = pressure drop in runner system;
μ = coefficient of viscosity;
L = length of runner or gate;
w = width of runner;
t = depth of runner or gate;
Q = quantity of material flowing through the runner per second.

Equations (6) and (7) may be written as

$\Delta p = k_1 L / d^4$ .. .. (6′)

$\Delta p = k_2 L / t^3$ .. .. (7′)

where k_1 and k_2 are constants for a particular runner and material under fixed condition.

With plastics material, the coefficient of viscosity is dependent upon the temperature, and in any injection process this will change because of the heat losses taking place within the runner system. This loss will be a direct function of the length of the runner and the speed of the injection. The heat losses will also tend to change the effective channel diameter d as the frozen skin adjacent to the surface of the runner becomes thicker at locations remote from the sprue.

From the preceding effects, it has been shown that for simplicity the changes in viscosity may be taken as a function of runner length, and the pressure loss in the runner system as a function directly proportional to the 1.25th power of the runner length. Thus

$\Delta p = k_1' L^{1.25}$ or $\Delta p = k_2' L^{1.25}$ .. .. (8)

Assuming that the inner diameter or dimensions are constant throughout the system, then

$k_1' = 2^7 Q \mu / \pi d^4$; and $k_2' = 2^5 Q \mu / wt^3$ .. .. (9)

It is also seen from these equations that the quantity of material flowing is directly proportional to the fourth power of the runner diameter; thus any small increase in this diameter results in a substantial increase in ease of flow.

EXAMPLE. If a runner diameter is increased from $\frac{1}{4}$ in. to $\frac{9}{32}$ in., find the percentage increase in material flow.

The ratio of the two diameters is 9 : 8;
hence the ratio $Q_2/Q_1 = 9^4/8^4 = 6561/4096$, and
the percentage increase in material flow is
$[(6561-4096)/4096] \times 100 = 60$ per cent.

With rectangular runners, the formula is only really closely applicable where the width of the runner w is not less than four times the depth t, but illustrates that the flow of material in these circumstances is proportional to the third power of the depth. Thus again any small increase in this dimension effects a considerable improvement in the flow. In cases where the width is less than $4t$, any increase in flow becomes progressively a function of the width as well as the depth. The formula again shows that the runner should be kept as short as possible, since the pressure drop is proportional to the 1.25th power of the runner length.

While minimum restriction to flow in the feed system should be the objective, a matt finish to the surface of the runners has some advocates. This is to retain the outer cooled skin of the flowing plastics material. This not only prevents it being swept away and replaced by other hot material, which is cooled in turn, but retains the layer which effectively insulates the hot inner material from the steel surfaces of the runner.

Gates. There are a number of forms of gate commonly used, and these are described in the following section. In addition, however, there are numerous forms of special gates adopted to suit particular mouldings. In general, however, these are all variations of the basic forms.

The point of gate location and its type require particular care and thought in the light of each specific moulding. It is good practice normally to feed into the thickest portion of the moulding, so that the material flows progressively from there to the thinnest sections. In addition, however, other flow-pattern factors have to be considered. These include flow and weld lines, gas traps, sinking, as well as considerations of visual acceptance and gate removal.

It is useful to refer to manufacturers' literature for special points on gates for specific moulding materials, since some forms of feed are more suitable than others, according to the flow characteristics of a given material.

Full Gates. The cross-sectional area of full gates is relatively large. Hence all types of thermoplastic material can be fed through this type, whereas not all are suitable for the more restricted types of feed. Full gates are particularly used for:

1. direct-gated (i.e. sprue-fed) mouldings on two-plate tools;
2. the feeding of high-viscosity materials such as acrylic and Kralastic, and often for the stiffer flow grades of polyolefin materials;
3. the gating of large deep-draw or thick and uneven mouldings, where

maximum utilization of pressure is necessary or where early gate freeze-off is not possible due to the need for extended packing pressure to prevent sticking.

This type of gate is usually applied as a direct sprue feed, as a full-edge gate on the outside of a component, or as an edge-gate feeding internally, on components such as frames—see Fig. 5.1, page 114. However, any non-restrictive gate is a full gate and can include disc, fan or ring gates, Full gates give a streamline flow with minimum shrinkage, bubbles and weld marks.

The direct sprue gate is very frequently employed on symmetrical components such as bowls and buckets, and also on rectangular boxes, radio cabinets, trays and doors. It has the advantage of giving a balanced tool design, since many of these components, if they were edge-gates on single-impression moulds, would result in a badly offset tool. With centre-fed moulds also, the direct-sprue gate allows the use of a two-plate mould instead of the three-plate tool necessary with a restricted centre feed. With the direct-sprue gate, however, it is necessary to cut off the sprue separately and, for a reasonable finish, to machine it as well; even then the mark of the gate can generally be clearly seen.

For free-flowing materials and small or medium moulds, relatively small diameters can be used; with stiff materials or large moulds and high injection rates, somewhat larger diameters are generally adopted.

In many cases, no cold well can be provided where the sprue gate enters directly into the moulding. This means that the first cold material that flows from the nozzle can enter the cavity and may cause scabs and surface flow marks, which are particularly objectionable in clear mouldings. Often, however, this is unavoidable and has to be accepted, although the effect is diminished by moulding technique.

One recommended method of eliminating or reducing the first cold slug of material and avoiding this mark is to use a reverse-taper injection nozzle, the nozzle itself being heated to its maximum operating temperature.

Edge gates are, of course, fed by runners. On a single-impression mould, one or several gates may be employed, according to the flow required. On multiple-impression tools, one gate is required for each cavity. The gates are commonly half-round, square or rectangular. The size will vary with the size and shape of the moulding and material, but a full gate is generally not less than the equivalent of $\frac{3}{32} \times \frac{3}{32}$ in. in area.

If the gate cross-sectional area or minimum dimension is too small, it may not be possible to control fully the sinking. If it is too large, the ram-forward time will be prolonged and the cycle time accordingly increased unnecessarily.

An example of a full standard edge gate is shown in Fig. 8.18. This is for a single-impression offset tool producing a circular dished dial in acrylic. The moulding weighs 59 grams and is produced on a four-ounce press. The short runner is of full-round profile connecting with a gate cut in the moving tool. The feed itself is of the reverse-taper type, to eliminate crazing into the moulding on degating, and in this case is of

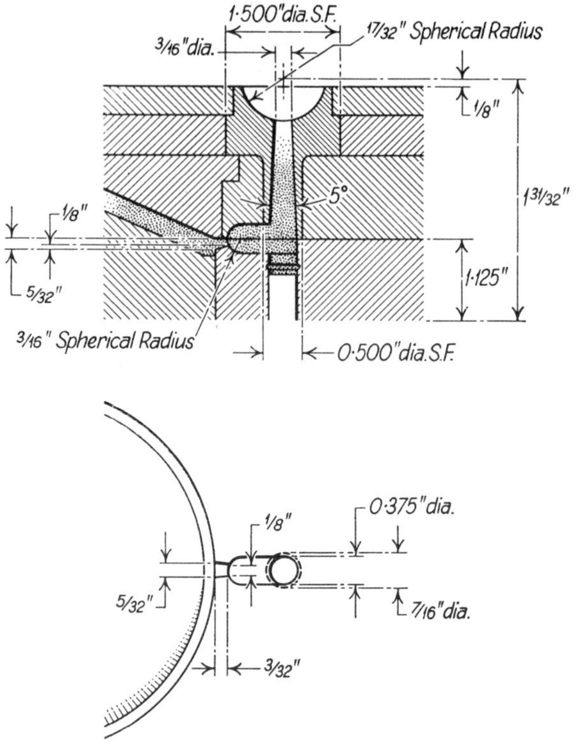

Fig. 8.18. Full-edge Gate for Acrylic Material.

generally square cross-section. The full true radius on the end of the runner as it approaches the gate is recommended, and will be noted. Streamlining the gate approach is not recommended.

Edge gates are fed either on the rim as in Fig. 8.19, or on the face as in

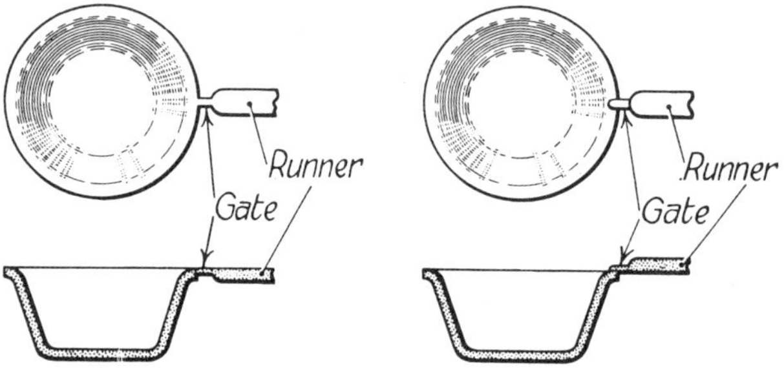

Fig. 8.19. [*Left*] Edge/Rim Gate. Fig. 8.20. [*Right*] Edge/Face Gate.

Fig. 8.20. The choice is usually one of function and appearance, or sometimes the avoidance of jetting. If the component fed on the face is viewed only from the front and the flange does not have to rest flush on another surface, this feed can merely be degated without the necessity of blending it into the moulding. With the rim-fed gate, however, polishing or finishing of the degated feed may be necessary to leave an acceptable appearance.

With edge gates, both full and restricted, it is desirable to radius the edges where the gate enters or joins the cavity (0.005 in. to 0.010 in.) to avoid splintering.

Restricted Gates. This type of gate is widely used as an edge or centre feed. It is particularly suitable for free-flowing materials, but less so for viscous materials such as Kralastic.

The chief advantages of the restricted gate are:

(a) the gate freezes off quickly after the material stops flowing and thus reduces the overall cycle time by reducing the ram forward period and the time necessary for a thick gate to cool;

(b) by the earlier freezing of the gate, packing is reduced with a corresponding lowering of stress in the gate area;

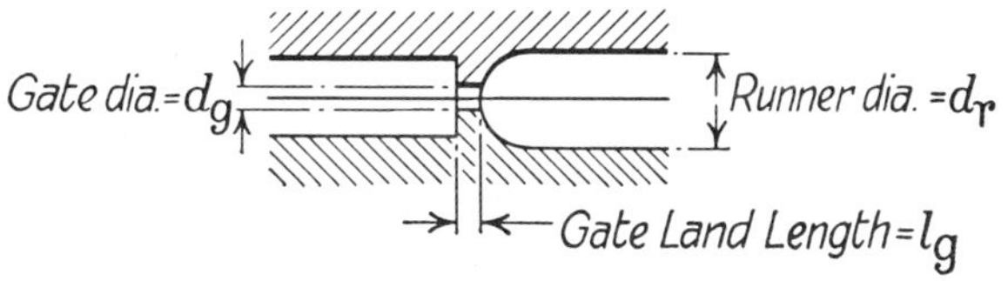

Fig. 8.21. Gate Proportions.

(c) the gate can be neatly trimmed off and hence improves the appearance of components without the extra finishing and polishing operations which may be necessary on a large gate.

Other factors are improved by the use of restricted gates with specific materials.

Restricted gates are normally between 0.03 in. and 0.06in. diameter, or correspondingly rectangular, and an approximate rule often adopted is that the diameter or thickness of the gate should be half that of the moulding section at the gate.

The land length of the gate is commonly made approximately equal to the gate diameter, or thickness, as indicated in Fig. 8.21. Excessive length causes too high a pressure drop, with consequent difficulty in filling the cavity.

The ratio d_r/d_g is made equal to 4.5, where d_r is the diameter of runner, and d_g is the diameter of the gate. This gives a ratio of runner area to gate area of approximately 20 to 1.

Frequently, the gate is of rectangular section, the same ratio being maintained. The width of this type of gate is frequently made about 2.5 times the depth.

EXAMPLE. Let the runner diameter be 0.25 in., then
$0.25/d_g = 4.5$, so that $d_g = 0.25/4.5 = 0.055$ in.
So that the full gate diameter $= \frac{1}{16}$ in., say.

For a rectangular gate:

area of a $\frac{1}{4}$ in. diameter runner $= 0.049$ sq. in.;

therefore the gate area at a 20 : 1 ratio is equal to

$0.049/20 = 0.0025$ sq. in.

If the gate depth $= t_g$ and the gate width $= w$, then the gate area $= t_g \times w$. Also, if $w = 2.5 \times t_g$, then the gate area is equal to $2.5\ t_g^2$, hence $t_g = \sqrt{(0.0025/2.5)} = 0.031$ in.

Whence, $w = 2.5 \times 0.031 = 0.0778$ in.

In the case of two-plate tools, the runner is cut into the face of one die plate, if a half-round or rectangular runner, or into both die plates if a full-round runner, and the runner terminates at the gate in a true full radius. The gate itself is generally cut only into one die plate, and hence is either of rectangular or half-round profile.

In three-plate tools, the same considerations apply to the runner in respect of the backplate and floating plate, but the gate will be full round.

Wherever possible, restricted gates should be located on the moulding in such a way as to ensure that the material flowing into the cavity meets some form of obstruction.* If this is not done, the material may spurt freely into the mould (i.e. "jetting": see Fig. 3.13A, page 91) which, in turn, can cause bad flow marks in the region of the gate on the final moulding. Two gate locations to reduce this effect are shown at (*X*) and (*Y*) in Fig.

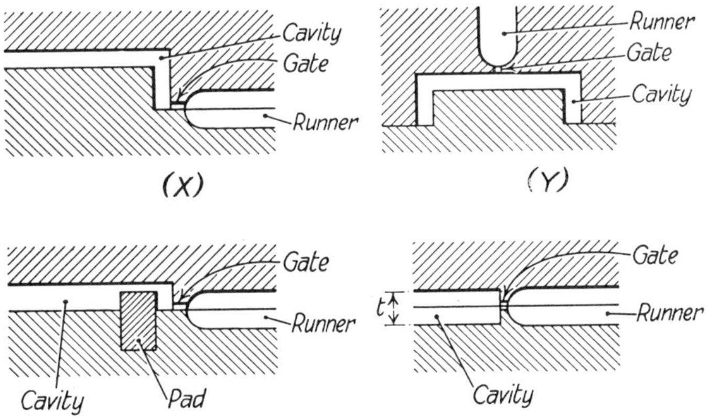

Figs. 8.22. (*X*) and (*Y*), [*Top*] Two Gate Locations to reduce Jetting. Fig. 8.23. [*Bottom Left*] Localised Pad to prevent Jetting. Fig. 8.24. [*Bottom Right*] Conditions favourable to Jetting.

---

*Occasionally this leads to erosion where the plastics material enters at high velocity and strikes the mould opposite the feed. This can be rectified by inserting a hardened steel pad at this point.

8.22, while the provision of a special pad is shown in Fig. 8.23. The tab gate illustrated in Fig. 8.34 is also used to avoid jetting. The worst conditions for jetting arise in the central feeding of a thick moulding where t is greater than $\frac{1}{8}$ in. (Fig. 8.24), and particularly with the high injection rates generated by restricted or needle gates.

Needle Gates. When the diameter of a restricted gate is below 0.03 in. it is commonly referred to as a needle feed, and is generally confined to free-flowing materials such as polystyrene. The minimum size is seldom less than 0.025 in. diameter. The land length should be made as short as possible and should not exceed the gate diameter. One of the most common uses of this type of gate is on mouldings which require feeding on a seen face and hence a small feed mark is essential. Very often they are associated with three-plate moulds. Highly restricted feeds of this type lead to turbulent flow with considerable orientation of the plastics material in the vicinity of the gate, especially in thin-walled mouldings.

This, in turn, leads to local weakness, resulting perhaps in cracking of

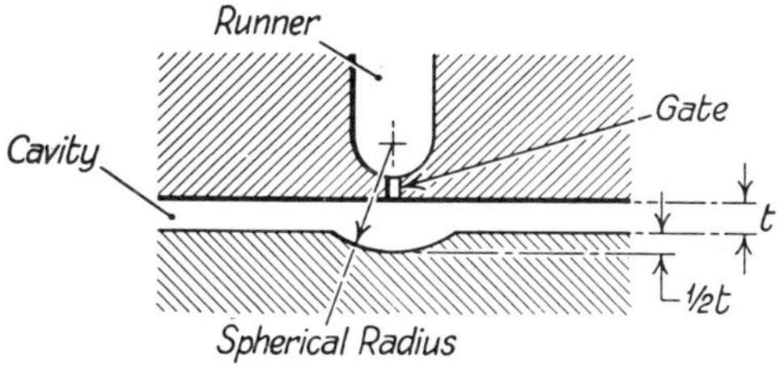

Fig. 8.25. Increase in Wall Thickness opposite to Gate to prevent Cracking.

the component, particularly in brittle materials. To offset this effect, it is desirable to increase the wall thickness immediately opposite the gate by an amount up to half the general thickness as indicated in Fig. 8.25.

Needle gates or small restricted gates can lead to difficulties on thick mouldings or mouldings with thick sections, since the freezing of the gate can prevent the application of packing pressure needed to offset sinking.

To prevent shattering of the component when the gate is broken off, again particularly with thin components in brittle materials, a reverse taper to the feed is employed, both on restricted gates and needle feeds. This ensures that the weakest part of the gate is adjacent to the runner. A full-round taper feed of this type is shown at (X) in Fig. 8.26, and a similar rectangular gate at (Y).

Automatic Degating. To facilitate production on high-speed moulds, self-shearing gates can be employed. On fully automatic moulds, this enables the mouldings to be produced ready for immediate use or packing, or, on fast cycles, enables the operator to attend the machine without having also to degate outside the press. If, however, the cycle is sufficient to carry out this operation, no economic advantage is gained, although automatic gating may be neater.

One of the simplest and most widely used methods is the submarine gate shown in Fig. 8.27. The gate is drilled at an angle from the runner to meet the moulding as shown. This produces a knife edge against which the feed is sheared, and if this edge is clean a very satisfactory cut-off can be obtained on the moulding. An illustration of a reassembled moulding with submarine gate is shown in Fig. 8.29.

If the gate is cut into the stationary half of the mould, as shown at *B* in Fig. 8.27, the moulding will move out on the moving tool as the mould opens, shearing off the gate at the knife edge. While this is occurring, the runner and gate remain stationary on the fixed half, and this is achieved by

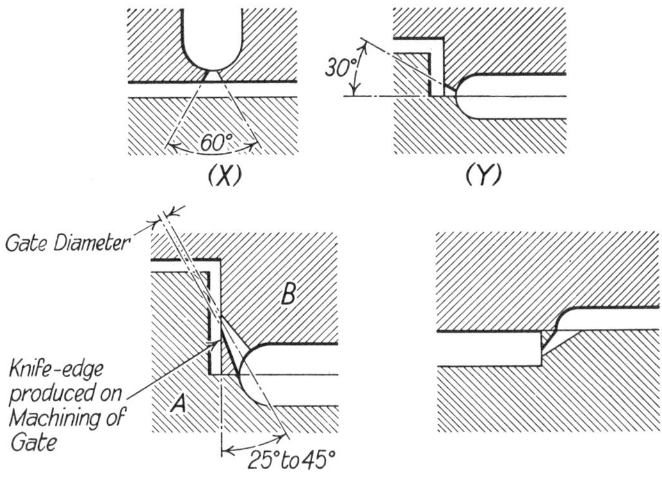

Fig. 8.26. [*Top*] Reverse-taper Gates: (*X*)—Fixed Tool; (*Y*)—Moving Tool. Fig. 8.27. [*Bottom Left*] Automatic Degating with Submarine Gate. Fig. 8.28. [*Bottom Right*] Jetting on Submarine Gate.

providing runner puller hooks close to the gate. When the moulding is clear, the runner system and gate are ejected by means of ejectors provided on the stationary tool.

When the gate is positioned on the moving side as shown at *A*, gate, runner and moulding go away together on the moving half. When the moulding and runner are finally ejected, the gate is sheared off. Moulding and runner are often ejected simultaneously, although prior ejection of the runner can be desirable.

The submarine gate has to deflect as it is ejected, and this fact and the angle of the knife edge required controls the angle at which the gate is cut.

Another automatic degating mould is shown in Fig. 6.33. It should be noted that if this type of gate is used on thick section mouldings in the

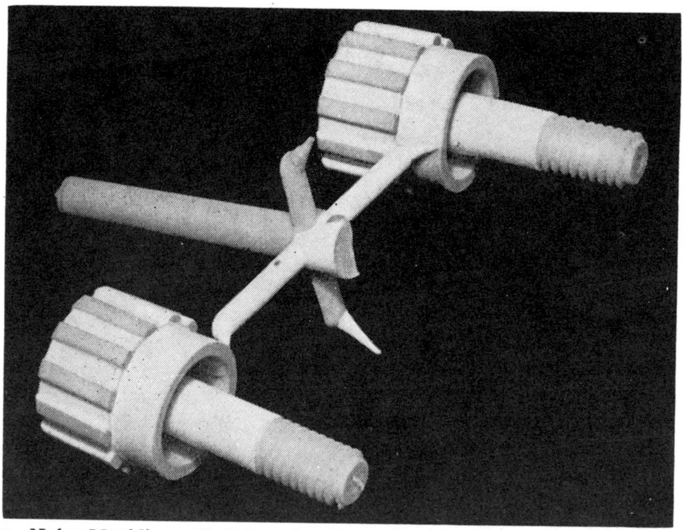

Fig. 8.29. Nylon Moulding with Submarine Gates. The Mouldings are normally separated from the Runners by Automatic Degating, but have been reassembled for Illustration. The Mould is a Two-impression Tool. The Blank Gates produce Two Preforms (not shown) at the same time. [*Courtesy: Ekco Plastics Ltd.*]

manner shown in Fig. 8.28, jetting will probably occur with resultant scab markings on the moulding.

There are several other methods of providing automatic degating, many of these being mechanically operated by means of finger cams.

Fan Gates. The kind of gate illustrated in Fig. 8.30 is often used to feed flat mouldings and mouldings that are both flat and thin, such as trays, scales, boxes and covers. The gate spreads the flow of material across the moulding and minimizes flow lines and lessens the effect of orientation. With some coloured materials, it reduces the possible objectionable flow pattern that emanates from a point gate.

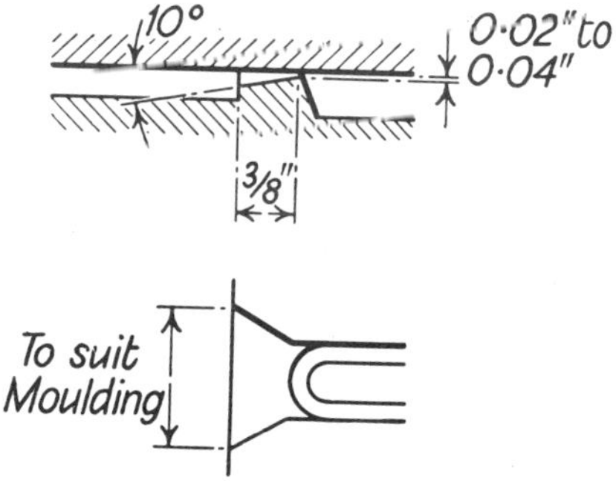

Fig. 8.30. Fan Gate.

The width of the fan does not, of course, necessarily anywhere approach the full width, except on narrow mouldings, but is commonly from $\frac{3}{8}$ in. to 1 in. Its cross-sectional area should not exceed that of the runner.

The film gate is an extension of the fan gate. In essence it consists of a runner parallel with the edge of the moulding, from $\frac{1}{8}$ in. to $\frac{1}{4}$ in. distant, and connected across to the cavity along its length by a gate which is from 0.005 in. to 0.03 in. thick. The gate thickness or its width may be varied to ensure even flow.

When the use of these gates is contemplated, the increased difficulty in degating should also be considered.

Disc Gates. In this type the feed is led to a cavity runner in the centre of an aperture formed in the moulding (often direct from the sprue) and connected to the mould cavity by a gate from 0.015 in. to 0.025 in. thick, completely around the periphery—see Fig. 8.31. Disc gates are used:

(i) When the moulding is symmetrical about the feed point, the

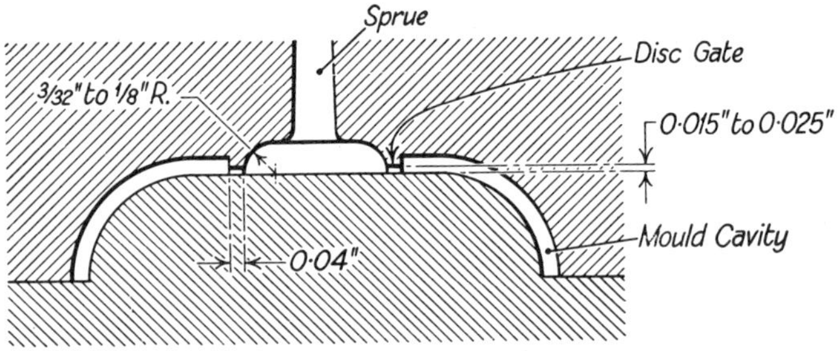

Fig. 8.31. Disc Gate.

continuous gate giving a uniform fill of the cavity. It is often applied to tubular mouldings or to flat or shallow annular forms.

(ii) To avoid weld lines from the aperture: if the moulding is fed from any other point, a weld line at the remote side of the hole is inevitable; by disc feeding from the centre of the hole itself this is eliminated. Venting is also facilitated.

(iii) On some large components, a disc feed in a convenient aperture can avoid the use of direct feed which may be visually unacceptable, or the construction of an offset tool which would be incurred by the use of a side feed.

The disc feed illustrated at Fig. 8.31 has its gate at the inside edge of the moulding wall thickness, whereas that illustrated in Fig. 8.32 is fed on the outside thickness. The feed is normally degated by punching out the disc, and the choice of this position depends upon the convenience of punching and the location requirements of the jig and moulding.

Fig. 8.32. Top-feed Disc Gate.

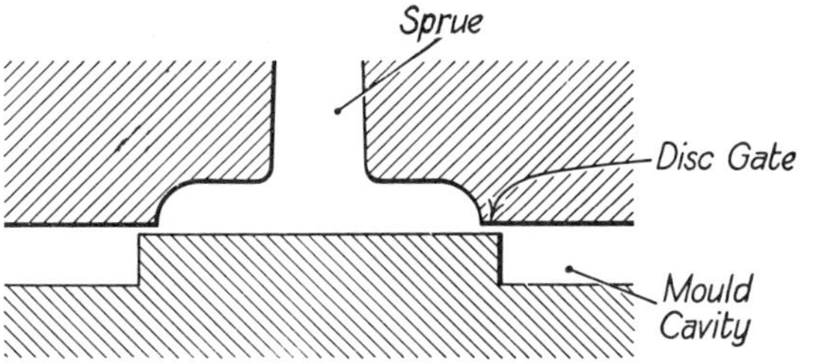

Ring Gates. These are used to produce tubular mouldings, as illustrated in Fig. 8.33. The material is led direct from the runner to the ring gate encircling one end of the moulding. From here it enters the cavity uniformly around the core pin. By this means, all air is forced to the bottom of the cavity and vented through the core pin, instead of becoming trapped, as it might if an edge gate were used. Similarly, the formation of a weld line is reduced. This gate has an advantage over the disc gate when the core pin is long and slender, since the core can be located at both ends for increased rigidity. It has the disadvantage, however, of more difficult degating in parting the mould from the feed. Various special forms of ring gate are often adopted to suit particular mouldings.

Tab Gates. These were specifically designed for acrylic materials in order to give more stress-free mouldings with greater optical clarity. The principle is that the material coming from the runner (Fig. 8.34), is directed through a restricted gate into the tab. From the tab, the material is led into the cavity. The restriction afforded by the gate causes a rise in temperature in the plastics material which is beneficial for moulding quality. By leading it into a tab at an angle to the gate, the material impinges on the opposite wall and then builds up into a smooth front which progresses into the cavity without the jetting that would arise if the gate itself led directly into the cavity.

The positioning of the restricted gate away from the moulding itself

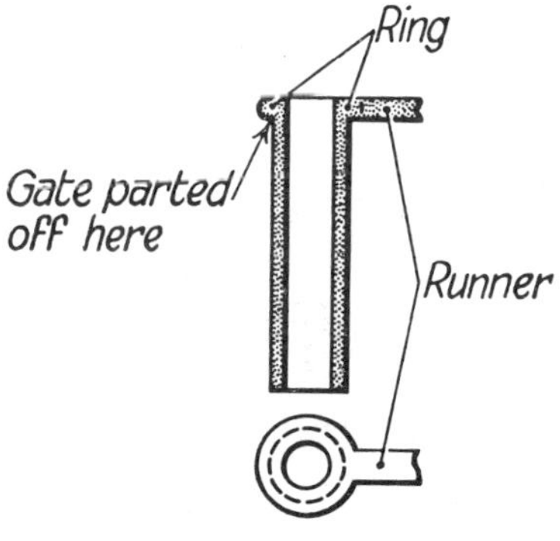

Fig. 8.33. Ring Gate.

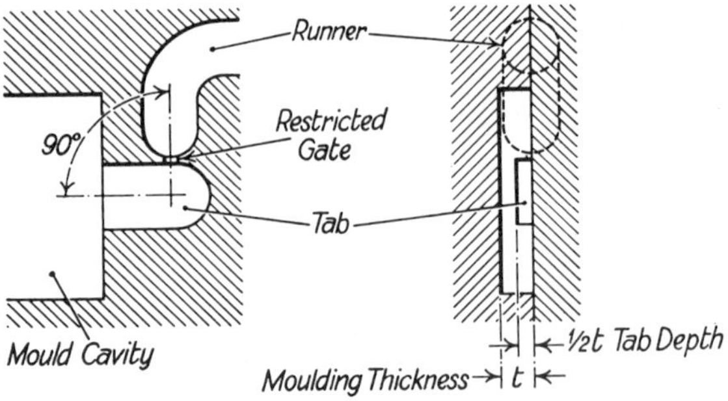

Fig. 8.34. Tab Gate.

also ensures that the high orientation generated by such a gate and remaining in this area after cooling will not be such a cause of weakness in the component. Additionally, any weakness from the notch effect inherent in a restricted feed is also excluded.

Gates are usually square or rectangular and the tab preferably rectangular, but sometimes half-round. The size of gate and of tab varies with the size of moulding. On mouldings for 4-oz. machines the gate may be 0.125 in. wide by 0.04 in. deep with a land length of 0.06 in. to 0.08 in., up to large mouldings with gates $\frac{3}{16}$ in. wide by $\frac{1}{8}$ in. deep and land length of $\frac{1}{16}$in. to $\frac{3}{16}$ in. The tabs themselves are from 0.5 in. long

Fig. 8.35. Tab Gate on Clear Acrylic Moulding. [*Courtesy: Ekco Plastics Ltd.*]

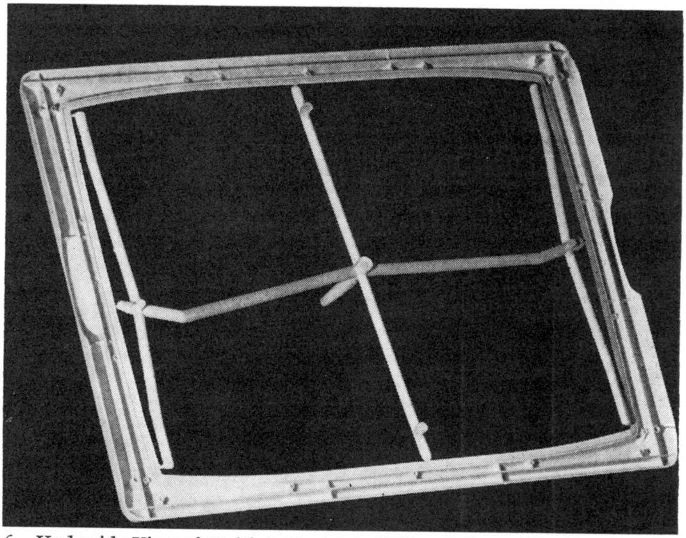

Fig. 8.36. Underside View of Multiple Restricted Edge Gates on Toughened Polystyrene Frame Moulding. Eight Ribs are employed to reduce Sinking on Bosses and Ribs. Weight of Moulding: 16 oz. [*Courtesy: Ekco Plastics Ltd.*]

and from 0.25 in. to 0.375 in. wide, and not less than half the moulding section in thickness, to 1.5 in. long by from 0.375 in. to 0.5 in. wide and not less than half the section in thickness. The tab is gated at the mid-point of its length, or at a point farther away from the cavity. Runners are preferably full-round and from $\frac{5}{16}$ in. to $\frac{3}{8}$ in. diameter. An illustration of a moulding with a tab gate is given in Fig. 8.35.

On some multi-impression tools, it is not always possible to gate into the tab at right angles. In such cases, the tab can be gated at a lesser angle, but the point of entry should be farther back. However, angles less than 90 deg. are not generally as efficient.

Gates for High-density Polythene. On moulds to be used with high-density polythene, it is preferred practice to increase the wall thickness of the moulding immediately opposite the gate by one-third—see Fig. 8.37. Increases in thickness up to one-half are sometimes adopted, but the consequent possibility of sinking must be borne in mind. Such increased thickness reduces the tendency to brittleness in the gate region arising from high orientation and packing stress.

This type of gate has advantages generally for large mouldings, since the button promotes uniform and symmetrical filling.

For small mouldings in polyolefins, gates of 0.03 in. to 0.06 in. can be used, but for larger or thick mouldings, gates are seldom less than 0.09 in. diameter. Due to distortion considerations, the actual location of the gates and the type of gate for these materials are important, and reference should be made to manufacturers' literature on these points.

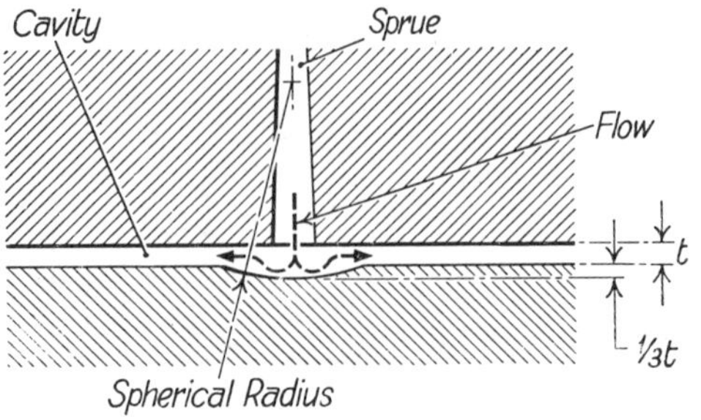

Fig. 8.37. Increase in Wall Thickness opposite Gate in High-density Polythene Mouldings.

Multiple Gates. These are used chiefly on large or flat-type mouldings where distortion difficulties may arise (e.g. high-density polythene), on moulds with long flow paths where premature cooling may occur (e.g. grilles), or on moulds where local sinking has to be countered.

In the latter case, the gates must be positioned adjacent to the thick sections that may sink, so that packing pressure may be applied to these sections. Even if a separate gate is not possible for every thick portion (e.g. on components with numerous bosses), the use of multiple gates will shorten the distance through which the packing pressure has to be applied. An example of multiple gates for this purpose is shown in Fig. 8.36.

If several gates are to be employed to facilitate the filling of the cavity or to reduce distortion, these should be placed, as far as possible, so that each will fill an equal amount. However, while the use of such gates has the advantage of allowing fast injection at lower temperatures and pressures with reduced warpage, they do have the disadvantage of causing extra flow lines and the possibility of gas pockets from entrapped air where the flow from the separate gates meet. The position of these weld lines should be considered when the gates are positioned, since it is desirable that they should be equidistant from the feed. If this be not done, then when the cavity is full, excessive packing will occur at the weld nearest the gate. If the gate is feeding to the edge of the component, this distance should be nearly the same as the distance to the weld lines set up by the flow from the same gate where it meets flow from other feeds. In general, this positioning is resolved in conjunction with the requirements of equal fill.

The gate sizes themselves should vary with the differences in runner length, and are determined as shown in a later section, on the basis that each is filling a separate cavity of the same size. The use of a multi-gate tool is illustrated in Fig. 5.10A, page 127.

Venting. As the plastics material is injected into the cavity, it must displace the air that is already there. If the pattern of flow in any mould does entrap any of this air, the compression will increase its temperature to such a degree as to cause a burn mark or even a hole. Alternatively, the surface may be covered with minute bubbles of dissolved air, giving a flake-like appearance. Such gas traps, with holes or burns, may occur on many tools, and are very likely to take place at points of flow junction on surfaces, in ribs or thick sections, or in blind spots at the end of flow such as the heads of bosses.

Either these marks must be accepted, the flow pattern altered, or special venting arrangements incorporated. Venting can take place between the tool parting faces, through the clearances of moving cores, down ejector pins or sleeves, or through special vent pins. Sometimes the fit of the parting faces is too close to allow the rapid egress of air, and the face of one half of the tool can be faintly grooved to a depth of 0.002 in. to 0.005 in. at the location where venting is required; for nylon, the minimum depth should be adopted. A similar problem often arises in venting down ejector pins, which can be solved by the addition of a slight flat to the pin.

It is important to consider the venting arrangements when the gate is positioned, since the regions of entrapped air will be those that are filled last. Venting should also be considered when the tool parting line is being decided, since some layouts may present a greater possibility of trapped air than others.

One of the advantages of constructing the cavity with a separate base such as shown in Figs. 6.4, 6.5, and 6.6., is the improved venting thereby obtained, since the air escapes down the joint lines of the blocks. However, if the fit of these is too close, then fine scratches, or grooves, can be added where required to facilitate the escape of air. These must not be deep enough to allow the entry of moulding material which may then break off from the moulding and obstruct the vent.

A special case relative to multi-gate tools is shown in Fig. 5.10c, page 129. In such a multi-feed tool, it is evident that the plastics material entering at each gate eventually converges, and may therefore entrap air and cause holes or burns. Two methods of preventing this are shown at (*d*) and (*e*).

With the equi-spaced gates shown, the point of convergence should be in the centre of the tool, and hence the air requires venting at this point. This condition is shown at (*d*) where the pin (8) is formed at one end into a vent pin of reduced diameter (9). At its other end, it forms a valve into which is cut a sprue hook. In operation, the entrapped air is vented up the clearance between the vent pin (9) and its mating hole. When the press opens, the sprue hook causes the whole pin to be pulled outwards until further movement is arrested by the cross pin. By means of this movement of the vented pin once every cycle, closure of the clearance by flash is prevented, and venting can continue on every moulding. The device is returned to its moulding position, as shown, by the valve head being pushed back by contact with the surface of the plate *A* when the tool closes.

Sometimes, the provision of vent arrangements must await initial sampling of the tool, so that the point of the entrapped air can be located. Should this point occur at any other than the central position, measures such as that indicated at (*e*) would have to be taken. This comprises a vent pin having a tapered lead which is mounted directly into the plate *A*. The pin is guided through the plate *B* by means of the tapered lead of the hole and is set with its parallel portion communicating with the mould cavity. When the tool opens, the pin is withdrawn from the plate *B*, and hence clears away any flash that may have occurred and which would have interrupted any future venting. This type of pin can be located wherever required in the mould.

Weld lines often contain entrapped air which, although not always leading to holes or burns, can cause a weakness at the weld, especially if the material is chilled after a long flow. Here again, venting may be found necessary to improve the strength of the bond.

Gate Balancing on Multi-impression Tools for the Same Component. Depending upon the runner layout, it is usual on multi-impression tools for the same component to find that, after the runner system has filled, the cavities nearest the sprue will fill first, with successive cavities being filled only after the resistance of the first cavities has reached equilibrium with the resistance of the gates of the next cavities from the sprue. This can lead to overpacking of some cavities, and shorts or sinking occurring in the outer cavities. This often happens and can be readily rectified only when the tool is actually running in the press. Thus, the gates of the outer cavities can be progressively increased in size until, as nearly as possible, simultaneous filling of all cavities is achieved.

Additionally, however, an effort can be made in the design stage to overcome some of this fault by designing the gates to present a degree of resistance that is proportional to a function of their distance from the sprue.

Variations of gate resistance can be obtained by two principal methods:

(a) maintaining all gate sizes identical but varying the gateland lengths;

(b) maintaining gate widths and land lengths constant but varying the gate depths.

Both these methods are equally convenient to effect during the initial machining of the runner system but (b) is easier to carry out when the tool is running on the press. Gates on multiple-impression tools are conveniently made rectangular, since width and depth are easily machined and measured. They are preferable also to a square gate, since balancing modifications on the latter are more critical, and may have to be carried out on both width and depth. Restricted-gate designs are used whenever possible, since these can be more readily adjusted for balanced flow.

It is imperative that all multi-impression tools have their gate dimensions controlled to a high degree, since in practice many troubles are experienced due to lack of accuracy. A tolerance of ± 0.002 in. on dimensions is advised.

Balancing of gates in the design stage by varying the land length can be considered along the following lines, although final adjustments by modifying the depth when the tool is running on the press will still be necessary.

If the cavities are to fill simultaneously, the resistances of all the runner and gate systems must be balanced. The pressure drop across each runner and gate system is equal to the sum of the pressure drop in the runner and the pressure drop at the gate.

The relationship between the pressure drop and the length of land at the gate may be taken as linear, because of its small proportions.

Thus, $\Delta P = \Delta p_r + \Delta p_g = k_r L_r^{1.25} + k_g L_g$

Therefore, $k_r L_{r1}^{1.25} + k_g L_{g1} = k_r L_{r2}^{1.25} + k_g L_{g2}$

From which,

$$L_{g2} = L_{g1} - (k_r/k_g)(L_{r2}^{1.25} - L_{r1}^{1.25}) \quad .. \quad .. \quad .. \quad .. \quad .. \quad (10)$$

Using round runners and gates,

$k_r/k_g = d_g^4/d_r^4 = 0.001$, for all practical purposes,

so that (10) becomes

$$L_{g2} = L_{g1} - 0.001(L_{r2}^{1.25} - L_{r1}^{1.25}) \quad .. \quad .. \quad .. \quad .. \quad .. \quad (11)$$

EXAMPLE. Let the runner lengths to three separate cavities be $L_{r1} = 1$ in.; $L_{r2} = 3$ in.; $L_{r3} = 5$ in.; with $L_g = 0.05$ in., $d_g = 0.04$ in., and $d_r = 0.25$ in.

Then,

$L_{g2} = 0.05 - 0.001(3^{1.25} - 1^{1.25}) = 0.05 - 0.001\ (2.95) = 0.047$ in.

and

$L_{g3} = 0.05 - 0.001(5^{1.25} - 1^{1.25}) = 0.05 - 0.001(6.48) = 0.044$ in.

With round gates, it is not generally found practical to balance the cavities by varying the diameter of the gates as the flow is too sensitive to this dimension, being directly proportional to the fourth power of the gate diameter.

Similarly, with rectangular gates it is to be preferred to vary the width of the gate rather than the depth or land length.

Round Runners with Rectangular Gates. In this case, for all practical purposes, $k_r/k_g = 4t_g^3/\pi d_r^4 L_g = 0.25$; then by equation (10)

$$1/w_{g2} = 1/w_{g1} = 0.25(L_{r2}^{1.25} - L_{r1}^{1.25}) \quad .. \quad .. \quad .. \quad .. \quad (12)$$

EXAMPLE. Let $L_{r1} = 1$ in.; $L_{r2} =$ 3in.; $L_{r3} = 5$ in.; $w_{g1} = 0.08$ in.; $d_r = 0.25$ in.; $L_g = 0.05$ in. Then, assuming that the depth of the gate is $t_g = 0.02$ in., we obtain by equation (12)

$$1/w_{g2} = 1/0.08 - 0.25(3^{1.25} - 1^{1.25}) = 11.76$$

hence $w_{g2} = 1/11.76 = 0.086$ in.

Similarly,

$$1/w_{g3} = 1/0.08 - 0.25(5^{1.25} - 1^{1.25}) = 10.88$$

hence $w_{g3} = 1/10.88 = 0.092$ in.

Therefore, the dimensions of the gates are:

1st cavity (initially assumed) $= 0.080$ in. $\times 0.02$ in. deep
2nd cavity .. .. .. $= 0.086$ in. $\times 0.02$ in. deep
3rd cavity .. .. .. $= 0.092$ in. $\times 0.002$ in. deep.

Gate Balancing on Multi-impression Tools for Components of Different Weights. To effect balanced gating under these conditions at the design stage is more difficult than the conditions for the previous example. It is not usually sufficient to modify the land lengths of the gates, and normal practice is to alter the gate diameters, or depths in the case of rectangular gates.

From equations (6) and (7),

$Q=\pi d_g^4 \Delta p/2^7 \mu L_g = k_1 d_g^4$, for round gates; and

$Q=w t_g^3 \Delta p/2^5 \mu L_g = k_2 t_g^3$, for rectangular gates.

Hence, using the equation for round gates,

$$Q_1/Q_2 = d_{g1}^4/d_{g2}^4 \qquad (13)$$

i.e.

$$d_{g2} = d_{g1}(Q_2/Q_1)^{\frac{1}{4}} \qquad (14)$$

Since Q is directly proportional to W, the weight of material,

$$d_{g2} = d_{g1}(W_2/W_1)^{\frac{1}{4}} \qquad (15)$$

where W_1 and W_2 are the respective weights of the components.

Similarly, for rectangular gates,

$$t_{g2} = t_{g1}(W_2/W_1)^{\frac{1}{3}} \qquad (16)$$

On some layouts, it can be arranged that the runner lengths to each cavity can be made similar, thus equalizing the pressure drop to each; an example of such a layout is given in Fig. 8.38. This is obviously advantageous, but should not be achieved at the expense of excessive lengths of runner or complex patterns. Apart from anything else, the waste material in such a feed system would be uneconomical.

From formulae (15) and (16) can be calculated the difference in size of depth of gate to compensate for changes in the quantity of material which must pass through into cavities of differing size. Variation in the size of gates can be minimized to some extent in practice by laying out the mould with the largest component nearest the sprue.

EXAMPLE. In a two-impression tool,

Cavity 1: $L_{r1}=1.5$ in.; $W_1=45$ grams;
$L_{g1}=0.06$ in.; $d_{g1}=0.06$ in.

Cavity 2: $L_{r2}=3.25$ in.; $W_2=105$ grams;
$L_{g2}=$? ; $d_{g2}=$?

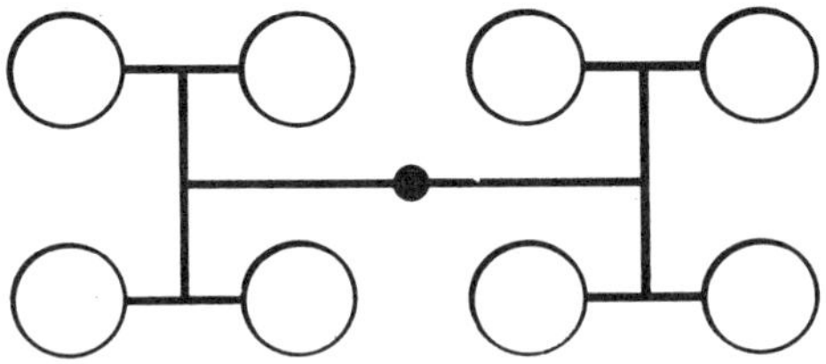

Fig. 8.38. Multiple-impression Tool with Equal Lengths of Runner to Each Cavity.

From equation (11)

$$L_{g2}=L_{g1}-0.001(L_{r2}^{1.25}-L_{r1}^{1.25})$$

$$=0.06-0.001(3.25^{1.25}-1.5^{1.25})=0.057 \text{ in.}$$

By equation (15)

$$d_{g2}=d_{g1}(W_2/W_1)^{\frac{1}{4}}=0.06(105/45)^{\frac{1}{4}}=0.074 \text{ in.}$$

hence the required dimensions are

$$L_{g2}=0.057 \text{ in. and } d_{g2}=0.074 \text{ in.}$$

APPENDIX A

BRIEF MOULD DESIGN CHECK LIST

APPENDIX A

BRIEF MOULD DESIGN CHECK LIST

PRESS

1. Is the weight of the moulding, runners and sprue within the shot capacity of the press?
2. Is the expected output well within the plasticising capacity of the press?
3. Is the clamping pressure of the press sufficient for the projected area of the mouldings and runners?
4. Will the mould pass between the press tie bars?
5. Do the clamping arrangements for the tool suit the platen bolt holes?
6. Is the closed thickness of the tool above the minimum and below the maximum required for the press?
7. Is the press opening stroke sufficient for the ejection and extraction of the moulding?

MOULDING

1. Is the flash line at the tool parting line position visually acceptable on the moulding?
2. Are the positions of any print lines arising from tool blocks, cores, ejectors, etc., visually acceptable on the moulding?
3. Is the gate position visually acceptable on the moulding?
4. Will the position of any flow or knit lines that may occur be acceptable in appearance?
5. Will any knit lines cause weakness in a critical area?
6. Will any heavy section in the moulding cause unacceptable sink marks?
7. Is the design free of any undercut areas that will prevent ejection?

TOOL

1. Is the tool parting line chosen the most efficient for mould operation and tool construction?
2. Have the punch and cavity been designed in the easiest manner for machining on the available equipment?
3. Will any slender blades or pins deform under cavity pressure or flow?

4. Is the cavity of adequate strength to resist internal cavity pressure?
5. Are all tool components exposed to side thrust from cavity pressure, located by solid steel and not only by screws or dowels?
6. Is the tool construction free from any chance of horizontal flash?
7. Does the ejector packing provide sufficient support to the die plate to prevent distortion under cavity pressure?
8. Can all parts of the tool be dismantled and separated in the event of tool breakdown or modification?
9. Are all necessary parts hardened?
10. Have all allowances for moulding shrinkage been provided?
11. Has sufficient moulding taper been allowed on all parts forming moulding surfaces?
12. Will the tool dimensions produce mouldings within component tolerances?
13. Will the moulding remain on the ejection side when the mould opens?
14. Is the ejection stroke sufficient to clear the moulding?
15. Have sufficient ejectors been provided to prevent sticking, cracking or distortion of the moulding?
16. Are the ejector and ejector bars sufficiently strong?
17. Is the ejector mechanism suitable for the particular press ejector system?
18. Has the ejector return mechanism been provided?
19. Have adequate guide pins been provided between the tool halves?
20. On split tools or moving cores is the opening movement provided by cams, cylinders, etc., sufficient to clear the undercuts on the component?
21. Are all inserts positively located for position and prevented from displacing themselves during tool closing or under plastic flow in the cavity?
22. On split tools or moving cores is the mechanism sufficiently foolproof to prevent damage by faulty operation?
23. On all splits and moving cores, is the cavity pressure resisted by solid steel locking faces and not by the split or core-operating cam or ram?
24. Have adequate cooling channels been provided?
25. Is the cooling too close or too distant from the mould surfaces?
26. Are the runners of sufficient size?
27. Has a sprue puller hook and sprue cold well been provided?
28. Are runner hooks and runner cold wells necessary?
29. On three-plate tools has sufficient opening been provided between the plates to allow extraction of mouldings and runner system?
30. Is the mould sufficiently vented?
31. On offset designs is the out-of-balance force too great?
32. Has an injection side platen location spigot been provided?
33. Do the spherical nose and orifice of the cylinder nozzle mate with the spherical seating and bore of the sprue bush?
34. On split tools can all ejectors be in the forward position without interfering with the closing of the splits?

If not—is mechanism provided to ensure ejectors are returned before the splits close?

35. Is sufficient clearance provided between ejector chains or links to withdraw the component through them without difficulty?
36. Have tool lifting eye bolts been provided?

APPENDIX B

SOLUTIONS OF SOME COMMON TRIGONOMETRICAL PROBLEMS IN MOULD DESIGN

APPENDIX B

SOLUTIONS OF SOME COMMON TRIGONOMETRICAL PROBLEMS IN MOULD DESIGN

Draft Angles

Referring to Fig. B.1, $t=h \tan \theta$, and the values of $\tan \theta$ for some usual values of θ are as follows:

θ	$\tan \theta$	θ	$\tan \theta$
00° 15′	0.00436	3° 00′	0.05241
00° 30′	0.00873	4° 00′	0.06993
1° 00′	0.01745	5° 00′	0.08748
2° 00′	0.03492		

Solution of Right-angled Triangle

Using the nomenclature indicated in Fig. B.2:

$$a=c \sin A, \text{ or } b \tan A, \text{ or } \sqrt{[c^2-b^2]}$$
$$b=c \cos A, \text{ or } a \cot A, \text{ or } \sqrt{[c^2-a^2]}$$
$$c=a \operatorname{cosec} A, \text{ or } b \sec A, \text{ or } \sqrt{[a^2+b^2]}$$

General Solution of Triangle

When all three sides are known (Fig. B.3), to find A, B or C:

$$\cos A=(c^2+b^2-a^2)/2bc$$
$$\cos B=(a^2+c^2-b^2)/2ac$$
$$\cos C=(a^2+b^2-c^2)/2ab$$

When two sides and the included angle are known (Fig. B.3) to find the opposite side:

$$a=\sqrt{[c^2+b^2-2bc \cos A]}$$
$$b=\sqrt{[a^2+c^2-2ac \cos B]}$$
$$c=\sqrt{[a^2+b^2-2ab \cos C]}$$

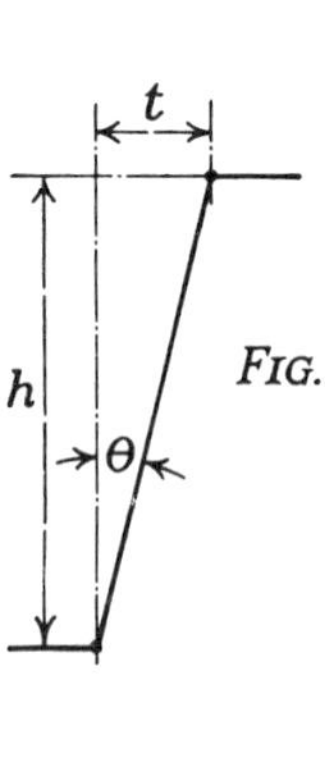

FIG. B.1

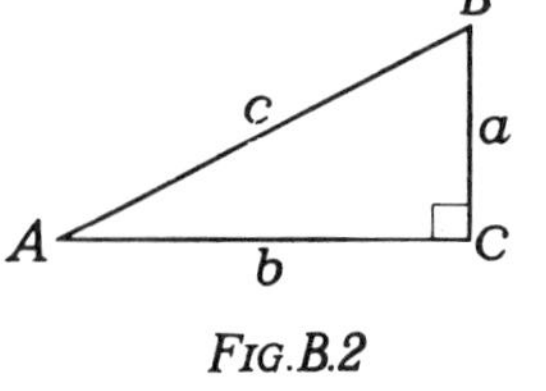

FIG. B.2

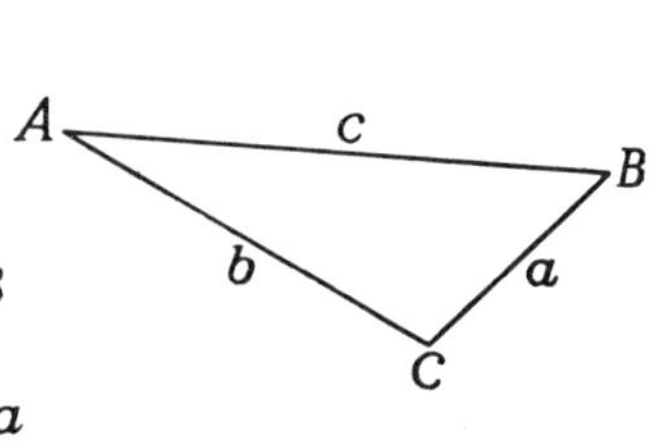

FIG. B.3

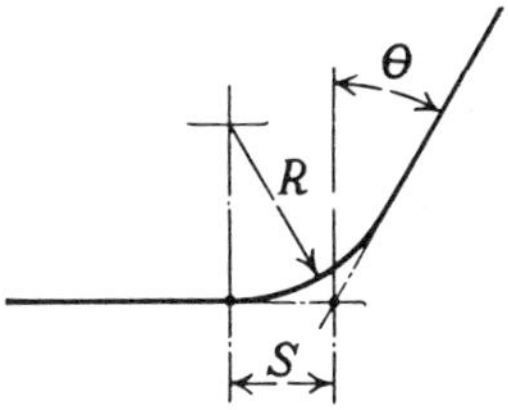

Fig. B.4

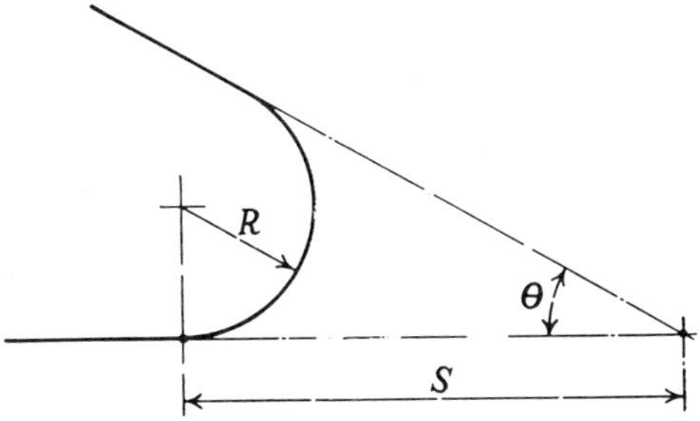

Fig. B.5

When one side and the opposite angle are known together with one other angle or side (Fig. B.3):

$$a/\sin A = b/\sin B = c/\sin C$$
$$a = b \sin A/\sin B;\ \text{or}\ c \sin A/\sin C$$
$$b = a \sin B/\sin A;\ \text{or}\ c \sin B/\sin C$$
$$c = a \sin C/\sin A;\ \text{or}\ b \sin C/\sin B$$

Radius-blending with Acute and Obtuse Angles

Obtuse Angle: Referring to Fig. B.4, if R and θ are known, then:

$$S = R \cot \tfrac{1}{2}(90° + \theta)$$

Acute Angle: Referring to Fig. B.5, if R and θ are known, then:

$$S = R \cot \tfrac{1}{2}\theta$$

Blend of Radius with Tapered Face (Fig. B.6)

When r, h, and θ are known:

$$q = r \sec \theta + (h - r) \tan \theta$$

When q, h, and r are known:

$$\theta = \tan^{-1}[q/(h-r)] - \sin^{-1}[r/\sqrt{\{(h-r)^2 + q^2\}}]$$

Blend of Tapered Face with Two Equal Radii (Fig. B.7)

When h, r and q are known, to find θ:

$$\theta = \tan^{-1}[q/(h-2r)] - \sin^{-1}[2r/\sqrt{\{(h-2r)^2 + q^2\}}]$$

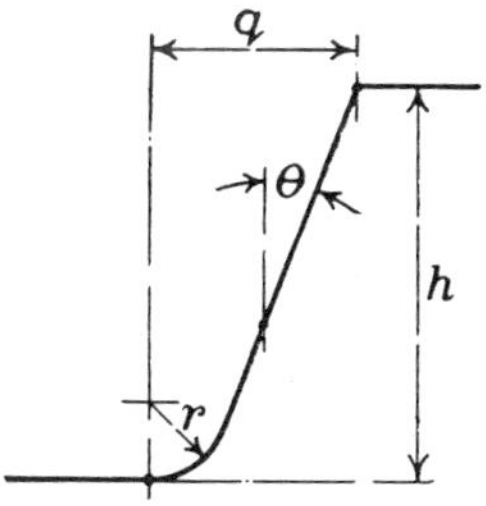

Fig. B.6

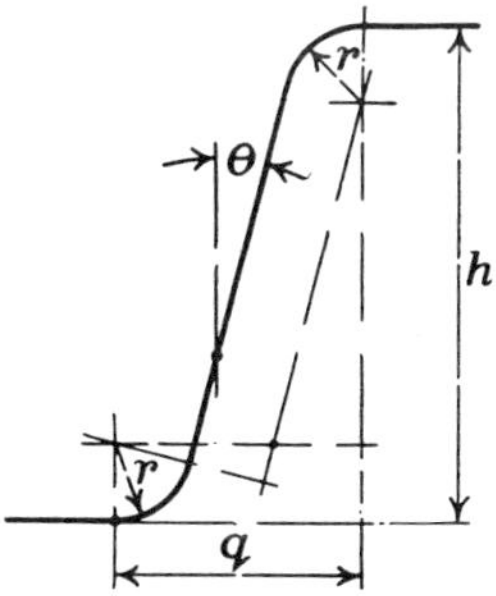

Fig. B.7

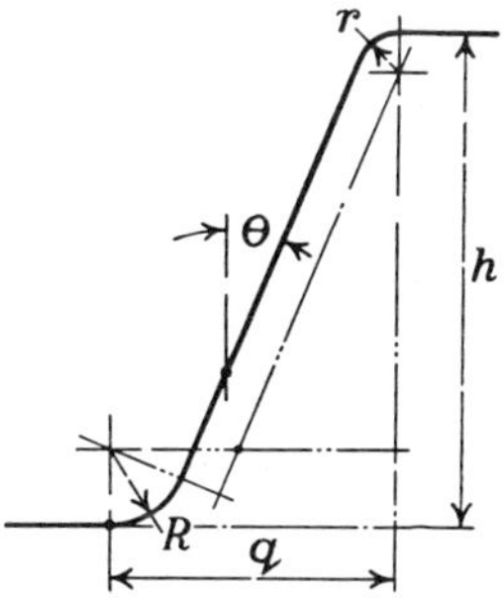

Fig. B.8

When r, h and θ are known, to find q:

$$q=2r \sec \theta+(h-2r) \tan \theta$$

When r, q and θ are known, to find h:

$$h\,(q-2r \sec \theta) \cot \theta+2r$$

Example. Let $r=0.5$ in., $q=2$ in., and $h=4$ in., then by the above formula:

$$\theta=\tan^{-1}[2/(4-2\times 0.5)]-\sin^{-1}[2\times 0.5/\sqrt{\{(4-2\times 0.5)^2+2^2\}}]$$
$$=\tan^{-1}[2/3]-\sin^{1}[\text{-}1/\sqrt{13}]$$
$$=\tan^{-1}0.66667-\sin^{-1}0.27735$$
$$=33^\circ\ 42'-16^\circ\ 6'$$
$$=17^\circ\ 35'$$
$$q=2\times 0.5\times \sec 17^\circ\ 35'+(4-2\times 0.5) \tan 17^\circ\ 35'$$
$$=\sec 17^\circ\ 35'+3 \tan 17^\circ\ 35'$$
$$=1.0490+(3\times 0.31690)$$
$$=1.9997 \text{ in. } (=2 \text{ in.})$$
$$h=[2-2\times 0.5 \sec 17^\circ\ 35'] \cot 17^\circ\ 35'+2\times 0.5$$
$$=0.9510\times 3.1556+1$$
$$=4.00097 \text{ in. } (=4 \text{ in.})$$

Blending of Tapered Face with Two Unequal Radii (Fig. B.8)

When h, r, and q are known, to find θ:

$$\theta=\tan^{-1}[q/(h-R-r)]-\sin^{-1}[(R+r)/\sqrt{\{(h-R-r)^2+q^2\}}]$$

When r, h, and θ are known, to find q:

$$q=(R+r) \sec \theta+(h-R-r) \tan \theta$$

When r, q and θ are known, to find h:

$$h=[q-(R+r) \sec \theta] \cot \theta+(R+r)$$

Radius passing through Three Points (Fig. B.9)

(i) When h and c are known, to find R:

$$R=(4h^2+c^2)/8h$$

(ii) When R and c are known, to find h:

$$h=R \pm\sqrt{[R^2-(\tfrac{1}{2}c)^2]}$$

(iii) When R and c are known, to find θ:

$$\theta=\sin^{-1} (c/2R)$$

(iv) When R and θ are known, to find l:

$$l=0.03490\ R\theta$$

Examples. Let $h=0.375$ in., $c=3$ in., then by the foregoing:

(i) $R=(4\times 0.375^2+3^2)/(8\times 0.375)$
$=9.5625/3$
$=3.1875$ in.

(ii) $h=3.1875\ \pm\sqrt{[3.1875^2-1.5^2]}$
$=0.375$ or 6.000 in. Obviously the second result is not applicable.
so $h=0.375$

(iii) $\sin^{-1} 3/(2\times 3.1875)=\sin^{-1} 0.47058=28^\circ\ 4'$

(iv) $l=0.0349\times 3.1875\times 28.0666$
$=3.12284$ in.

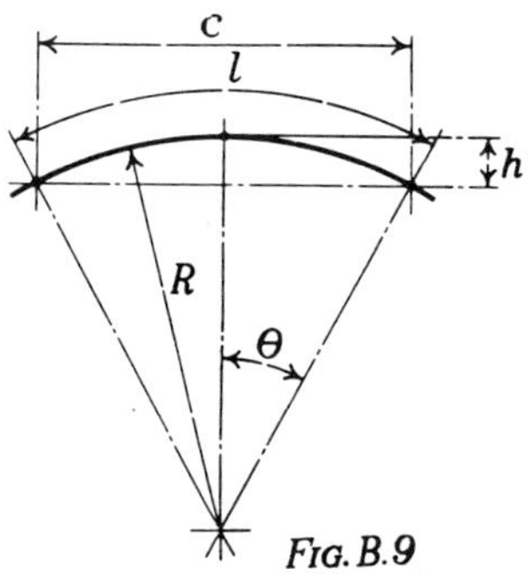

FIG. B.9

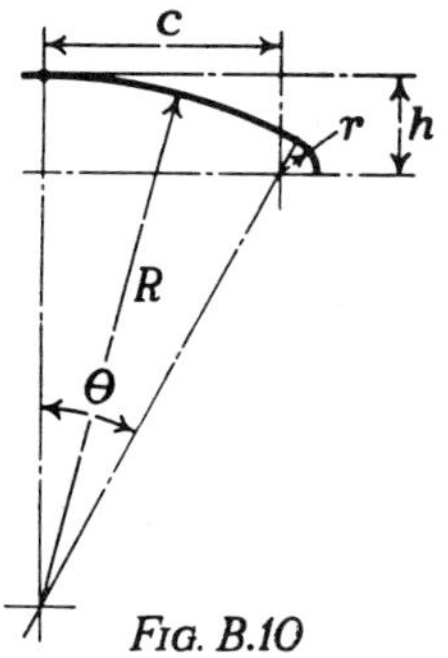

FIG. B.10

Blending of Two Radii (Fig. B.10)

(i) When r, c, and h are known, to find R:
$R=(c^2+h^2-r^2)/2(h-r)$

(ii) When R, c, and r are known, to find h:
$h=\sqrt{[(R-r)^2-c^2]}$

Examples. Let $r=0.25$ in., $c=1.5$ in., and $h=0.5$ in.

(i) $R=(2.250+0.250-0.0625)/(2\times0.25)$
$=4.875$ in.

(ii) $h=4.875-\sqrt{[(4.875-0.250)^2-2.250]}$
$=0.500$ in.

Blending of Three Radii, with Two of the Same Magnitude (Fig. B.11)

When c, h, R, and r are known, to find the position of the centre of

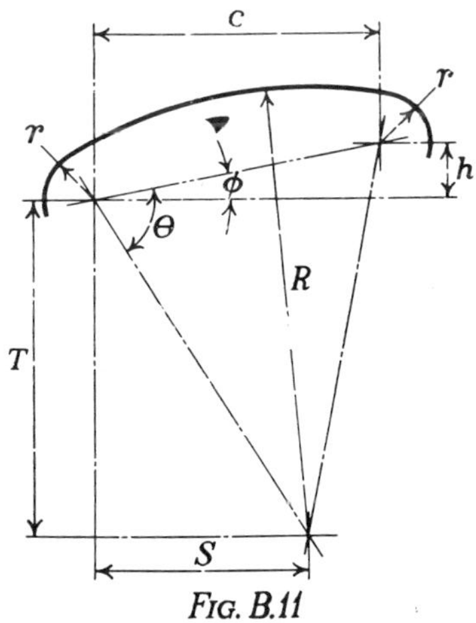

FIG. B.11

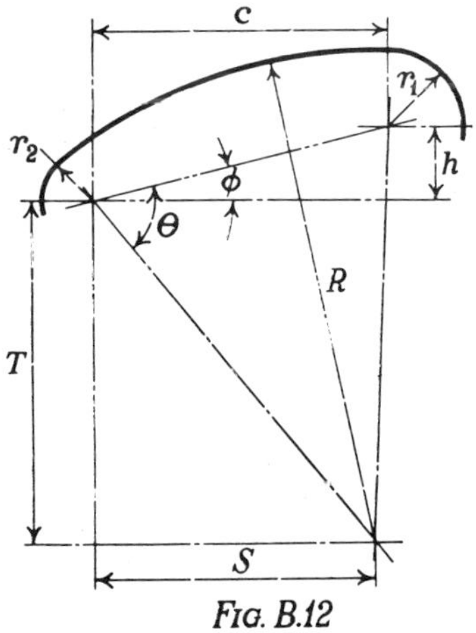

FIG. B.12

R, i.e. the dimensions T and S:

$$S=(R-r)\cos(\theta-\varphi)$$
$$T=(R-r)\sin(\theta-\varphi)$$

where $\cos\theta=\sqrt{(c^2+h^2)}/2(R-r)$

and $\tan\varphi=h/c$

Examples. Let $c=3$ in., $h=0.5$ in., $r=0.5$ in., and $R=5$ in.

Then, $\theta=\cos^{-1}[\sqrt{(9+0.25)}/2(5-0.5)=\cos^{-1}\ 0.33682$

Hence, $\theta=70°\ 19'$

Also $\varphi=\tan^{-1}[0.5/3]=\tan^{-1}\ 0.16667=9°\ 28'$

Therefore, $\theta-\varphi=60°\ 51'$, and consequently,

$S=4.5\cos 60°\ 51'=4.5\times 0.48710=2.1919$ in.

and $T=4.5\sin 60°\ 51'=4.5\times 0.87335=3.9301$ in.

Blending of Three Radii (Fig. B.12)

When h, c, r_1, r_2, and R are known, to find T and S:

$$S=(R-r_2)\cos(\theta-\varphi)$$
$$T=(R-r_2)\sin(\theta-\varphi)$$

Where $\theta=\cos^{-1}[\{h^2+c^2+(r_1-r_2)(2R-r_1-r_2)\}/\{2(R-r_2)\sqrt{(h^2+c^2)}\}]$

and $\varphi=\tan^{-1}[h/c]$

Examples. Let $h=0.5$ in., $c=3$ in., $r_1=0.75$ in., $r_2=0.5$ in., and $R=5$ in. Then, from above:

$\theta=\cos^{-1}[\{0.250+9+(0.250\times 8.750)\}/\{2\times 4.5\times\sqrt{9.250}\}]$

$=\cos^{-1} 0.41785=65°\ 18'$

$\varphi=\tan^{-1}[0.5/3]=\tan^{-1}0.16667=9°\ 28'$

Therefore, $S=4.5\cos 55°\ 50'=2.5272$ in.

and $T=4.5\sin 55°\ 50'=3.7233$ in.

APPENDIX C

SCHEMATIC EXAMPLES OF FEED AND EJECTION IN MOULDS, TAKEN FROM PRACTICE

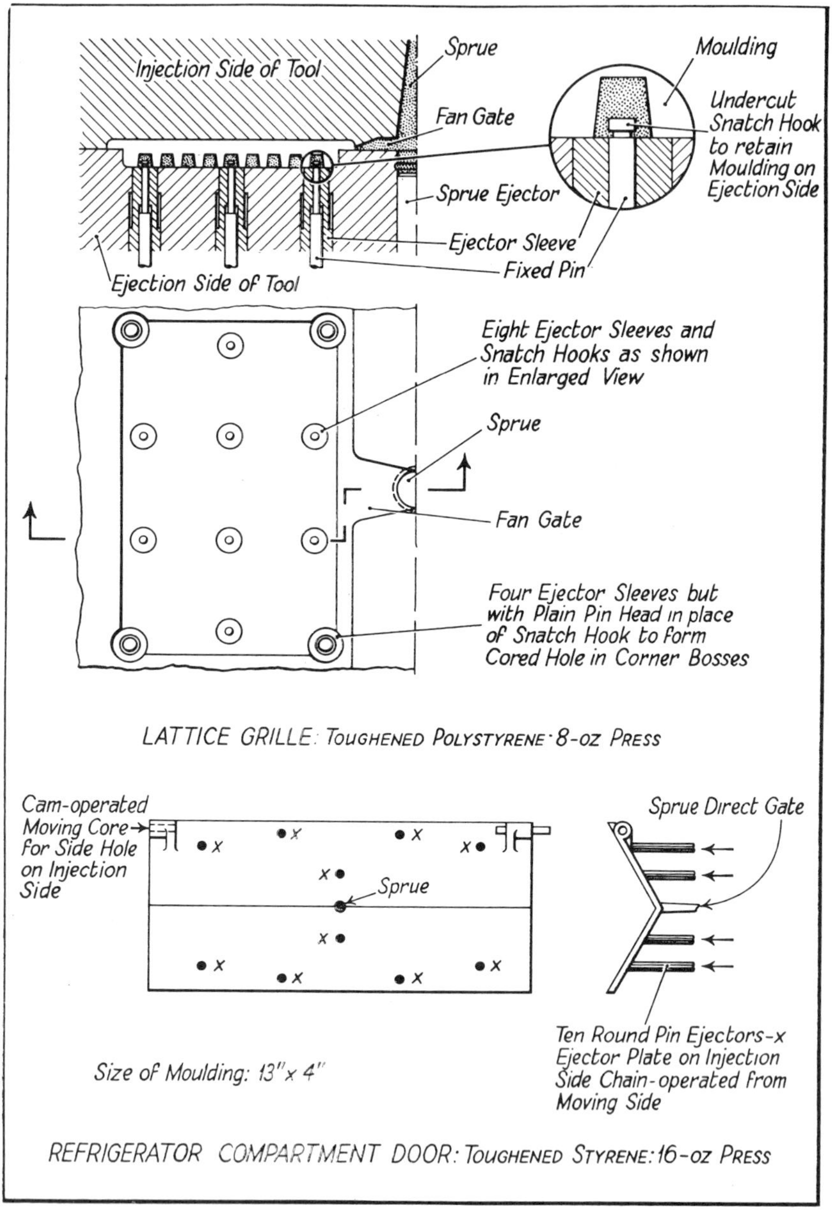
Injection Side of Tool
Sprue
Moulding
Fan Gate
Undercut Snatch Hook to retain Moulding on Ejection Side
Sprue Ejector
Ejector Sleeve
Fixed Pin
Ejection Side of Tool
Eight Ejector Sleeves and Snatch Hooks as shown in Enlarged View
Sprue
Fan Gate
Four Ejector Sleeves but with Plain Pin Head in place of Snatch Hook to form Cored Hole in Corner Bosses
LATTICE GRILLE: TOUGHENED POLYSTYRENE· 8-OZ PRESS
Cam-operated Moving Core for Side Hole on Injection Side
Sprue
Sprue Direct Gate
Ten Round Pin Ejectors-x Ejector Plate on Injection Side Chain-operated from Moving Side
Size of Moulding: 13" x 4"
REFRIGERATOR COMPARTMENT DOOR: TOUGHENED STYRENE: 16-OZ PRESS

FIG. C.1.

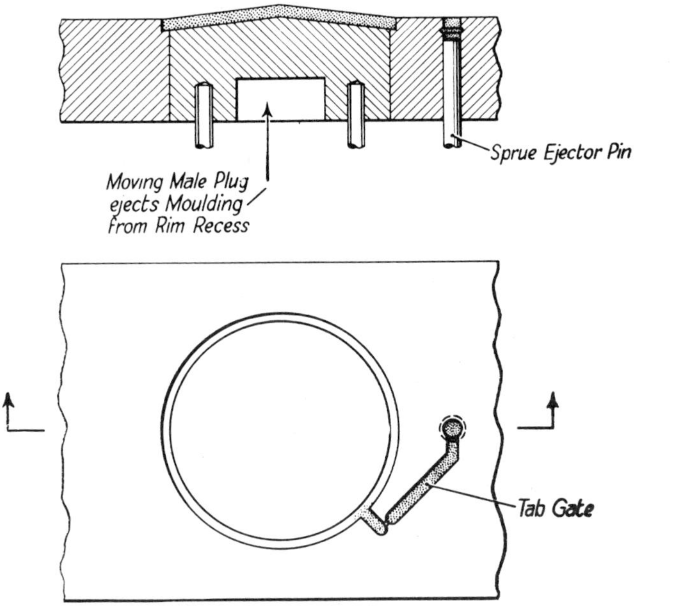

4-IN DIAMETER AUTOMOBILE SPEEDOMETER DIAL: CLEAR ACRYLIC: 8-OZ. PRESS

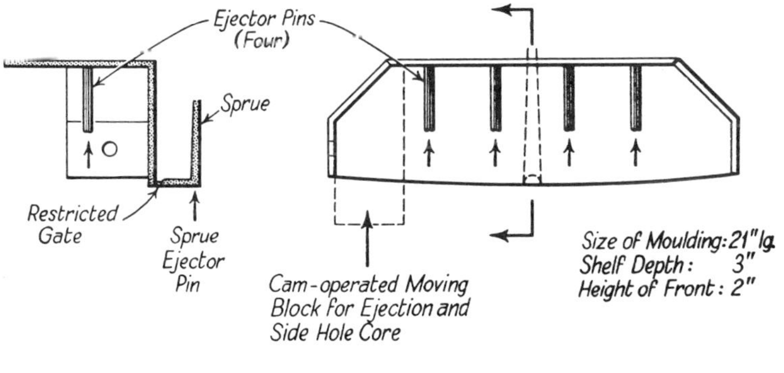

REFRIGERATOR SHELF AND FRONT: TOUGHENED POLYSTYRENE: 48-OZ. PRESS

FIG. C.2.

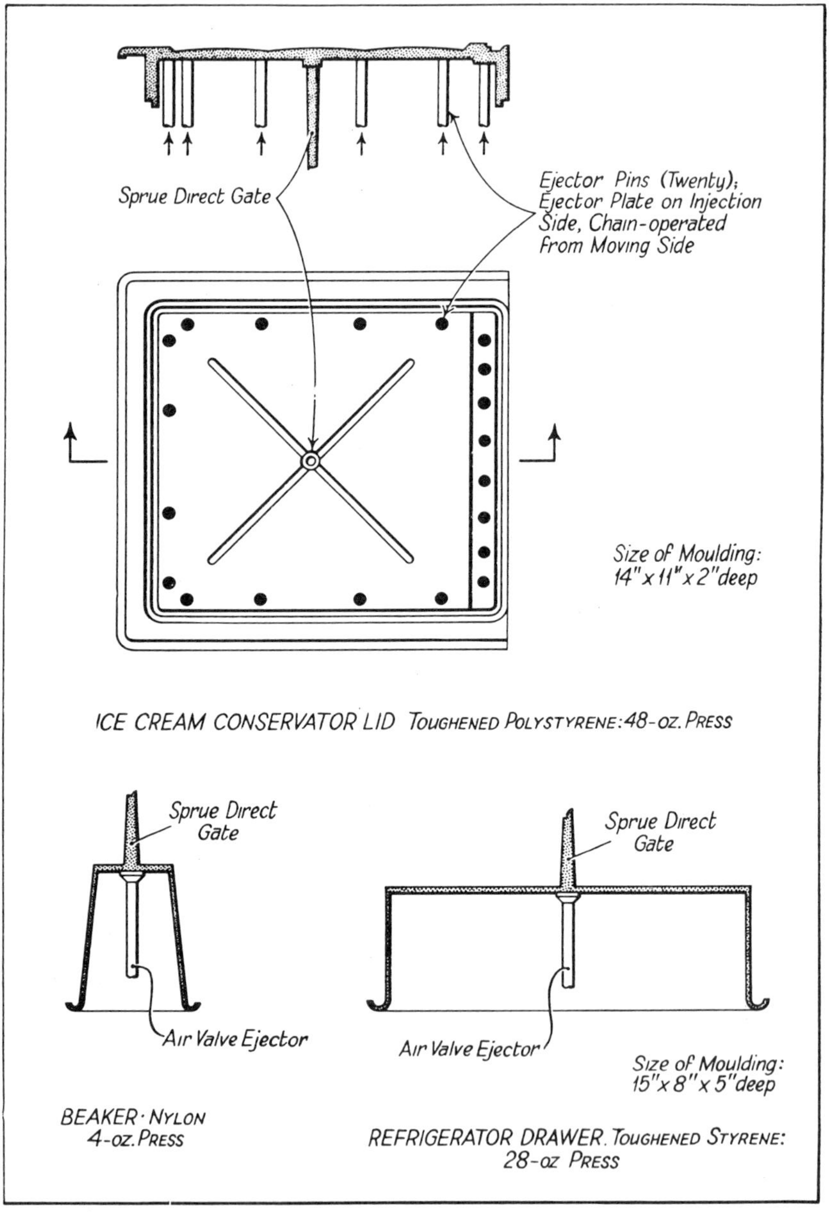
Sprue Direct Gate
Ejector Pins (Twenty); Ejector Plate on Injection Side, Chain-operated From Moving Side
Size of Moulding: 14" x 11" x 2" deep
ICE CREAM CONSERVATOR LID Toughened Polystyrene: 48-oz. Press
Sprue Direct Gate
Air Valve Ejector
BEAKER · Nylon 4-oz. Press
Sprue Direct Gate
Air Valve Ejector
Size of Moulding: 15" x 8" x 5" deep
REFRIGERATOR DRAWER. Toughened Styrene: 28-oz Press

Fig. C.3.

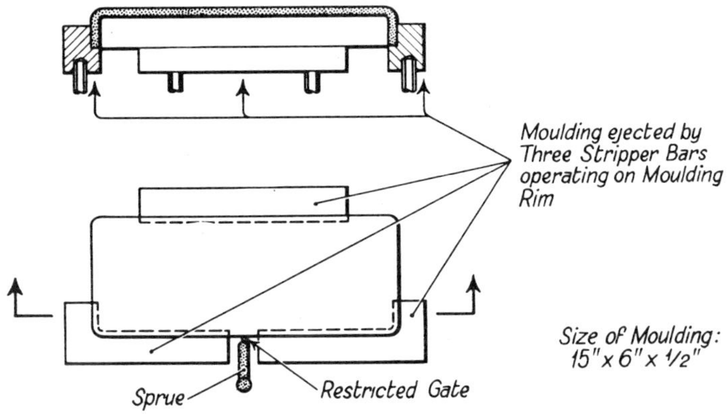

REFRIGERATOR EVAPORATOR DOOR · CLEAR POLYSTYRENE: 16-OZ. PRESS

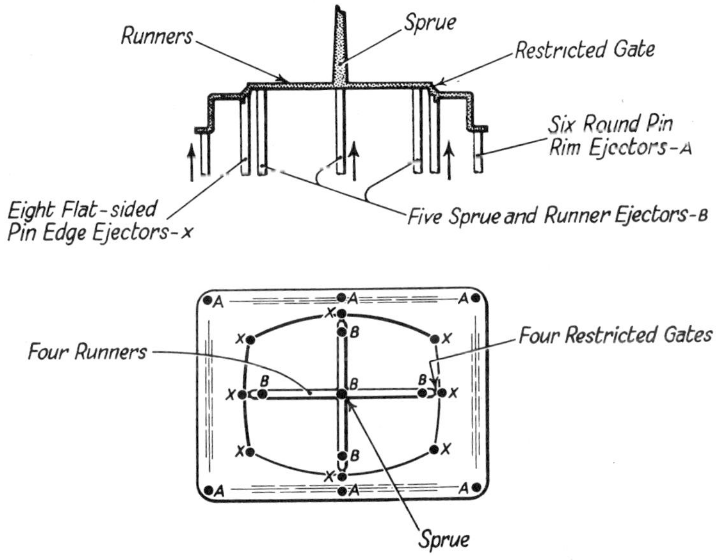

17-IN. TELEVISION SURROUND: TOUGHENED POLYSTYRENE: 28-OZ. PRESS

FIG. C.4.

APPENDIX D

PRINCIPAL CONDITIONS IN MOULDING DURING CYCLE

APPENDIX D

PRINCIPAL CONDITIONS IN MOULDING DURING CYCLE

	Stage of Cycle	Typical Results of Variation of Moulding Conditions	Action of Moulding Material
INJECTION TIME	Injection plunger or screw moves forward forcing plasticised material through nozzle.	Too short a cycle does not plasticise material sufficiently. Too low a cylinder temperature increases pressure losses. Material may oxidise or degrade if in cylinder too long at too high a temperature.	Approximately 20 per cent pressure loss compressing cold material. External lubricant on powder facilitates compression. Pressure squeezes air out from between granules which is vented back along cylinder. Heating increases volume of plastics but pressure decreases volume.
	Material enters sprue bush passing through runners to gate.		Cold material at front of plastics flow from nozzle is trapped in cold well. Outer skin of flowing material chilled on runner walls. Cold material at front of plastics flow in runners trapped in runner cold wells (if provided). Pressure loss due to viscous friction of approx. 10 per cent of original plunger pressure.

	Stage of Cycle	Typical Results of Variation of Moulding Conditions	Action of Moulding Material
INJECTION TIME	Material passes through gate and enters cavity.	Low material temperatures increase orientation in moulding.	If cold front of material not trapped, may freeze on mould face opposite gate, causing surface scab. Pressure loss through gate of 10–20 per cent of original plunger pressure depending on design. Steady streamline flow through full direct gates, turbulent flow with possible jetting from restricted gates. Restriction at narrow gate increases material velocity and thus reduces viscosity through gate. On other side, this velocity falls rapidly, increasing viscosity which together with cooling, seals gate when mould full. Small gates increase temperature of material but give higher orientation to the incoming material.

	Stage of Cycle	Typical Results of Variation of Moulding Conditions	Action of Moulding Material
INJECTION TIME	Material commences to fill cavity.	Too low an injection pressure or temperature will not fill mould. Too high an injection ram speed may hamper venting of mould, and cause excessive orientation in moulding. Too low a pressure or temperature can lead to flow lines or bad knit lines. High moulding temperatures decrease material viscosity, give faster filling less shear strain. Frozen strains are increased by the added cooling due to slow filling. At low material velocity rate, the shear rate of the material is also low and hence melt viscosity is high. At same temperature but high material flow rate viscosity falls. Thus modifying size of runners will not improve flow if material velocity is too low. Poor weld lines caused by too low injection pressure, injection speed, cylinder temperature or mould temperature.	Internal lubricant in material facilitates flow and ultimate mould release. Material at mould surfaces cools, increasing viscosity and decreasing flow velocity. Hotter internal material continues to flow at faster rate thus setting up shear stress and orientation through material from coolest outer skin to hottest at centre. These stresses are increased in thin sections due to the faster cooling rate arising from the greater surface-to-volume ration. Flow of material fastest along easiest path (e.g. thickest section). Flow path divides at pins or raised characters. Cold front of flowing material joins up with other flow paths leading to knit lines at junction. Air is displaced ahead of incoming plastics material—if this cannot escape may lead to burning or weak welds. Hottest material from cylinder is last to enter cavity. Heat leaves plastics as it enters cavity, increasing temperature of mould until removed by cooling water. As the plastics cools its viscosity increases and hence its flow velocity decreases. Greatest heat around spur and near gate due to constant inflow of fresh hot material. Uneven cooling of the mould causes uneven shear rates to be set up in the flowing material thus causing differing frozen strains in the moulding, leading to warpage and brittleness. Thus most cooling required near gate (hottest material) and least at extremities of mould (coolest material) to even out shear strains.

	Stage of Cycle	Typical Results of Variation of Moulding Conditions	Action of Moulding Material
INJECTION TIME	Cavity full. Packing pressure on.	Poor surface on moulding from too cold material or mould. Too high a temperature or pressure causes mould to flash.	Cavity under full pressure. As material in cavity commences to cool it shrinks and further material is forced in. This packing causes displacement of cooling, viscous material in gate area, causing molecular orientation and shear strain between cold outer skin and hot internal core. As material cools still further, continued packing causes thermo elastic deformation instead of plastic flow. This deformation causes further frozen strain in the component and reduces heat distortion of moulding.

	Stage of Cycle	Typical Results of Variation of Moulding Conditions	Action of Plastic Material
COOLING TIME	Gate freezes. Injection ram withdrawn.	If ram withdrawn before gate freezes some back flow of material out of cavity will occur. Surface shrinking or voids caused by too high a material temperature, too low a pressure, too short a ram forward time, or gate freezing too soon. Too long a packing time and/or too high an injection pressure can lead to residual mould pressure and high moulding weight. Alternatively, it may require longer cooling period to facilitate ejection by allowing shrinkage to offset higher cavity pressure, thus lengthening cycle. Too low an injection temperature leads to weak mouldings. Minimum shrinkage in moulding from low cylinder and mould temperature, maximum pressure, long packing time, low injection speed (cools as it fills). With male tools, long cooling time causes cold drawing of material thus reducing shrinkage but increasing internal stress. Warpage may be reduced in moulding by high cylinder and mould temperature (lower melt viscosity leading to lower shear strain and orientation and allowing more relaxation), low injection pressure and short packing time (also allowing less shear strain and more relaxation).	Material continues to cool and shrink. Cold outer layers retain frozen strain —hot inner core relaxes (thus thicker mouldings more strain free). Shrinkage is offset by residual elastic pressure within material. Excess shrinkage frees moulding from contact with female tool and grips male (setting up stress in moulding). Excess residual pressure seizes moulding between male and female tools causing difficult ejection. Cooling of heavy sections can lead to internal voids or surface sinking.

	Stage of Cycle	Typical Results of Variation of Moulding Conditions	Action of Plastic Material
COOLING TIME	Mould opens.	Too short a cooling time can lead to distortion by ejectors. Mould too hot lengthens cycle time. Sticking of moulding on male tool controlled by cooling rate, modulus of elasticity and tensile strength of material—if exceeded may cause cracking.	Ejectors move forward and under pressure eject moulding (vacuum may be generated between moulding and tool leading to distortion and resistance to ejection). Sprue withdrawn from sprue bush. Hot moulding continues to shrink in size on removal from press—internal stresses may lead to some distortion. With some materials moulding may continue to shrink at room temperature for some time. Shrinkage on to inserts sets up local stresses in bosses, etc.

INDEX